"The best ideas seem simple and obvious in re[...] simpler or more obvious than the principle of [...] when Darwin applied this 'simple and obvious' principle to phenotypic variation, we got natural selection, biology's biggest idea. Then, about one hundred years later, Skinner applied the same principle and gave us behavior analysis; a view of learning (among other things) that still has no serious explanatory rival in psychology. With this volume, David Sloan Wilson and Steven Hayes have assembled chapters by some of the most creative and rigorous minds in their respective disciplines. Their collective efforts offer an updated and elaborated account of these two big ideas, making clear along the way what only now—in retrospect—seems so simple and obvious: that *Evolution and Contextual Behavioral Science* are scientific sprouts off the same conceptual root."

—**James Coan, PhD**, professor at the University of Virginia

"This is a remarkable and unique volume that will likely start a paradigm shift in the behavioral sciences. Edited by the foremost leaders in their respective fields, David Sloan Wilson and Steven C. Hayes assembled a fascinating collection of chapters to derive a framework for understanding and changing human behavior. If Skinner and Darwin had been asked to write a book together after further developing their ideas, this would have been the one. *Evolution and Contextual Behavioral Science* will change the way you think about humans."

—**Stefan G. Hofmann, PhD**, professor in the department of psychological and brain sciences at Boston University

"The human mind is different from that of all other animals because of our cognitive abilities. It has given rise to extraordinary cultures both good and bad, to cooperation, language, art, science, and medicine but also war, slavery, factory farming, and the enjoyment of violence. The key to understanding how the human mind works is an understanding of its evolved functional motives and competencies, and their contextual phenotypic organization. What nature prepares us with, our social niche grows, patterns, and choreographs. This volume brings together two of the world's outstanding and leading exponents on both dimensions, the evolved nature of mind and the contextual choreographies of mind, partly but not only through language. Together this range of contributors provide a fascinating and scholarly set of writings with the focus to integrate and cross-fertilize these two bodies of scientific investigation. Together they take a deep dive into the link between nature and culture. It's a must-read for anybody interested in this fundamental analysis of the human condition."

—**Paul Gilbert, OBE, PhD**, is professor in the department of psychology at the University of Derby, and author of *Human Nature and Suffering* and *The Compassionate Mind*

EVOLUTION

&

CONTEXTUAL

BEHAVIORAL

SCIENCE

AN INTEGRATED FRAMEWORK *for* UNDERSTANDING,

PREDICTING & INFLUENCING HUMAN BEHAVIOR

Edited by

DAVID SLOAN WILSON, PhD

STEVEN C. HAYES, PhD

CONTEXT PRESS

An Imprint of New Harbinger Publications, Inc.

Publisher's Note

Distributed in Canada by Raincoast Books

Copyright © 2018 by David Sloan Wilson and Steven C. Hayes
Context Press
An imprint of New Harbinger Publications, Inc.
5674 Shattuck Avenue
Oakland, CA 94609
www.newharbinger.com

Cover design by Amy Shoup

Acquired by Catharine Meyers

Indexed by James Minkin

Library of Congress Cataloging-in-Publication Data on file

20   19   18

10   9   8   7   6   5   4   3   2   1   First Printing

Contents

Behavioral and Physical Health

Small Groups

Psychopathology and Behavior Change

Dialogue on Psychopathology and Behavior Change 312

Foreword

I think the most important contribution I've made to the behavioral sciences is to introduce Steven C. Hayes and David Sloan Wilson to each other.

As contextual behavioral scientists, Steve and I go back a long way. As long as twenty-five years ago, we were discussing the need to extend the contextual behavioral way of thinking beyond psychology. A little over ten years ago, I encountered David. I was interested in talking to him about his evolutionary account of religion, *Darwin's Cathedral*. He told me about a new book he had published, *Evolution for Everyone*. As I read it, I kept saying to myself, *This is how I think!* But then I would encounter a mention of B. F. Skinner's work and go, *Ah…no.*

As I got to know David, though, I realized that he was not only a contextualist but a scientist open to new ideas and eager to bring together people who shared his evolutionary perspective. I told Steve that he had to meet this guy and brought them together at a meeting I organized in Denver. They have been learning from each other ever since, and this volume is one important result of their collaboration.

It comes at a critical time. Over the past 250 years, science and capitalism have combined to transform the world in ways that would seem miraculous to people living in the nineteenth century. Yet the miracles of our progress are accompanied by great dangers. We have produced nuclear weapons that continue to proliferate and threaten horrific destruction and international terrorist organizations whose growth and harm has been enabled by the Internet. Our steadily increasing material wealth has been accompanied by increasing levels of economic inequality. And despite solid scientific understanding of the looming, catastrophic consequences of climate change, our exploration, production, and use of fossil fuels continues to expand.

This volume attests to the fact that science can be done within a framework that is guided by a set of explicit values. The community of scientists writing in this book are focused on ensuring the well-being of every person. They are doing this through the pursuit of basic principles that we can use to evolve more nurturing societies.

As this volume shows, the combination of evolution and contextual behavioral science produces a science of substantial scope—addressing issues as diverse as animal learning, human symbolic learning, adolescent development, psychopathology, empathy and emotions, evolutionary mismatch, and human health, and the selection of group and organizational practices.

It is also a science of considerable depth, addressing evolution at the genetic, epigenetic, behavioral, and symbolic levels—and fostering research on the complex interactions among these processes.

Every facet of human behavior addressed in this book is relevant to the project of evolving societies with greater well-being and better prospects for preventing the catastrophes that endanger us. The fundamental challenge we face is whether we can promote prosocial and communitarian values, behaviors, and organizational practices in the face of contingencies that have selected for wealth accumulation and self-aggrandizement, especially over the past fifty years.

As this volume attests, we are making progress in understanding—and influencing—symbolic thought and communication. The contextual analysis of the behavior of individuals has reached a stage where we understand the kinds of environments that people need to thrive.

We are making substantial progress in treating psychological and behavioral problems and guiding adolescent development so that lifelong problems are prevented.

We are beginning to understand the ways in which evolutionary mismatch threatens modern humans' health and we are defining what is needed to address these mismatches.

We are making progress in understanding emotion and the relational processes involved in people developing the empathy that enables them to cooperate with others.

We are learning how to cultivate psychological flexibility in individuals and organizational flexibility in groups and institutions.

We need much more work, however, on creating contingencies in larger society that select groups and institutions to benefit society at large—not just their members and investors. In all these areas, we as scientists are still far from knowing how to translate what we have learned into widespread adoption. This book is an important foundation for this vital next step. May it help you to advance your own efforts to build a world in which most everyone is thriving, and future generations are ensured a safe and healthy existence.

—Anthony Biglan, PhD
author of *The Nurture Effect: How the Science of Human Behavior Can Improve Our Lives and Our World*

CHAPTER 1

Evolution and Contextual Behavioral Science

David Sloan Wilson

Binghamton University

Steven C. Hayes

University of Nevada, Reno

This volume seeks to integrate two bodies of knowledge that have developed largely independently during the last half-century. The first is evolutionary science (ES), which already provides a unifying theoretical framework for the biological sciences and is increasingly being applied to the human-related sciences. The second is contextual behavioral science (CBS), which seeks to understand the history and function of human behavior in the context of everyday life, where behaviors actually occur, and also to *influence* behavior in a practical sense. The term CBS is relatively new, but it is a modern face of functional and contextual views of psychology that developed around learning theorists such as B. F. Skinner and that trace themselves back to the functional views of pioneers such as William James and John Dewey.

This volume is an outgrowth of a personal and professional relationship between the two editors that is over ten years old. Anthony Biglan, a major prevention scientist and a former president of the Society for Prevention Research, introduced us, and the three of us found our conversations to be extraordinarily productive in a number of areas. Along with Dennis Embry, a scientific entrepreneur who markets evidence-based behavioral change practices, we wrote a lengthy article titled "Evolving the Future: Toward a Science of Intentional Change," which was published with about two dozen commentaries in the premier academic journal *Behavioral and Brain Sciences* (Wilson, Hayes, Biglan, & Embry, 2014). That article can be regarded as the first milestone in the integration of ES and CBS that this volume seeks to advance further; thus, it is especially fitting for Tony to have written the foreword for this volume.

The histories of these two bodies of knowledge are thoroughly entwined. The purpose of this intoductory chapter is to trace the historical roots of both bodies of knowledge so that we can explore the utility of the framework that emerges from their reintegration without triggering disruptive associations from the past. When bodies of knowledge become isolated from each other, it is not only for mundane reasons (e.g., scientists organized into separate societies with few opportunities to interact) but may be due to misunderstandings that made it hard to work effectively together or to configurations of ideas that resist mixing, even when they are brought together and understood. Creation of a more unified framework requires a careful examination of these different kinds of disconnections, including their histories, and a willingness to adopt new configurations, understandings, and approaches.

A brief examination of history will also help make sense of the main purposes of this volume: to continue to move CBS under the umbrella of ES, and to help ES see how work with contextually oriented behavioral scientists can foster both the development of its own principles and their application to practical affairs. Rather than the sequential relationship that is typically imagined, where the "big questions" are pursued for their own sake inside a hope that they may someday lead to practical applications, this volume envisions a parallel relationship, where the best basic science is simultaneously most relevant for understanding, predicting, and influencing positive change in the real world. This kind of "reticulated" development strategy is a vigorous feature of CBS and has been central to the cooperative efforts of the editors of this volume. After reviewing the relevant history of ES and CBS, we will touch on current possibilities and on the organization of this volume and its specific purposes.

A Brief History of Evolutionary Thought in Relation to CBS

There can be no doubt that in Darwin's mind, his theory of natural selection applied to the length and breadth of humanity in addition to the natural world. Over twenty years before the publication of *The Origin of Species*, Darwin wrote in his private notebook (1837/1980): "Origin of man now proved. Metaphysics must flourish. He who understands baboon would do more toward metaphysics than Locke." Darwin was not the only evolutionist of his period, and his contemporaries were even more focused on the improvement of the human condition. Herbert Spencer had a towering reputation and his evolutionary cosmology, while largely forgotten today, was built around the betterment of humanity. It was Spencer, not Darwin, who caused the word "environment" to become frequently used and who wrote extensively about the organism-environment relationship (Wilson & Pearce, 2016).

In America, a small group of intellectuals that included Oliver Wendell Holmes, Charles Sanders Peirce, William James, and John Dewey developed a philosophical school of thought called Pragmatism that was inspired by evolutionary theory. Their story is beautifully told by Louis Menand (2001) in his book *The Metaphysical Club*, which was awarded the Pulitzer Prize for history. The central insight of Pragmatism is that ideas are not disembodied facts but rather tools of survival that can only be understood in the context of people living their everyday lives. As James put it in his 1907 book *Pragmatism*, there is "no difference in abstract truth that doesn't express itself in a difference in concrete fact and in conduct consequent upon that fact, imposed on somebody, somehow, somewhere, and somewhen."

Pragmatism provides the philosophical foundation for a conception of basic and applied science taking place in parallel rather than in sequence. It was exemplified by Dewey, who created the Laboratory School at the University of Chicago, which doubled as a school for children and a laboratory for studying the educational process.

Even though the study of evolution in the biological sciences and in relation to human affairs started out intermingled, the two aspects developed along separate tracks for most of the twentieth century. The reasons are complex, but one major factor was the emergence of the science of genetics. Darwin formulated his theory in terms of variation, selection, and a resemblance between parents and offspring (heredity), which was an easily documented fact even if the underlying mechanisms were obscure. He struggled to identify the mechanism(s) of heredity for his whole life without success and without realizing that his contemporary, Gregor Mendel, had a key to the answer with his experiments on peas. When Mendel's work was rediscovered in the early twentieth century, genes quickly became regarded not only as *a* mechanism of inheritance, but as *the only* mechanism of inheritance. In other words, the study of evolution became highly genecentric, as if the only way that offspring can resemble their parents is by sharing genes. This is patently false—offspring can resemble their parents by sharing the same cultural traits, for example, but models of cultural change as an evolutionary process with its own mechanisms of inheritance weren't developed from within evolutionary biology until the last quarter of the twentieth century. Thus, with their narrow focus on genes, most evolutionary biologists ceded the study of nongenetic change to other disciplines (Jablonka & Lamb, 2006).

It is widely thought that evolutionary theory earned a bad name for itself in the social sciences by being used to justify inequality—Social Darwinism. The real history is more complex and interesting, as recounted for a general audience in a special edition of the online journal *This View of Life* (https://evolution institute.org/this-view-of-life/) titled "Truth and Reconciliation for Social Darwinism" (Wilson & Johnson, 2016). The actual term "Social Darwinism" is largely used as a pejorative to describe any laissez-faire policy that permits the rich to take advantage of the poor. Hardly anyone calls himself a Social Darwinist, and those accused of Social Darwinism seldom use Darwin's (or even Spencer's) theory to justify their views. Progressive thinkers such as Dewey, who were clearly influenced by Darwin, are never called Social Darwinists. Confusingly, the use of the term "Social Darwinism" as a boogeyman in the social sciences and humanities coexists side by side with the scholarship that can easily dispel the myth (Hodgson, 2004).

Academic disciplines such as cultural anthropology (Wilson & Paul, 2016) and sociology (Wilson & Schutt, 2016) had their own reasons to declare independence from biology (especially of the gene-centric variety), and even from psychology, in order to establish their own autonomy. In psychology, William James comes across as strikingly modern as an evolutionary psychologist, and he and his contemporary, James Mark Baldwin, developed ideas about learning as a guiding force for genetic evolution that were far ahead of their time. For the most part, however, these early developments in psychology were eclipsed by the advent of behaviorism and learning theory, which dominated academic psychology during the middle of the twentieth century. Understanding the history of behaviorism is critical for creating a framework that unifies ES and CBS, as we will now show.

One reason for the advent of behaviorism during the first half of the twentieth century was because learning theorists thought—with justification—that the scientific progress of psychology was being slowed by a focus on hypothesized internal mental events and introspection as a research method. Behaviorists did not object to the study of thinking and reasoning per se. Skinner explicitly rejected the idea that behavioral psychology needed to limit itself to overt behavior (Skinner, 1945), and it is not generally known that even classical behaviorism developed creative methods for the study of such topics as problem solving and reasoning. For example, John B. Watson, the father of American behaviorism, developed a "talk aloud" procedure to study thinking, an approach that was rediscovered by cognitive science as part of "protocol analysis" and is now regularly used in the cognitive sciences to understand the processes underlying problem solving (Ericsson & Simon, 1993). For the most part, however, behaviorists used direct observations of nonhuman animals rather than human introspection as their basis for the development of principles.

Superficially, these features and several others should have brought ES closer to behavioral thinking and vice versa. Both are monistic and when they consider behavior, they focus on actions in the context of history and circumstances. The functional wing of behavioral thinking was based on Pragmatism as a philosophy of science (Hayes, Hayes, & Reese, 1988), and major theorists such as Skinner explicitly embraced evolutionary thought. Consider this paragraph drawn from one of his most important articles, "Selection by Consequences" (Skinner, 1981, p. 501):

> Selection by Consequences is a causal mode found only in living things, or in machines made by living things. It was first recognized in natural selection, but it also accounts for the shaping and maintenance of behavior of the individual and the evolution of cultures. In all three of these fields, it replaces explanations based on the causal modes of classical mechanics. The replacement is strongly resisted. Natural selection has not made its case, but similar delays in recognizing the role of selection in other fields could deprive us of valuable help in solving the problems which confront us.

In this passage, Skinner is saying that there is something about open-ended individual and cultural change that is *like genetic evolution*. In all three cases, there is *variation* in behaviors, the different behaviors have different *consequences*, and there is a *selection* of the behaviors that adapt the organism to its environment. All three are evolutionary processes that rely on "blind variation and selective retention," as the social psychologist Donald Campbell was fond of putting it (Campbell, 1960).

Skinner's phrase "replaces explanations based on the causal modes of classical mechanics" requires exegesis. Consider the mundane observation that many desert-living species are sandy colored to conceal themselves from their predators and prey. The evolutionary explanation is clear enough: individuals varied in their coloration in the past, those that blended into their background survived and reproduced better, and the variation was heritable, resulting in the well-camouflaged organisms that we see today. Notice that we can make this prediction for all kinds of desert-living species, such as snails, insects, amphibians, reptiles, birds, and mammals. Each of these taxonomic groups have different physical exteriors (snail shells are made of calcium carbonate, insect exoskeletons are made of chitin, and so on) coded by different genes. What enables us to understand the properties of desert-living species without requiring any knowledge about their physical makeups? The answer is heritable variation. The extent to which the physical makeup of organisms results in heritable variation—that is, the extent to which we can ignore the physical makeup and make predictions on the basis of the molding action of the environment. This applies as well to the molding action of the environment on behaviorally flexible individuals and cultures, which is what Skinner meant by his claim that "selection by consequences" replaces explanations based on the causal modes of classical mechanics.

Evolutionary biologists commonly employ a reasoning strategy called "natural selection thinking" or "the adaptationist program," which begins by assuming that heritable variation exists for all traits and that enough time has elapsed for adaptations to have evolved by natural selection. These assumptions are not defended as true, but are treated as conveniences so that the question "What would be the properties of the organism if it were well adapted to the environment?" can be asked. This reasoning strategy results in hypotheses that provide good starting points for empirical inquiry, when the messy details of inheritance mechanisms and populations out of evolutionary equilibrium come into play. In short, evolutionary biologists make their own "blank slate" assumption about heritable variation as a useful heuristic, which should be kept in mind when we consider "blank slate" assumptions about the capacity of individuals and cultures to change. In all these respects, behaviorism is deeply informed by evolutionary theory and stands on solid ground.

Given all of that, why the disconnection? The differences were largely over analytic strategy—differences that could not be overcome until both sides were further developed.

Although there are many important differences among various types of learning theory, all these scientists sought the development of behavioral principles that had high precision (concepts could be applied unambiguously) and high scope (concepts could be applied across a range of situations) and that did not contradict well-established and well-supported theories in other scientific domains (that is, they had coherence across levels of analysis, or "depth"). The scientific strategy deployed to find these principles was initially laboratory-based and experimental, rather than examining animal behavior over time in its natural context in an observational way. Psychobiological and ethological researchers objected strongly to this strategy, often seeing it as an evolutionarily ignorant assumption that principles identified in the lab *had* to generalize regardless of the biological makeup of particular organisms. Pinker stated this objection clearly when he claimed that "behaviorists believed that behavior could be understood independently of the rest of biology, without attention to the genetic make-up of the animal or the evolutionary history of the species" (2002, p. 20). Many other well-known evolutionary theorists have made similar comments (e.g., Gould & Marler, 1987; de Waal, 2001). Ethological researchers were also concerned that the attempt to find generally applicable principles that applied across tips of evolutionary branches was bound to miss key features of behavior that had been fitted through evolution to specific historical or situational features. Putting a rat in an operant chamber, they thought, would tell you very little about the rat's natural behavior; nor would studying a rat in a barren environment be likely to tell you much about human behavior.

For their part, behaviorists understood that worry. Many behaviorists (Watson and Skinner included) cut their academic eye teeth on psychobiological research (for example, one of Skinner's first three publications was on tropisms in ants). They forged ahead anyway, feeling that behavioral principles with high precision, scope, and depth that generalized across tips of evolutionary branches might exist, and if so they might be more readily identified without an initial focus on the complexity of field-based observations. It is often forgotten that learning theory initially was a branch of comparative psychology, and that a great deal of early learning research examined the generalization of learning principles such as reinforcement or discriminative control across species. The behaviorist strategy was a risky strategy, and could have failed, but it turned out that many of these principles do in fact apply to virtually all species that evolved during the Cambrian period or later. Behaviorists also soon realized they had the backup of application to human affairs to ensure that their principles were important, even if those principles were not developed initially by field studies.

We can think of these differences in light of Tinbergen's four key questions about any phenotypic trait (1963), including behavior. Every trait with a functional basis must be understood in functional terms. It will have a physical mechanism, and it will develop. It also will have a history. Tinbergen used the word *evolution* primarily to describe the history question, but evolution can be expanded to consider all four questions (e.g., Wilson et al., 2014). Behaviorists examined development and function within short time frames as a place to start, with only a nod to physical mechanism and history; mainstream evolutionists thought history and function over longer time frames would work better as a beginning, but they were skeptical about the central relevance of development and function over shorter time frames, and again with only a nod to physical mechanisms.

The deep connections between behaviorism and evolutionary theory we have noted might have worked themselves naturally into a new era of integration as these fields matured to consider questions as yet unaddressed, if it had not been for a small handful of historical events. Cognitive approaches, which largely replaced behaviorism in basic psychology during the last part of the twentieth century, were more mechanistically based. The central metaphor of information-processing models was the mind as an all-purpose computer—thus the need to understand its circuitry. A key phrase was "beyond the information given" (Bruner,

1973), which meant that the mind must have a great deal of inborn knowledge in order to process information from the environment, an insight that was denied, or at least greatly downplayed, by the behaviorist tradition.

Evolutionary psychology (Barkow, Cosmides, & Tooby, 1992) harnessed and redirected this change in Zeitgeist by arguing that the mind is not a single *all-purpose* computer but rather a large number of special-purpose computers or modules that emerged by genetic evolution to solve specific problems of survival and reproduction that had been encountered in the ancestral environment. While evolutionary psychologists disagreed with the cognitivists about the details in the organization of the mind, they shared a common disdain for the strategies of behavioral psychology. Tooby and Cosmides (1992) made a sharp contrast between evolutionary psychology, on one hand, and the "Standard Social Science Model" on the other, which includes behaviorism and other so-called "blank slate" traditions in the social sciences, including anthropologists such as Margaret Mead and Clifford Geertz. Steven Pinker (2002) added his considerable weight with his book *The Blank Slate*.

The effect of these developments was that evolutionists were largely unaware of the openness to evolutionary thinking that existed in behaviorist approaches. Furthermore, in the gene-centric era, there were few reasons for evolutionists to seek out knowledge of learning principles. When behaviorism was largely exorcized from basic academic psychology, it didn't go extinct but remained alive and well in the applied behavioral sciences, where the need to influence behaviors in the real world had a moderating influence on the shifting winds of academic fashion. To mainstream evolutionists, however, behaviorism was increasingly invisible, and used more as a whipping boy than as a focus of serious consideration. Skinner's analogy between contingencies of reinforcement and contingencies of survival or cultural selection was widely rejected because genes and memes lived and died across lifetimes, while behaviors were strengthened within lifetimes. For their part, behaviorists recognized the importance of evolution but saw little room for their own practical and research agenda at the evolution science table.

There it stood. For decades.

And then things began to change, and this volume became possible.

In ES, what has changed is that the hugely greater mechanistic knowledge afforded by an analysis of the genome paradoxically weakened simple genetic explanations of phenotypic traits. The central role of development was pulled into ES in the form of "Evo-Devo" research (West-Eberhard, 2003), while the increasing intellectual strength of multidimensional and multilevel extended evolutionary approaches (e.g., Pigliucci & Muller, 2010) began to create a renewed interest in the relevance of learning and behavior to niche creation, niche selection, and

8

genetic regulation and accommodation. Two landmark books in this change are *The Symbolic Species*, by Terrence Deacon (1997), and *Evolution in Four Dimensions*, by Eva Jablonka and Marion Lamb (2006). We are proud that both Deacon and Jablonka are included among the authors in this volume.

Jablonka and Lamb identify no fewer than four mechanisms of inheritance: (1) genetic; (2) epigenetic (involving transgenerational changes in gene expression rather than gene frequency); (3) forms of social learning found in many species; and (4) forms of symbolic thought that are distinctively human. To these we could add the adaptive component of the immune system, with the creation and selection of antibodies as a fifth evolutionary process with its own mechanism of inheritance.

For its part, the behavioral tradition too found that it needed to turn to modern evolutionary ideas such as multilevel selection to deal with the failure of classical behavioral principles to explain human language and cognition (Hayes & Sanford, 2014). It became clear empirically that knowledge of how behavioral principles impacted gene expression through epigenetic mechanisms was now of central importance (e.g., Dias & Ressler, 2014). Evolutionary ideas increasingly were used to guide basic behavioral research programs (Hayes, Sanford, & Chin, 2017).

The elementary realization that there is more to evolution than genetic evolution has led to a renaissance of thinking among evolutionary scientists on topics that previously were almost exclusively the domain of the social sciences and humanities. It is also increasingly recognized that organisms are systems for turning environment and behavior into biology. Thus, a two-headed arrow is now in place between fields that study evolutionary mechanisms and those that study behavior with a functional and contextual approach.

The vertebrate immune system provides an especially helpful analogy for thinking about our evolved system for behavioral change (Wilson et al., 2014). The immune system includes a so-called innate component, which consists of highly automated defenses that don't change during the lifetime of the organism, and a so-called adaptive component, which includes the formation of approximately 100 million different antibodies and selection of those that successfully bind to antigens. The terms are a bit confusing, because the innate component is highly adaptive and the adaptive component includes mechanisms that are highly innate. In any case, if we think of our evolved mechanisms for behavioral change as including both an innate and an adaptive component, then the modular view of evolutionary psychologists and the open-ended view of the so-called Standard Social Science Model can be seen to be compatible rather than being argued against each other (Wilson, 2017).

An example can be found in taste aversion. When it was originally encountered, it was viewed as an entirely new form of learning (e.g., Garcia, Lasiter, Bermudez-Rattoni, & Deems, 1985) that directly challenged or even falsified the strategy of learning theory (e.g., Seligman, 1970) because it so dramatically violated existing parameters of behavioral principles. As features of classical conditioning were examined and proved to apply to this case, it became more common to think of it as an evolutionarily established modification of the temporal and stimulus parameters impacting classical conditioning (e.g., Revusky, 1971). Without principles of classical conditioning, that coherent explanation could not have been reached. At the same time, the fact that only specific cues that differed across species could acquire functions via taste aversion required evolutionary insight to be understood. The modern consensus appears to be that the phenomenon applies to cues that are food related within the ecological niche of specific species. Both learning science and evolution science were used and were necessary to help understand the overall phenomenon. There is a larger point that examples such as this can make in hindsight. While "selection by consequences" thinking can be insightful by itself, it can become limited by a lack of knowledge of physical mechanisms and the failure to examine development and function in a way that expands across context and evolutionary history. Similarly, a gene-centric view of evolution grossly undervalues the role of learning and behavior across a vast range of evolutionary topics. Thus:

> Everyone was wrong, and progress requires movement on all sides. Evolutionists need to consult the human behavioral sciences and humanities respectfully—to discover what these disciplines know about learning and symbolic systems. Scientists and scholars from the human behavioral sciences and humanities will benefit by thinking about their work as inside the orbit of evolutionary theory (Wilson et al., 2014, p. 401).

In short, there is much to be learned and taught on all sides—but only if disciplinary boundaries can be overcome and the many different configurations of ideas can be related to each other. That is the purpose of this volume.

About This Volume

In this book, we explore the view that functional and contextual approaches in behavioral science need to be reintegrated with evolution science for both fields to advance at the proper clip in order to create a better framework for understanding, predicting, and influencing human behavior. The structure of the volume explores that idea in an innovative way. The volume pairs up teams of evolutionists and behavioral scientists around a specific topic relevant to the relationship of these two approaches. Authors were given a broad topic such as "learning" or "language and symbolic communication" and asked to select a specific focus for their paper within that topical area in any way they saw fit. Those papers then became chapters of this book. Teams were selected because of their prominence in evolutionary perspectives or in contextual behavioral science perspectives. There was no restriction of the specific approaches or issues, and in fact, each team was ignorant of the other team's work until their chapter was completely written. Each side then read each other's chapter and held a recorded conversation, moderated by the editors of the volume, about their similarities and differences. The transcript of that conversation was then edited down to a compact dialogue that follows each pair of chapters.

This extraordinary arrangement allows the reader to see each field as the experts do, and then to see the experts searching for avenues of connection and reintegration. Because there was little attempt to restrict the approach taken to a large topic area, even the very issues chosen for explication provided a kind of assessment of the degree of overlap and integration that already existed. This volume is designed to be equally accessible to both communities, with the reader having a front row seat witnessing a historical reintegration—one that is taking place among the experts in real time through their essays and conversations.

As editors, we tried to highlight this process more than to dictate its form. We did, however, consider the importance of studying behavior in the context of everyday life, where the behavior actually takes place, in selecting teams and topics. Both fields agree on this emphasis at the level of assumptions, but not necessarily at the level of research strategy. Evolutionary approaches take the view that species can only ultimately be understood in relation to past and present environments. That is why field studies figure so prominently and why most evolutionists insist that laboratory research must be informed by field studies to avoid the risk of being uninterpretable. Because contextual behavioral science takes prediction *and influence* with precision, scope, and depth to be its model of understanding (that is, its truth criteria), they agree that laboratory studies need to be read with a grain of salt until they are applied to the study of human behavioral and cultural diversity in the context of real-world problems. For these reasons, we deliberately selected teams that took context and human impact seriously, in either of these two forms. As human-related evolutionary research begins to place the same emphasis on field studies as research on other species, the idea of doing basic and applied research in parallel nests nicely into that overall approach, facilitating another point of connection between ES and CBS.

Although CBS is a newly labeled branch of the applied behavioral sciences, it is formulated to be a modern face of behavioral psychology. Rather than attempting to characterize its state of play, we will let the chapters in this volume tell that story. The history that we have reviewed shows why CBS relies in large part on basic learning principles drawn from the animal lab, but the reader will also see that its modern development has led to an empirically vigorous theory of symbolic thought called relational frame theory (RFT), which informs its practical methods for influencing behavior, including acceptance and commitment therapy/acceptance and commitment training, or ACT. (For a comprehensive review of the CBS research program, see Zettle, Hayes, Barnes-Holmes, & Biglan, 2016.)

One asymmetry that comes through in this volume is that contextual behavioral scientists have thought more about evolutionary science over the last decade than vice versa. The Association for Contextual Behavioral Science (ACBS), founded in 2005, has about 8000 members across the world and chapters in 27 countries and 18 languages. Evolutionists have regularly been plenary speakers at the annual meetings, and there is an evolution science Special Interest Group. Thus, at this moment, the relatively compact CBS community has learned more about evolution than the much larger and more diffuse evolutionary community has learned about CBS. Even counting fellow travelers, there are likely fewer than 25,000 scientists and professionals with a CBS orientation worldwide, and most are likely to have seen the flow of evolutionary articles and presentations in their areas over the last decade.

The same is not true of evolution scientists, but that too is changing. Many of the issues examined in this volume have been touched on in the online magazine *This View of Life* (TVOL), which provides in-depth coverage of ES for the general public and interested professionals, and which highlights links to the contemporary scientific literature (https://evolution-institute.org/this-view-of-life/). *TVOL* is managed by David as part of his nonprofit organization, The Evolution Institute (EI: https://evolution-institute.org). ACBS is an organizational supporter of EI, and CBS authors have contributed to *TVOL* articles. These efforts constitute additional important resources for the reunification of ES and CBS.

Thus, this volume is a snapshot of an ongoing conversation with a clear trajectory: the reintegration of evolution science and a functional contextual approach to the study of behavior into a comprehensive framework for addressing human complexity. We are happy to share the excitement that we have felt about the integration of ES and CBS with a larger audience in the pages of this groundbreaking volume.

References

Barkow, J. H., Cosmides, L., & Tooby, J. (1992). *The adapted mind: Evolutionary psychology and the generation of culture.* Oxford: Oxford University Press.

Bruner, J. S. (1973). *Beyond the information given.* Oxford, UK: Norton.

Campbell, T. D. (1960). Blind variation and selective retention in creative thought and other knowledge processes. *Psychological Review, 67,* 380–400.

Darwin, C., & Herbert, S. (Ed.). (1980; originally written ca. 1837). *The red notebook of Charles Darwin.* Ithaca, NY: Cornell University Press.

Deacon, T. W. (1997). *The symbolic species.* New York: Norton.

de Waal, F. B. M. (2001). *The ape and the sushi master: Cultural reflections of a primatologist.* New York: Basic Books.

Dias, B. G., & Ressler, K. J. (2014). Parental olfactory experience influences behavior and neural structure in subsequent generations. *Nature Neuroscience, 17,* 89–96.

Ericsson, K. A., & Simon, H. (1993). *Protocol analysis: Verbal reports as data.* Boston: MIT Press.

Garcia, J., Lasiter, P. S., Bermudez-Rattoni, F., & Deems, D. A. (1985). A general theory of aversion learning. *Journal of the New York Academy of Sciences, 443,* 8–21.

Gould, J. L., & Marler, P. (1987). Learning by instinct. *Scientific American, 256,* 74–85.

Hayes, S. C., Hayes, L. J., & Reese, H. W. (1988). Finding the philosophical core: A review of Stephen C. Pepper's *World Hypotheses. Journal of the Experimental Analysis of Behavior, 50,* 97–111. doi:10.1901/jeab.1988.50–97

Hayes, S. C., & Sanford, B. (2014). Cooperation came first: Evolution and human cognition. *Journal of the Experimental Analysis of Behavior, 101,* 112–129. doi:10.1002/jeab.64

Hayes, S. C., Sanford, B. T., & Chin, F. (2017). Carrying the baton: Evolution science and a contextual behavioral analysis of language and cognition. *Journal of Contextual Behavioral Science, 6,* 314–328. doi:10.1016/j.jcbs.2017.01.002

Hodgson, G. M. (2004). Social Darwinism in anglophone academic journals: A contribution to the history of the term. *Journal of Historical Sociology, 17,* 428–463. doi:10.1111/j.1467 –6443.2004.00239.x

Jablonka, E., & Lamb, M. (2006). *Evolution in four dimensions: Genetic, epigenetic, behavioral, and symbolic variation in the history of life.* Cambridge, MA: MIT Press.

James, W. (1907). *Pragmatism: A new name for some old ways of thinking.* New York: Longmans, Green & Co.

Menand, L. (2001). *The Metaphysical Club: A story of ideas in America.* New York: Farrar, Straus & Giroux.

Pigliucci, M., & Muller, G. B. (2010). *Evolution, the extended synthesis.* Cambridge, MA: M.I.T. Press.

Pinker, S. (2002). *The blank slate: The modern denial of human nature.* New York: Viking.

Revusky, S. (1971). The role of interference in association over a delay. In W. K. Honig & P. H. R. James (Eds.), *Animal memory.* New York: Academic Press.

Seligman, M.E. (1970). On the generality of the laws of learning. *Psychological Review, 77,* 406–418.

Skinner, B. F. (1945). The operational analysis of psychological terms. *Psychological Review, 52,* 270–277. doi:10.1037/h0062535

Skinner, B. F. (1981). Selection by consequences. *Science, 213*, 501–504.

Tinbergen, N. (1963). On aims and methods of ethology. *Zeitschrift für Tierpsychologie* [*Journal of Animal Psychology*], *20*, 410–433.

Tooby, J., & Cosmides, L. (1992). The psychological foundations of culture. In J. H. Barkow, L. Cosmides, & J. Tooby (Eds.), *The adapted mind: Evolutionary psychology and the generation of culture*. Oxford: Oxford University Press.

West-Eberhard, M. J. (2003). *Developmental plasticity and evolution*. New York: Oxford University Press.

Wilson, D. S. (2017). Evolutionary psychology and the standard social science model: Regaining the middle ground. *This View of Life*. Retrieved from https://evolution -institute.org/article/evolutionary-psychology-and-the-standard-social-science-model -regaining-the-middle-ground/?source=tvol

Wilson, D. S., Hayes, S. C., Biglan, A., & Embry, D. (2014). Evolving the future: Toward a science of intentional change. *Behavioral and Brain Sciences, 37*, 395–460.

Wilson, D. S., & Johnson, E. M. (2016). Truth and reconciliation for Social Darwinism. *This View of Life*. Retrieved from https://evolution-institute.org/wp-content/uploads/2016/11 /2Social-Darwinism_Publication.pdf

Wilson, D. S., & Paul, R. (2016). Cultural anthropology and cultural evolution: Tear down this wall! A conversation with Robert Paul. *This View of Life*. Retrieved from https:// evolution-institute.org/article/cultural-anthropology-and-cultural-evolution-tear -down-this-wall-a-conversation-with-robert-paul/

Wilson, D. S., & Pearce, T. (2016). Was Dewey a Darwinian? Yes! Yes! Yes! *This View of Life*. Retrieved from https://evolution-institute.org/article/was-dewey-a-darwinian-yes-yes-yes -an-interview-with-trevor-pearce/

Wilson, D. S., & Schutt, R. (2016). Why did sociology declare independence from biology (and can they be reunited)? An interview with Russell Schutt. *This View of Life*. Retrieved from https://evolution-institute.org/article/why-did-sociology-declare-independence -from-biology-and-can-they-be-reunited-an-interview-with-russell-schutt/

Zettle, R. D., Hayes, S. C., Barnes-Holmes, D., & Biglan, T. (Eds.). (2016). *The Wiley handbook of contextual behavioral science*. West Sussex, UK: Wiley-Blackwell.

CHAPTER 2

The Contextual Science of Learning: Integrating Behavioral and Evolution Science Within a Functional Approach

Michael J. Dougher
University of New Mexico

Derek A. Hamilton
University of New Mexico

At a very general level, there are two interacting but epistemologically distinct approaches to the psychological study of learning: mechanistic (structural) and functional. The former, which is clearly predominant and includes most contemporary cognitive and neuroscience approaches, focuses on the structure, organization, and functions of the internal mechanisms that are presumed to underlie or drive learning (e.g., Bechtel, 2008; Domjan, 2010; Ormrod, 2012). The latter focuses on the relations between observed changes in an organism's behavior and correlated or contingent environmental events (e.g., Catania, 2013; De Houwer, Barnes-Holmes, & Moors, 2013; Donahoe & Palmer, 1994). Researchers from these two traditions interact around data, but they differ substantially in their underlying assumptions, theories, experimental methods, and interpretations of experimental results.

Within contemporary evolution science, there has been a movement away from an exclusive reliance on natural or genetic selection as an explanation of phenotypic adaptation and an increasing interest in the roles played by other inheritance systems, including epigenetics, learning processes, symbolic or language processes, and culture (e.g., Bolhuis, Brown, Richardson, & Laland, 2011;

Caporael, 2001; Jablonka & Lamb, 2014; Papini, 2002). Obviously, these systems overlap considerably with those at the heart of the learning and behavioral sciences. In addition, functional approaches to learning share with evolution science a commitment to selectionist models of causality and a focus on *functional* adaptations to environments. Given these epistemological, theoretical, and content commonalities, we take the position here that both the learning and evolution sciences would be well served by adopting an integrated, functional-contextual approach to the study of learning and behavior. In what follows, we lay out the distinctions between mechanistic and functional approaches to learning via the example of their different approaches to spatial learning, discuss previous attempts to integrate learning and evolution, and conclude with a call for an integrated contextual science of learning and evolution.

Two Approaches to the Study of Learning

Modern *mechanistic* approaches to the science of learning can be seen as intellectual descendants of the positivistic, hypothetico-deductive, neo- or methodological behaviorism that characterized experimental psychology in the middle of the twentieth century. Donahoe and Palmer (1994) label these approaches collectively as inferred-process approaches: Behavioral data serve as indicators or sources of inference about the cognitive and/or neurophysiological mechanisms that are hypothesized to underlie learning. Within this approach, experiments, typically involving groups of organisms, are conducted to test hypotheses deduced from theories about underlying learning mechanisms. The data are normally averaged across subjects (for generalizability), and correspondence, in the form of predicted outcomes, serves as the criterion by which hypotheses (and parent theories) are determined to be supported or falsified.

In contrast, modern *functional* approaches to the science of learning are largely descendants of the "Skinnerian tradition" and fall under the rubric of behavior analysis or contextual behavioral science (CBS). Rather than being seen as an indicator of hypothesized learning mechanisms, the behavior of whole organisms in relation to its historical and immediate environmental context is taken to be the subject of study in its own right, and explanations (causes) of behavior are to be found in the reliably observed functional relations between behavior and relevant environmental contexts. Within this approach, the basic units of analysis are environment-behavior relations, where each component of the relation is mutually defined by its functional relation to the other. That is to say, behavior is defined in terms of its environmental antecedents and consequences, and these, in turn, are defined in terms of their observed effects on behavior. Because prediction *and* influence serve as the criteria by which functional contextual theories are evaluated, experiments tend to involve individual subjects, and experimental methods are designed to demonstrate behavioral control with maximal internal validity. Generalizability of results is achieved through replication across participants or contexts, not through group designs per se.

A recent paper by De Houwer and colleagues (2013) nicely illustrates the difference between mechanistic and functional approaches to the science of learning. From the dominant mechanistic perspective, learning is typically defined as changes in an organism resulting from experience. These authors find that definition problematic for several reasons and, instead, suggest a functional definition of learning as an ontogenetic adaptation. The emphasis here is not on the organism or its internal mechanisms but on the observed functional regularities between behavior and environments. On this view, different types of learning, e.g., classical and operant conditioning, depend not so much on different organismic mechanisms or structures as on different types of environmental regularities and their relation to behavior. The environmental regularities involved in classical conditioning (i.e., regularities in the ordered co-occurrence of relevant antecedent stimuli) are different from those that are involved in operant conditioning (i.e., regularities in the relations between behavior and relevant consequent stimuli). While the differences between mechanistic and functional approaches are more philosophical than empirical, we argue that functional approaches are better suited for integration with modern evolution science.

Mechanistic and Functional Approaches to the Study of Spatial Learning

To illustrate the differences between mechanistic and functional approaches as they have been applied in learning research beyond simple operant and classical conditioning, we turn to spatial learning and navigation. Because spatial learning is both complex and relatively consistent across species, it was a central focus of experimental psychology in its earliest days and has factored heavily in our understanding of the neural bases of mammalian learning and memory over the past fifty years.

After the early work of Watson (1907) and others that focused on the basic sensory processes required for spatial learning, emphasis shifted to a debate regarding what is learned in spatial learning arrangements (see, e.g., Knierim & Hamilton, 2011). This was perhaps best exemplified in the classic debate between Tolman (1938) and Hull (1943) regarding the acquisition of cognitive maps as opposed to the acquisition of response tendencies or learned associations. O'Keefe and Dostrovsky's (1971) discovery of place cells in the hippocampus (of rats) led to the formulation of the highly influential cognitive mapping theory of hippocampal function proposed by O'Keefe and Nadel (1978). At the heart of this view, the function of the hippocampus is to create and maintain a spatial representation of environmental stimuli and their relations (a cognitive map) that is *independent* of the observer's (organism's) perspective and behavior. Their provocative prediction that cognitive maps are wholly formed and updated rapidly (i.e., non-incrementally and without regard to behavioral contingencies) motivated several laboratories to investigate whether well-established learning phenomena such as blocking and overshadowing would occur in spatial learning tasks (Mackintosh, 2002). For example, Chamizo, Aznar-Casanova, and Artigas (2003) investigated overshadowing in a virtual spatial navigation task for humans. After efficient navigation to a goal location had been learned, removal of some environmental stimuli disrupted navigation performance, leading these authors to question the cognitive mapping account of spatial leaning.

Although these (and other) results would seem to support the learning account over the cognitive mapping account, it is important to point out that both accounts are mechanistic, in that navigation is explained by an appeal to hypothesized mechanisms that are the products of learning: cognitive maps versus associations. The specific functions of relevant environmental stimuli in relation to navigating behavior are neither identified nor included in the explanation of that behavior.

In response, Hamilton, Rosenfelt, and Whishaw (2004) conducted an experimental analysis of the stimulus functions operating through the entire duration of spatial navigation trials. Results showed that distinct segments of navigation toward a goal location depended on different environmental stimuli, with the initial selection of a direction based on more global features of the environment and subsequent/terminal aspects of the trip based on stimuli located at or near a goal. Thus, fine-grain functional relations are operating in spatial navigation tasks, demonstrating the importance of functional analyses for both mechanistic and functional accounts of behavior.

If inferred mechanisms are to be used as explanations of behavior, then it is critical that they be defined in terms of empirically determined functional relations, lest they lead to the neglect of important functional relations. Once these functional relations are discovered, the field can consider the incremental value of incorporating inferred mechanisms in behavioral explanations, especially when those mechanisms are themselves defined in terms of those same functional relations. As they are commonly conceptualized, however, mechanistic accounts of behavior generally do not sufficiently emphasize environment-behavior relations that are the bases of functional theories of learning. With respect to evolution science, there would seem to be a clear potential for alignment between questions about the functions of spatial behaviors (including learning) in this respect and their adaptive value.

Learning and Evolution

While mechanistic and functional approaches to the study of learning differ in fundamental ways, both have long acknowledged that learning itself is an evolutionary adaptation and both have attempted to integrate evolution into their thinking about learning. Nevertheless, as Bolles and Beecher (1988) report in the preface of their edited volume, *Evolution and Learning*, the two fields have advanced largely independently of each other, and when there are points of interaction they are often in the form of suggestions that an advancement in one of the fields be considered for the development of the other. Typically, however, when psychologists include evolutionary considerations in their theories of learning, evolution scientists pay little attention (see also Ginsburg & Jablonka, 2010). The same holds when evolution scientists include psychological findings in their views on learning. Added to this is the tendency for both evolution and learning scientists to make uneducated assumptions about the other tradition. For example, psychologists have been criticized for their imprecise and sometimes erroneous use of the term "adaptation" (Beecher, 1988), and evolution scientists have been criticized for adhering to an assumption of continuity between human and non-human behavior in the face of compelling differences between humans and non-humans in their higher-order relational capabilities (Penn, Holyoak, & Povinelli, 2008). Further, as Donahoe (2012) notes in his review of *Evolution Since Darwin: The First 150 Years* (Bell, Futuyma, Eanes, & Levinton, 2010), despite the well-recognized importance of environment-behavior relations in natural selection and the fact that a change in an animal's morphology is often preceded by a change in behavior, essentially nothing about the scientific study of behavior appears in the book.

Mechanistic Approaches to Learning and Evolution

As would be expected, when psychologists incorporate evolutionary concepts into their theories of learning and behavior, they tend to do so in ways that reflect their epistemological inclinations. Accordingly, those from mechanistic traditions look to evolutionary processes, mostly genetic selection, to explain the development or occurrence of the biological or cognitive mechanisms (modules) that underlie the behavior of interest (e.g., Dickins, 2003; Domjan, 2010; Klein, Cosmides, Tooby, & Chance, 2002).

Since the late 1960s, mechanistic approaches to the study of learning have been significantly influenced by an adaptationist view that sees learned behavior as part of an organism's biological equipment that allows for a better fit with its environment. Papini (2002) collectively refers to this position as the "ecological view," which can be contrasted with the "general-purpose view" that sees learning as open-ended and generalized across species. The ecological view was prompted to a large extent by the "biological constraints on learning" movement driven by discoveries in taste-aversion learning, avoidance learning, autoshaping, and foraging that seemed to refute the traditional view of learning as a general, open-ended process and to support the idea that learning mechanisms are adaptations shaped by natural selection to achieve the best solutions to specific environmental problems. This view was epitomized in Seligman's (1970) influential critique of general learning processes across species and his call for a focus on specific learning mechanisms defined by certain learning tasks. Once these specific, biologically relevant processes and mechanisms were identified, mechanistic learning scientists tested the extent to which they could be modified through various conditioning procedures (e.g., Domjan & Hollis, 1988).

Although not specifically focused on learning, an approach that has come to be known as evolutionary psychology (EP) provides a developed example of the ecological view of mind and behavior. While the range of topics subsumed under EP is quite broad and eclectic (see Caporael, 2001), the term is commonly associated with the work of a specific group of researchers (e.g., Barkow, Cosmides, & Tooby, 1992; Buller, 2005; Buss, 1999; Pinker, 2002; Symons, 1979). These researchers take the position that human cognitive abilities are the result of a set of specific information-processing modules designed by natural selection to solve adaptive problems faced by our hunting and gathering ancestors. EP is characterized by two overarching perspectives: (1) an opposition to general process explanations of complex human behavior in favor of "massive modularity"; and (2) a reliance on inclusive fitness or natural selection as the primary driver of the evolution of these modules.

Others have questioned the empirical and conceptual adequacy of EP as a general platform for evolution science (e.g., Bolhuis et al., 2011; Caporael, 2001; Papini, 2002; D. S. Wilson, Hayes, Biglan, & Embry, 2014). We find these arguments convincing. For example, D. S. Wilson and colleagues (2014) point to the interactions of the innate and adaptive components of the immune system as a good analogy for understanding the roles of the innate and adaptive components of human learning and behavior. That analogy suggests that the distinction between the ecological (massive modularity, elaborate innateness) and general systems views of learning and behavior is a false dichotomy, given that differential environmental pressures can lead both to innate preparedness and cross-species differences in learning as specific modifications of general learning processes. These modifications themselves result in nongenetic mechanisms of inheritance capable of rapidly adapting organisms to their current environments in a domain-general fashion. In the end, EP is an inadequate platform for an integrated contextual science of evolution and learning because it fails to accommodate the increasing evidence (discussed in more detail below) for humans' susceptibility to such nongenetic mechanisms of inheritance as learning, language, and culture, and fails to recognize these as drivers of evolution.

Behavior Analysis and Evolution

Perhaps ironically, there have been fewer attempts to formally integrate evolutionary science and behavior analysis. This is ironic because from its very beginning, behavior analysis classified itself as a branch of biology (Donahoe, 1996; Skinner, 1974), and selection-based causality is a central position common to both fields. Early on, Skinner claimed the behavior of organisms to be a single field of inquiry in which both phylogeny and ontogeny must be taken into account. He acknowledged that behavior could not be understood, predicted, or controlled without understanding organisms' instinctive patterns, evolutionary history, and ecological niches, and he portrayed operant conditioning as an ontogenetic analogue of natural selection (Skinner, 1938, 1974, 1981; Donahoe, 2012; see also Morris, Lazo, & Smith, 2004). Prominent behavior analysts have since advocated for more integration of evolutionary science in behavior analysis (e.g., Baum, 2012; Catania, 2013; Donahoe, 2012; Donahoe & Palmer, 1994; Hayes & Sanford, 2014; Rachlin, 2014; D. S. Wilson et al., 2014), and in 1987 a dialogue between E. O. Wilson and B. F. Skinner ended with a call from both for a more integrated approach to the study of behavior and human nature (reprinted in Naour, 2009).

One example of the incorporation of evolutionary ideas into behavior analysis is Baum's (2012) paper, wherein he redefines reinforcement as the contingent shaping of the allocation of behavior that produces phylogenetically important events and associated stimuli. Another example is Rachlin's "teleological behaviorism" (2014). Rachlin argues that the typical analytic focus in behavior analytic research is too temporally restricted and suggests that behavior analysis borrow from evolutionary biology a more temporally extended view of selection. Instead of a focus on discrete acts, he calls for an emphasis on more complex, extended patterns of behavior that are selected and organized over longer time scales.

Susan Schneider's book, *The Science of Consequences: How They Affect Genes, Change the Brain, and Impact Our World* (2012), attempts to integrate the research across a number of disciplines, focusing on the use of consequences to alter behavior, genes, brains, and environments. She takes an "interacting systems" approach and attempts to organize this multidisciplinary work under the collective heading of a science of consequences. The book well demonstrates the impressive power of selection by consequences at virtually every level of analysis from neurons to cultures, and makes or at least implies a strong case for integration and consilience across disciplines that take a selectionist approach. However, despite these efforts and the acknowledged commonalities between evolution science and behavior analysis (see also Monestès, 2016), there are still relatively few instances of true collaboration or integration between behavior analysis and evolution science.

Several factors may have played a role in keeping the development of the two fields relatively independent. One is that despite Skinner's views, neither he nor other behavior analysts gave developments in evolution sufficient weight in understanding operant behavior. Some of that may have had to do with the sheer amount of work involved in discovering and developing the principles of operant conditioning via laboratory analyses. That was necessary to establish clear experimental control, but the lack of ecological validity in this work may have contributed to its relative neglect by evolution scientists and behavioral ecologists. Relatedly, the focus within behavior analysis on laboratory analyses of behavior may have signaled an indifference to evolution-based biological constraints on learning and an acceptance of discredited stimulus equipotentiality and reinforcer transituationality assumptions that rendered the data from operant laboratories as too contrived to be of interest to evolution scientists. In addition, up until recently, evolution scientists tended to see behavior as the phenotypic expression of the genotype, and learning was seen as a dependent variable rather than an explicative or causal variable. Moreover, the received view of the Skinnerian position, despite many published declarations to the contrary, is that it is radically environmentalist and that behavior analysis appeared to be unconcerned with issues of mind and complex behavior (Kokko & Jennions, 2010), which is precisely what human evolutionists were trying to explain. Finally, behaviorism had come to be associated in the media with the behavior modification movement and the use of *Clockwork Orange*–like procedures and punishment to modify unwanted behavior. Not exactly the kind of public image with which most scientists would want to affiliate.

Contextual Behavioral Science and Evolution

Although we have characterized behavior analysis as a coherent functional approach to the study of learning and behavior, the field is not entirely unified in its underlying epistemology, defined subject matter, units of analysis, views on language and cognition, or criteria for acceptable explanations of behavior (e.g., Hayes & Sanford, 2014; Hayes, Barnes-Holmes, & Roche, 2001; K. G. Wilson, 2016). While some behavior analysts embrace a fully contextual epistemology, take a molar analytical approach, emphasize the study of language and cognition, and incorporate verbal processes as controlling variables in their theories of behavior, others retain a more traditional realist epistemology, take a molecular analytic approach, focus on discrete operant behaviors in controlled situations, and eschew as mentalistic explanations of behavior that appeal to verbal processes. In light of these differences, there have been attempts within behavior analysis to situate it within a larger conceptual framework and to expand its breadth by integrating within it academic disciplines with similar conceptual perspectives. By far the most successful of these is the recent attempt to situate behavior analysis within a larger contextual behavioral science (Hayes, Barnes-Holmes, & Wilson, 2012; Zettle, Hayes, Barnes-Holmes, & Biglan, 2016).

A full description of CBS is beyond the scope of this chapter, but Hayes and colleagues (2012, p. 2) provide a concise definition:

> [CBS] … is grounded in contextualistic philosophical assumptions, and nested within multi-dimensional, multi-level evolution science as a contextual view of life, it seeks the development of basic and applied scientific concepts and methods that are useful in predicting and influencing the continually embedded actions of whole organisms, individually and in groups with precision, scope and depth.

For present purposes, it is useful to unpack and clarify some of the terms in this definition.

First is the reference to contextualistic philosophical assumptions. Although behavior analysis emphasizes environmental controlling variables, the notion of context is more developed and elaborated within CBS. Here, context includes an organism's history (both ontogenetic and phylogenetic), its current situation, and the entire range of potentiated consequences both immediate and delayed that are contingent on its behavior. Context is inseparable from the behavior of interest and is itself part of the behavioral stream. Accordingly, contexts themselves are units of selection. As Monestès (2016) contends, "In a larger evolutionary context, considering the effects of consequences on the organism should go hand in hand with consideration of the environmental modifications that result from the organisms' behavior" (p. 104).

Another defining characteristic of contextualistic philosophy is the criterion by which theories and explanations are considered to be true. The truth criterion for contextualism is effective action or the extent to which a scientific account affords prediction and, where possible, influence. Prediction and influence are scientific goals that CBS shares with behavior analysis, but a behavioral science and technology of influence (change) at both the individual and group level is a very explicit and fundamental goal of CBS. It is also a primary reason for nesting it within evolution science. Evolution science is the science of change and, as such, provides a natural overarching structure for the learning and behavioral sciences.

Note in this definition the added objectives of prediction and influence with precision, scope, and depth. In this case, precision refers to the extent to which a given set of scientific concepts and principles are unambiguous, that is, the range of relevant determining variables and their relations to each other are well developed and specified. Scope refers to the range of contexts to which those scientific concepts apply, and depth refers to the range of fields of analysis or disciplines to which they usefully apply. As an example, in both the learning and evolution sciences, selection as a scientific concept has been shown to have both precision and depth. The extent to which selection processes are central to both fields speaks to its scope.

The second point to note is that CBS is nested within multidimensional, multilevel evolution science. Nesting CBS within evolution science is a straightforward commitment to selection as a central, unifying causal process. Driving this commitment is the increasing evidence that natural (genetic) selection is just one inheritance mechanism driving human evolution (e.g., Caporael, 2001; Corning, 2014; Cziko, 1995; Ginsburg & Jablonka, 2010; Jablonka & Lamb, 2014; Monestès, 2016; West-Eberhard, 2003). In our view, the clearest and strongest case for a multilevel, multidimensional approach in evolution science has been put forth by Jablonka and Lamb (2014), who claim that four inheritance systems (dimensions) must be considered in any adequate account of human evolution: genetic, epigenetic, behavioral, and symbolic (language). As Cziko (1995), Caporael (2001), and others point out, what makes a theory of behavior evolutionary is not just that it is based in biology or on genetic selection, but rather that it relies on the integration of three interacting principles of change: variation, selection, and retention (transmission). Across evolutionary research domains, the focus on variation, the conditions for selection, and the mechanisms of retention may differ, but neuronal selection, selection in the immune system, and learning are all considered examples of evolution. As Caporael (2001) contends, the "gene's eye view" of evolution is too narrow to accommodate recent evidence regarding the roles of epigenetics, learning, and socio/cultural factors in human evolution. From the CBS perspective, evolutionary processes are operating simultaneously and similarly at every level of analysis (gene, cell, individual, culture), and all relevant selection systems or dimensions (genetic, epigenetic, behavioral, symbolic, and cultural) contribute to evolutionary development.

Finally, CBS calls for the analysis of the contextually defined behavior of whole organisms, not isolated systems, mechanisms, or disembodied organs. Again, this is in line with the views of several modern evolution scientists who are also calling for a new evolution science. Caporael (2001), for example, argues that "the process of evolution is situational, relational, and embodied in specific contexts. Multilevel selection calls attention to *organisms'* interactions with their environments and to phenotypical development" (p. 5, italics added). And: "The situatedness of activity makes the unit of analysis the organism-in-setting, where the focus is on the relation between entity and context, or person and setting" (p. 6). Further, Bateson (2013) contends that "while the behavior of whole animals can be informed by knowledge of the underlying mechanisms, the process of reassembly can only be conducted at the level of the whole organism." He goes on to state that "referring to genes as being adapted to the environment no longer makes any sense. Adaptation is at the level of the phenotype" (p. 8).

Given the central role of selection in both evolution and learning science and the increased emphasis on learning, language, and cultural influences on human evolution, it is clear why CBS wants to situate the behavioral sciences under the cover of evolution science. There is an inherent reciprocal relation between the two disciplines. Learning, symbolic behavior, and culture are themselves evolutionary processes and should be understood and investigated in the larger context of evolution science. In addition to being inheritance systems in their own right, learning, symbolic behavior, and culture clearly influence evolution, and a deeper understanding of these processes is critical to advancing evolution science.

A Contextual Science of Learning

A contextual science of learning situated within evolution science would have much to gain and much to offer. Given the increasingly recognized importance of behavioral, symbolic, and cultural dimensions of evolution, it is clear that a developed understanding of human learning and language is critical to the advancement of human evolution science. Indeed, if Ginsburg and Jablonka's (2010) contention that contingency learning is one of the key factors that drove the so-called "Cambrian explosion," then it follows that operant conditioning itself is a key factor of evolution. Conversely, if learning and language processes are themselves evolutionary adaptations, then it is critical that they be understood within an evolution science context.

An excellent example of a potential contextual behavioral science contribution to evolution science is the increasingly supported contextualistic theory of language and cognition called relational frame theory (RFT) (Hayes, et al., 2001; Dymond & Roche, 2013). A full description is beyond the scope of this chapter, but the core of RFT is based on a special learned-response repertoire called derived relational responding. What differentiates derived relational responding from other operant behavior is that once relating repertoires are acquired, they allow for the contextually controlled derivation of untrained relations among stimuli and the untrained acquisition of stimulus functions by those stimuli. In short, it provides an empirically supported functional account of the acquisition and development of symbolic behavior. Monestès (2016) offers an excellent description of how RFT provides a compelling platform for the experimental analysis of Jablonka and Lamb's (2014) fourth dimension in evolution, language, and symbolic behavior.

On the flip side, a clear example of the benefits to contextual learning science afforded by an integration with evolution science can be seen in a recent paper by D. S. Wilson and colleagues (2014), who attempt to link derived relational responding to evolutionary concepts. With respect to human symbolic behavior and its role as an inheritance system, these authors liken the role of symbolic (derived relational) networks to that of a genotype. They label these symbolic networks "symbotypes" to emphasize the comparison where, once established, every symbotype leads to a suite of behaviors that potentially influence survival and reproduction in the same way that genotypes produce phenotypes. Like genotypes, symbotypes survive and evolve based on the behavior they produce in the real world. Also like genotypes, the elements of symbotypes can recombine exponentially, resulting in thousands of derived relations from just a few directly trained verbal relations. This process results in an almost infinite capacity for open-ended behavioral and cultural change. These authors go on to outline how the conceptual integration within CBS of findings from a disparate range of behavioral science disciplines can result in an effective science of intentional change.

Another example of the benefits of imbedding learning in contemporary evolution science is provided by Hayes and Sanford (2014). Although RFT was first described as operant behavior resulting from many multiple exemplars, it soon became clear that an adequate account of the development of relational responding itself required an account of the selection pressures that would produce the genetic differences that allow multiple exemplar training to produce the effects it does, and also an account of the emergence of a verbal community that could provide that training. Appealing to recent developments in evolution science, especially those concerned with between-group competition and multilevel selection, these authors conclude that intraspecies cooperation is an evolutionary prerequisite for the development of human language and cognition. It is only in the context of a cooperative species that operant contingencies that produce derived relational repertoires could be effective.

As stated earlier, the goals of CBS are to develop a multilevel, multidimensional science characterized by precision, scope, and depth. Within both learning and evolution science there is already a good deal of precision and depth. The integration of learning and evolution sciences would add to the scope of both. To be sure, there will be obstacles to overcome in developing an interdisciplinary approach to an integrated science of learning and evolution, not the least of which are intradisciplinary and institutional expectations and practices. But the consilience and theoretical interdependence that could result seems well worth the effort.

References

Barkow, J. H., Cosmides, L., & Tooby, J. (Eds.). (1992). *The adapted mind: Evolutionary psychology and the generation of culture.* Oxford: Oxford University Press.

Bateson, P. (2013). Evolution, epigenetics and cooperation. *Journal of Biosciences, 38,* 1–10.

Baum, W. B. (2012). Rethinking reinforcement: Allocation, induction, and contingency. *Journal of the Experimental Analysis of Behavior, 97,* 101–104.

Bechtel, W. (2008). Mechanisms in cognitive psychology: What are the operations? *Philosophy of Science, 75,* 983–994.

Beecher, M. D. (1988). Some comments on the adaptationist approach to learning. In R. C. Bolles & Beecher, M. D. (Eds.), *Evolution and learning.* Hillsdale, NJ: Lawrence Erlbaum.

Bell, M. A., Futuyma, D. J., Eanes, W. F., & Levinton, J. S. (2010). *Evolution since Darwin: The first 150 years.* Sunderland, MA: Sinauer Associates.

Bolhuis, J. J., Brown, G. R., Richardson, R. C., & Laland, K. N. (2011). Darwin in mind: New opportunities for evolutionary psychology. *PloS Biology, 9.* doi:10.1371/journal.pbio .1001109

Bolles, R. C., & Beecher, M. D. (1988). *Preface.* In R. C. Bolles & M. D. Beecher (Eds.), *Evolution and learning.* Hillsdale, NJ: Lawrence Erlbaum.

Buller, D. J. (2005) *Adapting minds: Evolutionary psychology and the persistent quest for human nature.* Cambridge, MA: MIT Press.

Buss, D. M. (1999). *Evolutionary psychology.* Boston: Allyn & Bacon

Caporael, L. (2001). Evolutionary psychology: Toward a unifying theory and a hybrid science. *Annual Review of Psychology, 52,* 607–628.

Catania, A. C. (2013). *Learning* (5th ed.). Cornwall-on-Hudson, NY: Sloan Publishing.

Chamizo, V. D., Aznar-Casanova, J. A., & Artigas, A. A. (2003). Human overshadowing in a virtual pool: Simple guidance is a good competitor against locale learning. *Learning and Motivation, 34,* 262–281.

Corning, P. A. (2014). Evolution "on purpose": How behavior has shaped the evolutionary process. *Biological Journal of the Linnean Society, 112,* 242–260.

Cziko, G. (1995). *Without miracles: Universal selection theory and the second Darwinian revolution.* Cambridge, MA: MIT Press.

De Houwer, J., Barnes-Holmes, D., & Moors, A. (2013). What is learning? On the nature and merits of a functional definition of learning. *Psychonomic Bulletin and Review, 20,* 631-642.

Dickins, T. (2003). General symbol machines: The first stage in the evolution of symbolic communication. *Evolutionary Psychology, 1,* 192–209.

Domjan, M. (2010). *The principles of learning* (7th ed.). Boston: Cengage.

Domjan, M., & Hollis, K. L. (1988). Reproductive behavior: A potential model system for adaptive specialization in learning. In R. C. Bolles & M. D. Beecher (Eds.), *Evolution and learning.* Hillsdale, NJ: Lawrence Erlbaum.

Donahoe, J. W. (1996). On the relation between behavior analysis and biology. *Journal of the Experimental Analysis of Behavior, 19,* 71–73.

Donahoe, J. W. (2012). Reflections on behavior analysis and evolutionary biology: A selective review of *Evolution since Darwin: The first 150 years. Journal of the Experimental Analysis of Behavior, 97,* 249–260.

Donahoe, J. W., & Palmer, D. C. (1994). *Learning and complex behavior.* Boston: Allyn & Bacon.

Dymond, S., & Roche, B. (2013). *Advances in relational frame theory: Research and applications.* Oakland, CA: Context Press.

Ginsburg, S., & Jablonka, E. (2010). The evolution of associative learning: A factor in the Cambrian explosion. *Journal of Theoretical Biology, 266,* 11–20.

Hamilton, D. A., Rosenfelt, C. S., & Whishaw, I. Q. (2004). Sequential control of navigation by locale and taxon cues in the Morris water task. *Behavioural Brain Research, 154,* 385–397.

Hayes, S. C., Barnes-Holmes, D., & Roche, B. (2001). *Relational frame theory: A post-Skinnerian account of human language and cognition.* New York: Plenum.

Hayes, S. C., Barnes-Holmes, D., & Wilson, K. G. (2012). Contextual behavioral science: Creating a science more adequate to the challenge of the human condition. *Journal of Contextual Behavioral Science, 1,* 1–16.

Hayes, S. C., & Sanford, B. T. (2014). Cooperation came first: Evolution and human cognition. *Journal of the Experimental Analysis of Behavior, 101,* 112–129.

Hull, C. L. (1943). *Principles of behavior: An Introduction to behavior theory.* New York: Appleton-Century-Crofts.

Jablonka, E., & Lamb, J. (2014). *Evolution in four dimensions: Genetics, epigenetics, behavioral and symbolic variation in the history of life* (2nd ed.). Cambridge, MA: MIT Press.

Klein, S. B., Cosmides, L., Tooby, J., & Chance, S. (2002). Decisions and the evolution of memory: Multiple systems, multiple functions. *Psychological Review, 109,* 306–329.

Knierim, J. J., & Hamilton, D. A. (2011). Framing spatial cognition: Neural representations of proximal and distal frames of reference and their roles in navigation. *Physiological Review, 91,* 1245–1279.

Kokko, H., & Jennions, M. D. (2010). Behavior ecology: The natural history of evolutionary biology. In M. A. Bell, W. F. Eanes, & D. F. Futuyma (Eds.), *Evolution since Darwin: The first 150 years*. Sunderland, MA: Sinauer.

Mackintosh, N. J. (2002). Do not ask whether they have a cognitive map, but how they find their way about. *Psicologia, 23*, 165–185.

Monestès, J.-L. (2016). A functional place for language in evolution. In R. D. Zettle, S. C. Hayes, D. Barnes-Holmes, & A. Biglan (Eds.), *The Wiley handbook of contextual behavioral science*. West Sussex, UK: Wiley-Blackwell.

Morris, E. K., Lazo, J. F., & Smith, N. G. (2004). Whether, when and why Skinner published on biological participation in behavior. *The Behavior Analyst, 27*, 153–169.

Naour, P. (2009). *E. O. Wilson and B. F. Skinner: A dialogue between sociobiology and radical behaviorism*. New York: Springer.

O'Keefe, J., & Dostrovsky, J. (1971). The hippocampus as a spatial map: Preliminary evidence from unit activity in the freely-moving rat. *Brain Research, 34*, 171–175.

O'Keefe, J., & Nadel, L. (1978). *The hippocampus as a cognitive map*. Oxford: Oxford University Press.

Ormrod, J. E. (2012). *Human learning* (6th ed.). London: Pearson.

Papini, M. R. (2002). Pattern and process in the evolution of learning. *Psychological Review, 109*, 186–201.

Penn, D. C., Holyoak, K. J., & Povinelli, D. J. (2008). Darwin's mistake: Explaining the discontinuity between human and nonhuman minds. *Behavioral and Brain Sciences, 31*, 109–130.

Pinker, S. (2002). *The blank slate: The modern denial of human nature*. New York: Viking.

Rachlin, H. (2014). *The escape of the mind*. New York: Oxford University Press.

Schneider, S. M. (2012). *The science of consequences: How they affect genes, change the brain, and impact our world*. Amherst, NY: Prometheus Books.

Seligman, M. E. (1970). On the generality of the laws of learning. *Psychological Review, 77*, 406–18.

Skinner, B. F. (1938). *The behavior of organisms*. New York: Appleton-Century-Crofts.

Skinner, B. F. (1974). *About behaviorism*. New York: Random House.

Skinner, B. F. (1981). Selection by consequences. *Science, 213*, 501–504.

Symons, D. (1979). *The evolution of human sexuality*. New York: Oxford University Press.

Tolman, E. C. (1938). The determiners of behavior at a choice point. *Psychological Review, 45*, 1–41.

Watson, J. B. (1907). Kinesthetic and organic sensations: Their role in the reaction of the white rat to the maze. *Psychological Monographs, 8*, 1–101.

West-Eberhard, M. J. (2003). *Developmental plasticity in children*. New York: Oxford University Press.

Wilson, D. S., Hayes, S. C., Biglan, A., & Embry, D. (2014). Evolving the future: Toward a science of intentional change. *Behavioral and Brain Sciences, 37*, 395–460.

Wilson, K. G. (2016). Contextual behavioral science: Holding terms lightly. In R. D. Zettle, S. C. Hayes, D. Barnes-Holmes, & A. Biglan (Eds.), *The Wiley handbook of contextual behavioral science*. West Sussex, UK: Wiley-Blackwell.

Zettle, R. D., Hayes, S. C., Barnes-Holmes, D., & Biglan, A. (Eds.). (2016). *The Wiley handbook of contextual behavioral science*. West Sussex, UK: Wiley-Blackwell.

CHAPTER 3

Classical and Operant Conditioning: Evolutionarily Distinct Strategies?

Zohar Z. Bronfman
Tel-Aviv University

Simona Ginsburg
Open University of Israel

Eva Jablonka
Tel-Aviv University

Over a century of scientific research has been devoted to investigating associative learning (AL). At a very early stage in this endeavor, a fundamental distinction was drawn between classical (Pavlovian) and operant/instrumental (Skinnerian/Thorendikian) learning. Although the nature (and merit) of this distinction was debated during the first half of the twentieth century, it eventually became one of the most basic tenets of AL studies, taken for granted by many contemporary scholars. We present a brief history of the distinction, examine its status from an evolutionary perspective, and attempt to determine whether the two types of conditioning can indeed be regarded as distinct learning systems. We use three criteria, based on suggestions made by learning and memory theorists (Tulving, 1985; Sherry & Schacter, 1987; Schacter & Tulving, 1994; Foster & Jelicic, 1999), that can be considered to be jointly sufficient for drawing a distinction between different systems of learning and memory, and apply them to the classical/operant distinction. Our analysis suggests that, especially when viewed from an evolutionary perspective, it is not clear that classical and operant conditioning should be considered as distinct learning and memory systems. We conclude by offering alternative, complementary distinctions between forms of AL that are based on the notion of *learning breadth*: the degree to which learning encompasses complex stimuli, actions, and values. We suggest that two progressive evolutionary changes involving changes in control hierarchy occurred during the evolutionary history of AL.

The Distinction Between Classical and Operant Learning: Early and Current Arguments

At the very end of the nineteenth century and the first half of the twentieth century, associative learning became a field of experimental study and rigorous formulation. Ivan Pavlov in Russia (Pavlov, 1927) and Edward Thorndike and B. F. Skinner in the United States (Thorndike, 1898; Skinner, 1938) studied and documented the manner by which animals associate stimuli, actions, and values as a consequence of their ontogenetic experience, in order to adapt their behavior to the environment.

Pavlov's theory, known today as classical (or "respondent") conditioning, entails the formation of an association between a conditioned stimulus (CS) and an unconditioned stimulus (US). A US is defined as a stimulus that elicits a reflex response (an unconditioned response, UR); for example, the smell of food innately elicits salivation. Conversely, the sound of a metronome does not elicit salivation prior to learning and is therefore considered a CS. The CS-US association is formed when the US repeatedly follows the CS, usually in close temporal proximity. Due to the formation of the CS-US association, a conditioned response (CR) is acquired: the next presentation of the CS will elicit the CR, even in the absence of the US. Thus, the organism has learned to respond to the CS (e.g., to salivate) as if the US were about to arrive.

Thorndike's and Skinner's theory, known today as operant conditioning (also termed Thorndike's law of effect, or instrumental conditioning), describes situations in which the probability for eliciting a certain action changes as a function of its reinforcing history: actions that were followed by a positive (or negative) outcome will be more (or less) likely to occur in the future, under similar circumstances. For example, a rat's tendency to press a lever will either increase or decrease if this act is repeatedly followed by delivery of food or electric shock, respectively. Skinner (1938), who developed this approach, claimed that operant (emitted) behavior is shaped by reinforcement. He argued that a stimulus (e.g., a flash of light signaling that pushing the lever may produce food) comes to control an operant if the response (pushing the lever) is reinforced, and that complex behavior is the result of a sequence of stimuli, so that a discriminative stimulus not only sets the occasion for subsequent behavior, but can also reinforce a behavior that precedes it.

Although operant and classical associative learning are differently operationalized and defined, in the early days of their study, several theorists, including Pavlov and Skinner, attempted to clarify the relations between them. Pavlov and several other Russian scientists regarded the two types of conditioning as based on the same Pavlovian principles, assuming that there are bi-directional excitatory and inhibitory connections between the neural centers controlling stimuli (CS and US) and responses (CR and UR) that differ in their relative strength and temporal coincidence (reviewed in Gormezano & Tait, 1976). Hence, in some cases, an innately elicited UR can lead to the stimulation of a reflex, thus implementing instrumental learning. Some American psychologists, on the other hand, explained Pavlovian conditioning as a case of instrumental/operant conditioning that occurs when the consequences of the US are reinforced by the CR (Kimmel, 1965; Smith, 1954; see also review in Gormezano & Tait, 1976).

Another position held by many psychologists regarded the two modes of learning as based on two distinct processes. For example, in a 1935 paper, Skinner distinguished between operant and Pavlovian conditioning. He argued that the essence of Pavlovian conditioning is "the substitution of one stimulus for another, or, as Pavlov has put it, signalization. It prepares the organism by obtaining the elicitation of a response before the original stimulus has begun to act." In operant learning, however, there is no signalization: "the organism selects from a large repertory of unconditioned reflexes[1] those of which the repetition is important with respect to certain elementary functions and discards those of which it is unimportant." The conditioned response in this latter case does not prepare for the reinforcing stimulus, but rather produces it.

Although subject to many qualifications (see Rescorla & Solomon, 1967), during the decades that followed, the distinction between classical and operant learning that was suggested by Skinner became more acceptable. This is evident from the literature discussing the evolution of associative learning, where it is usually assumed that classical and operant conditioning are distinct systems, and that (the more complex) operant learning evolved from classical conditioning (Wells, 1968; Razran, 1971; Moore, 2004; Perry, Barron, & Cheng, 2013).

There was also recognition of the fundamental similarities shared by the two types of conditioning (e.g., Donahoe, Burgos, & Palmer, 1993). Both types of AL

---

1 Therefore, according to Skinner, a new reflex may be created in Pavlovian conditioning, since practically any stimulus may be attached to the reinforcement. Conversely, operant conditioning is not a device for increasing the repertoire of reflexes; the response continues to be elicited by the one stimulus with which it began.

entail: (1) perception of a stimulus or stimuli that initiate the process (CS in classical conditioning; the internal state of the exploring animal and the object upon which it acts in operant conditioning); followed by (2) behavior (UR in classical conditioning; the reinforced behavior in operant conditioning); and a signal of (3) value (carried either by the US or the reinforcer) that determines the salience of the elicited response. Moreover, it was also clear that distinguishing operant and classical conditioning does not imply that the two processes do not interact. Attempts to study the interactions, along with experiments showing that classical conditioning modifies instrumental learning, were an important part of learning research (e.g., Rescorla & Solomon, 1967).

What, then, is the status of this distinction? Clearly, there are interesting pragmatic differences between the typical types of responses, reinforcements, and behavioral outcomes of operant and classical learning. But are these differences like those of learning two languages, say Japanese and Hebrew, which entail, no doubt, many differences at several levels (behavioral, neural) but seem, nevertheless, to belong to a single cognitive-functional domain? Or are the differences more like those between the ability to learn to sing and the ability to learn to navigate? Or perhaps more like learning that is based on episodic versus procedural memory or like the difference between nonassociative and associative learning? Below, we examine the status of the distinction between operant and classical conditioning and argue that they are neither distinct learning systems (like declarative and procedural memory-based learning or non-AL and AL) nor specialized types of learning like navigation and song learning, but are different facets of a single AL system.

Criteria for Distinguishing Between Learning and Memory Systems and Their Application to the Operant/Classical Distinction

Cognitive theorists (Tulving, 1985; Sherry & Schacter, 1987; Schacter & Tulving, 1994; Foster & Jelicic, 1999) suggested that distinguishing between learning and memory systems (such as procedural and declarative memory) should be based on individually necessary and jointly sufficient functional-evolutionary criteria. Building on their proposal, we suggest the following three criteria:

1. *Functional distinctiveness*: Distinct forms of learning should adhere to different functional principles and rely on dissociable neural structures. Operationally, functional distinctiveness can be inferred via double dissociations (see also Dunn & Kirsner, 1988).

2. *Taxonomic distinctiveness*: Members of some animal taxa should exhibit one form of learning while members of other animal taxa should exhibit the other (if the assumption that secondary loss had occurred is deemed unlikely). In addition, the distinction should yield explicit predictions regarding the learning abilities of as yet untested organisms.

3. *Adaptive-evolutionary distinctiveness*: Distinct forms of learning should have distinct evolutionary adaptive rationales that can account for qualitative variations in learning abilities between members of different taxa. For example, bird-song learning and navigation learning are specialized forms of learning that cannot be reduced to one another (they "solve" different selection pressures; Sherry & Schacter, 1987). The learned information is assumed to be distinct in each case.

Based on these criteria, we now turn to analyze the classical/operant distinction.

Are classical and operant learning functionally distinct?

The two forms of learning share fundamental learning principles. As discussed above, in both forms, the reinforcement should follow the CS or the action (e.g., Skinner, 1938; Ferster & Skinner, 1957). In addition, both forms are governed by the prediction-error learning principle, according to which only unpredicted reinforcement gives rise to learning (Kamin, 1968, 1969; Rescorla & Wagner, 1972; Schultz, Dayan, & Montague, 1997; Clark, 2013).

In spite of these similarities, there are some empirical reports of dissociations between the two forms of learning, at several levels of biological organization. Studies in rodents showed that lesions in the orbitofrontal cortex impair Pavlovian conditioning, while lesions in the prelimbic cortex interfere with operant conditioning (Ostlund & Balleine, 2007). At the cellular level, at least in *Aplysia*, studies reveal that both forms of conditioning are accompanied by different effects on the membrane excitability of a critical neuron (B51) that is involved in the two types of conditioning (Brembs, Lorenzetti, Reyes, Baxter, & Byrne, 2002; Lorenzetti, Mozzachiodi, Baxter, & Byrne, 2006; Mozzachiodi & Byrne, 2010).

Nonetheless, the studies showing dissociations between classical and operant learning await replication, and there are too few such studies to allow any significant conclusion, especially given that null-effects (presumably reporting lack of dissociations) tend not to be published. For comparison, hundreds, if not thousands, of papers have addressed the issue of whether other forms of learning—recollection and recognition—are dissociable (reviewed in Yonelinas, 2002). Thus, future studies must address this important question, and provide much needed data about whether classical and operant conditioning are indeed functionally dissociable.

Perhaps one of the greatest difficulties in distinguishing between operant and classical learning is that both take place during the same learning episode (even in humans; see Staats, Minke, Martin, & Higa, 1972). The difficulty can be partially overcome only if one takes special measures to isolate the effects of predictive sensory stimuli from the associated reinforcing effects of the animal's actions. Indeed, a prominent study (Brembs & Plendl, 2008) took such measures using the flight simulator paradigm (Wolf & Heisenberg, 1991). In this paradigm, a single *Drosophila* fly is glued to a small hook and attached to a torque meter, thus "flying" stationarily in the center of a cylindrical panorama. Due to this design, the experimenter has full control over the perceptual stimuli to which the fly is exposed (e.g., blue/green color), as well as over the relation between the fly's movements (e.g., turn left/right) and the reinforcement it receives (e.g., heat).[2] Hence, under these unique circumstances, it is possible to study the two types of learning separately. Brembs showed that learning about the external perceptual stimuli (a proxy of classical conditioning) and learning about one's own movements (a proxy of operant conditioning) were differentially impaired by genetic manipulations of adenylyl cyclase and protein kinase C, respectively (Brembs & Plendl, 2008). On the basis of these studies, Brembs (2011) suggests that instead of the traditional distinction based on the processes involved in each type of learning, a classification based on what is learned should be adopted: "Today, we can propose a terminology to better distinguish what is learned (stimuli or behavior) from how it is learned (by classical or operant conditioning). We define self-learning as the process of assigning value to a specific action or movement. We define world-learning as the process assigning value to sensory stimuli. While only world-learning occurs in classical conditioning experiments, both processes may occur during operant conditioning" (Colomb & Brembs, 2010, p. 142).

We believe that the distinction suggested by Brembs and colleagues is useful, although in most conditions, including most classical conditioning experiments, animals learn both about the world and their own behavior, a point that was stressed by early psychologists (reviewed in Rescorla & Solomon, 1967). Consider a case of classical conditioning when a particular cue (e.g., a particular smell) that predicts the presence of a prey elicits in the animal a reflex biting reaction (the

---

2 Note, however, that even in the absence of changes in external stimuli, internal perception (termed interoception) may vary.

UR). However, the actual reflex response must be tailored to the specific prey that the cue predicts (its size, its texture, etc.). The animal learns both which cue predicts the prey and the modified UR (the CR) that the cue elicits. More generally, when the CR is not identical to the UR (it is a modification of it that is specific to the eliciting CS), which is very often the case when the UR is a locomotor-pattern, learning the CR is part of what the animal learns.

Hence, we believe that the assumption that classical conditioning requires only world-learning does not apply in many natural conditions where both self- and world-learning co-occur, although the animal's dependence on world- and self-learning may vary in important ways. Future studies should clarify whether and how the distinction between world- and self-learning applies to organisms other than the fruit fly, and under which ecological conditions such dissociations are functionally significant.

Another consideration that is relevant to the distinction between self (usually motor-behavior based) and world (perception based) learning is the codependence between perception and motor response: even when an animal is still and the object of perception is stationary, perception involves motor action. Except for some special conditions, animals do not passively perceive their world. In general, they perceive and explore the world through active movements of the body and/or its sensory organs, with motor and sensory variables being players of equal importance (Ahissar & Vaadia, 1990; Ahissar & Arieli, 2001). The idea that perception is passive is probably one reason for the common assumption that (active) operant conditioning evolved from (the more passive) classical conditioning (Moore, 2004; Perry et al., 2013). The notion that the sensorimotor system is evolutionarily entangled is compatible with the increasingly influential conception of the sensorimotor system as a unified system, where inputs—external (exteroceptive), action-oriented (proprioceptive), and internal (interoceptive)—are selected by the animal according to their compatibility with fitness-promoting states (Clark, 2013).

Are classical and operant learning taxonomically distinct?

We are unaware of any organism in which both types of learning were sought that possesses one type of learning and not the other. For example, the nematode, whose learning abilities are relatively simple, shows both forms of learning (Rankin, 2004). Similarly, *Aplysia* has been shown to possess both classical (Carew, Walters, & Kandel, 1981) and operant (Brembs et al., 2002) learning. We believe, though, that the question of such taxonomic independence requires further study, since current knowledge about the learning abilities of most animal phyla is very poor. Any observation of animals that can learn only via operant learning or only through classical conditioning will support the hypothesis that operant and classical conditioning are distinct learning systems, at least in some taxa.

Do the two forms of conditioning meet distinct evolutionary challenges?

Associative learning seems to be a domain-general adaptation *par excellence* (Macphail, 1982; Heyes, 2012), and it is difficult to define distinct "evolutionary challenges" to which operant and classical AL specifically rise, since both appear to meet most challenges. Clearly, both classical and operant learning rise to the general challenge of matching the most appropriate action to specific (ontogenetically encountered) circumstances, by means of associating actions, circumstances, and value. For example, consider a wolf that, during foraging, learns that the smell coming from the bush is usually followed by a hopping rabbit crossing the shrub. The next time the smell is encountered, the wolf will expect a rabbit, and will therefore prepare for an onslaught (e.g., leap forward, increase heart rate), as per classical conditioning. This preparation will render the wolf more likely to capture the rabbit. Thus, by associating the smell (CS) with the hopping rabbit (US), the wolf can maximize its response to the rabbit (which carries, as a prey, a salient positive motivational value). This scenario also comprises operant aspects—the wolf must choose the most appropriate action for making the onslaught most efficient (e.g., leap versus crawl). The value of the chosen action (and of competing ones) will be determined by the history of the wolf's previous operant conditioning (to the context, rabbits, etc.), where each action was more or less likely to result in achievement of the goal (in this case, capturing the rabbit).

Nonetheless, on a finer-grained scale, it could be argued (as per Skinner's hypothesis) that operant conditioning meets the challenge of actively changing the organism's state (e.g., from hungry to satiated), while classical conditioning entails the ability of the organism to better prepare itself for future events (which it cannot control). However, as we stressed earlier when discussing self- and world-learning, under natural, ecological conditions, these two "challenges" systematically coexist, and the system is likely to have evolved to reflect this entanglement.

We conclude that it is not clear that the distinction between operant and classical conditioning fulfills all the criteria necessary for deducing that the two types of AL are distinct learning systems. There are a few studies that report dissociations between operant and classical learning, yet these are too few, given the importance of this question. Very little is known about the learning abilities of relatively simple organisms, yet to the best of our knowledge, those that were studied show both forms of conditioning. Finally, it is far from clear that the evolutionary challenges that classical and operant learning "solve" are distinct, especially in natural ecological settings, in which both world- and self-learning usually co-occur.

A Complementary Evolutionary Classification

The distinction between operant and classical conditioning is still not scientifically established after a century of research into AL. Does this mean that AL, which first appeared in the early Cambrian (Ginsburg & Jablonka, 2010a, 2010b), remained essentially unchanged ever since, and that the only difference in learning among animals is to be attributed to changes in the sensory, motor, and motivational systems, as Macphail (1982, 1987) argued? Should we focus, alternatively, on domain-specific adaptations in learning and abandon the attempt to discern domain-general adaptations (Shettleworth, 1998; see review by Papini, 2002)? Or can we adopt a third option—an evolutionary-transition-oriented approach (Maynard Smith & Szathmary, 1995)—and seek qualitative progressive change within the framework of the domain-general ability for learning by association? We adopt this third alternative, which, we maintain, can elucidate the nature of the relations between AL and other cognitive capacities and clarify questions concerning the distribution of AL in the animal world. We suggest two qualitative stages leading to increasingly broad (flexible) AL, which include both classical/world learning and operant/self learning:

1. Limited AL, in which only preexisting (evolved) reactions can be modified by experience. Limited AL relies on the evolution of temporal signaling, resulting in a transition from general sensitization, where the order of the stimuli presented does not affect the response, to learning dynamics, which lead to a stronger and more enduring response when the CS precedes the US (also known as alpha conditioning) or when the reinforcer follows exploratory activity. An example of alpha conditioning was described by Carew and colleagues (1981). They paired a weak touch to the *Aplysia* siphon, which elicited a feeble reflex motor response of gill-withdrawal (a CS), with a subsequent strong electric-shock to the tail, which resulted in strong gill-withdrawal (a US). This CS-US sequence of pairing led to the weak siphon stimulation yielding a strong gill withdrawal (CR), significantly stronger than a random pairing between the two stimuli, or reverse pairing, when the strong stimulus preceded the weak one (i.e., US-CS

sequence). Similarly, Brembs and colleagues (2002) showed that *Aplysia* can undergo limited operant conditioning by learning to increase its propensity for a certain innate motor behavior (biting) when this behavior is immediately followed by a positive reinforcing stimulus (electric stimulation of reinforcing dopamine neurons; see also Cook & Carew, 1986). The simplest types of updating, based on the mismatch between (or the surprising effect of) incoming signals and pre-existing networks states, are apparent at this level and manifest in blocking and reafference. Blocking (Kamin, 1969) is the impairment of formation of associations if, during the conditioning process, a CS preceding a US is presented together with another CS that has already been associated with, and is fully predictive of, the unconditioned stimulus. As noted earlier, this dependence on the surprise value of the CS is termed the prediction error principle (Rescorla & Wagner, 1972; Schultz et al., 1997; Schultz, 1998; Schultz & Dickinson, 2000). Reafference is the modulation of sensory processing by motor command, so that the sensory effects of one's own action can be distinguished from those coming from the world (reviewed in Crapse & Sommer, 2008). Reafference is necessary for all large moving animals, while blocking has been reported in *Aplysia californica*, Planarians, and moth larvae (Pszczolkowski & Brown, 2005). Limited AL, which generally includes association between any elemental (noncomposite) stimulus and behavior, is presumably quite common, although our knowledge about learning in many animal phyla is at present very scant.

2. Progressive evolution led to Unlimited Associative Learning, in which a compound, non-reflex-eliciting input comprising a specific conjunction of several features (e.g., a blue, sunflower-like large shape) or a specific non-reflex sequence of actions (e.g., pressing a key and then moving left) can be associated with value and support second-order conditioning. Thus, the evolution of unlimited AL has gone hand in hand with the evolution of object recognition and categorization, as well as with gist perception. Moreover, it has been repeatedly argued that binding necessitates top-down attention (Treisman & Gelade, 1980), and we suggest that unlimited AL involves increasingly higher levels of bi-directional processing and control: updating of past neural representations in the light of new surprising information occurs through mechanisms that go beyond lateral inhibition of non-reinforced sensory processing. In other words, the emergence of

full-fledged unlimited AL involved the evolution of the recall of bound stimuli through perceptual fusion and top-down attention that relied on multilevel hierarchical predictive coding (Rao & Ballard, 1999; Friston, 2005; Clark, 2013). Lastly, and as we elaborate below, unlimited AL entails minimal consciousness since it presupposes the functional and structural attributes deemed essential to minimal consciousness by philosophers, psychologists, and neurobiologists (Bronfman, Ginsburg, & Jablonka, 2016a, 2016b).

As we argue below, these forms of AL comply with the above criteria and can be regarded as distinct learning-memory systems, although, as is inevitable with most evolution-based distinctions, gray areas, which defy classification, exist between these transitions.

Why should our suggested distinction between limited and unlimited associative learning matter in a theoretical or a practical sense? While there are, we believe, many functional implications to the suggested distinction between limited and unlimited AL, we wish to focus here on one (major) implication, which is both theoretical and practical. We have argued in detail elsewhere (Bronfman et al., 2016a) that the evolutionary transition from limited AL to unlimited AL marks the evolutionary transition to sentience, or basic consciousness. In a nutshell, our argument is that the enabling system of unlimited AL—the sum total of mechanisms and dynamics underlying it—exhibits all the properties that are deemed individually necessary and jointly sufficient for basic consciousness, and that the distribution of unlimited associative learning among animals is compatible with independent suggestions based on the functional anatomy of the brain advanced by neurobiologists, cognitive scientists, and evolutionary biologists to explain minimal consciousness (Merker, 2007; Barron & Klein, 2016; Feinberg & Mallatt, 2016).

If UAL is indeed a plausible evolutionary marker of minimal consciousness, our suggestion bears important implications. First, second-order conditioning of compound stimuli and actions is expected only when the animal is conscious/aware. This is supported by the finding that masked, unconsciously perceived, and novel CSs, such as pictures of exotic flowers or mushrooms, do not give rise to conditioning, while masked angry faces do (Öhman & Soares, 1993). Studies of this type need to be extended to test for second-order conditioning of novel and already-learned compound stimuli and action in both humans and animals (whose ability to become conditioned to masked stimuli of different types has not been investigated). Such studies in animals from different taxonomic groups can provide insights regarding the distribution of minimal consciousness in the animal kingdom. At the more theoretical level, the study of functional models of unlimited AL can aid in further characterizing and reverse-engineering the mechanisms underlying minimal consciousness (discussed in Bronfman et al., 2016a, 2016b).

Are limited AL and unlimited AL functionally distinct?

Unlimited AL depends on the addition of new learning principles and controlling neural structures to the limited AL system. Limited AL relies on the signaling of temporal contiguity and elementary predictive coding, while unlimited AL relies—in addition—on perceptual fusion and categorization and multilevel hierarchical predictive coding. At the anatomical level, we expect limited AL to require a brain with integrating circuits (minimally, circuits of interneuron intervening between the sensory units and sensory and motor units, such as the circuits apparent in the cerebral ganglia of *Aplysia*). Unlimited AL requires, in addition, specialized learning-supporting structures like the midbrain ventral tegmental area in mammals (which encompasses dopaminergic neurons that are necessary for the signaling of the prediction error) and the central complex in insects, as well as organized regions responsible for memory, perceptual fusion, and pattern completion and separation, such as the mushroom bodies and hippocampus in insects and mammals respectively (Bronfman et al., 2016b).

Are limited AL and unlimited AL taxonomically distinct?

Existing evidence strongly suggests that limited AL and unlimited AL are taxonomically distinct (although the studies on the learning abilities of different animals have not been tailored to our criteria). There are organisms that possess only limited AL (e.g., *Aplysia*, Planaria), while a subset exhibits unlimited AL as well (all vertebrates, many arthropods, and some mollusks; Bronfman et al., 2016b). We suggest that limited AL started evolving during the early Cambrian era as body size and the correlated lifespan increased.

Do the two forms of AL meet distinct evolutionary challenges?

The two forms of learning seem to enable solutions to increasingly complex conditions. Limited AL allows "action recombination": for the first time, different innate responses (and the circuits underlying them) can be associated with each other via learning, thus enriching the behavioral repertoire of the animal and generating a new combinable functional module. Moreover, as illustrated by blocking and reaference effects, animals can disregard irrelevant information, an important adaptation for animals that live in an inherently changing world. By learning only conjunctions, unlimited AL also solves the overlearning problem: while with limited AL individual stimuli that comprise the learned object (e.g., a large, dark, rapidly moving object) will each trigger the learned response (Ginsburg & Jablonka, 2010b), an animal having unlimited AL can fine-tune its responses and respond only to the conjunction of these stimuli, to the compound and fused precept, thus avoiding overresponsiveness to specific (and nondiagnostic) features. We suggest that the evolution of recall of bound features, which involves a fusion of the various constituents of the percept, and that coevolved with hierarchical predictive coding, may have been driven by the problems of overlearning. We proposed that the evolution of these forms of AL was a factor in the Cambrian explosion (Ginsburg & Jablonka, 2010a, 2010b), with limited AL present already during the early Cambrian, and unlimited AL evolving during later Cambrian stages.

Our classification of AL seems to conform to all the criteria that were suggested to distinguish between learning and memory systems. In addition to these qualitative evolutionary changes, during the last 540 million years of animal evolution, which involved several major adaptive radiations and many specializations in different animal taxa, the extent and scope of limited AL and unlimited AL may also have been modulated in a taxon-specific manner according to the sensory, motor, and motivational constraints and affordances experienced by members of different species in their specific niches. Progressive learning evolution did not, of course, end with the evolution of unlimited AL. The evolution of episodic memory in several groups and the evolution of symbol-based learning in humans were additional qualitative, progressive evolutionary changes. The discussion of their evolution is, however, beyond the scope of this chapter (but see chapters 4 and 5 of the present volume).

Summary and Conclusions

Following a century of effort, the distinction between classical and operant conditioning remains an open question. We used three functional and evolutionary criteria to examine this distinction, and found that there is little evolutionary justification for it: we are unaware of any organism that possesses one form of learning but not the other, and there is little reason to believe that under natural, ecological conditions the two forms of learning occur separately. Nonetheless, several studies that showed partial dissociation between the two forms of learning support the distinction to a limited extent. The pioneering work of Brembs and colleagues (Brembs & Plendl, 2008; Colomb & Brembs, 2010; Brembs, 2011) is a promising avenue for future exploration of these dissociations.

We offered an alternative, complementary classification of AL, which is motivated mainly by evolutionary considerations. We suggested that AL became increasingly more flexible since it first emerged during the early Cambrian; we proposed a distinction between two major levels of associative learning—limited and unlimited—that are hierarchically and evolutionarily related, but nevertheless functionally and evolutionarily distinct. Our classification offers testable behavioral predictions, and can be used to guide future comparative research on the evolution of learning and memory systems. Furthermore, if validated, it may shed further light on the functional organization associated with each learning system and the neural structures underlying them.

References

Ahissar, E., & Arieli, A. (2001). Figuring space by time. *Neuron, 32*, 185–201.

Ahissar, E., & Vaadia, E. (1990). Oscillatory activity of single units in a somatosensory cortex of an awake monkey and their possible role in texture analysis. *Proceedings of the National Academy of Sciences, 87*, 8935–8939.

Barron, A. B., & Klein, C. (2016). What insects can tell us about the origins of consciousness. *Proceedings of the National Academy of Sciences, 113*, 4900–4908.

Brembs, B. (2011). Spontaneous decisions and operant conditioning in fruit flies. *Behavioural Processes, 87*, 157–164.

Brembs, B., Lorenzetti, F. D., Reyes, F. D., Baxter, D. A., & Byrne, J. H. (2002). Operant reward learning in Aplysia: Neuronal correlates and mechanisms. *Science, 296*, 1706–1709.

Brembs, B., & Plendl, W. (2008). Double dissociation of PKC and AC manipulations on operant and classical learning in Drosophila. *Current Biology, 18*, 1168–1171.

Bronfman, Z., Ginsburg, S., & Jablonka, E. (2016a). The evolutionary origins of consciousness: Suggesting a transition marker. *Journal of Consciousness Studies, 23*, 7–34.

Bronfman, Z., Ginsburg, S., & Jablonka, E. (2016b). The transition to minimal consciousness through the evolution of associative learning. *Frontiers in Psychology.* doi.org/10.3389/fpsyg.2016.01954

Carew, T. J., Walters, E. T., & Kandel, E. R. (1981). Classical conditioning in a simple withdrawal reflex in Aplysia californica. *Journal of Neuroscience, 1*, 1426–1437.

Clark, A. (2013). Whatever next? Predictive brains, situated agents, and the future of cognitive science. *Behavioral and Brain Sciences, 36*, 181–204.

Colomb, J., & Brembs, B. (2010). The biology of psychology: Simple conditioning? *Communicative and Integrative Biology, 3*, 142–145.

Cook, D. G., & Carew, T. J. (1986). Operant conditioning of head waving in Aplysia. *Proceedings of the National Academy of Sciences, 83,* 1120–1124.

Crapse, T. B., & Sommer, M. A. (2008). Corollary discharge across the animal kingdom. *Nature Reviews Neuroscience, 9,* 587–600.

Donahoe, J. W., Burgos, J. E., & Palmer, D. C. (1993). A selectionist approach to reinforcement. *Journal of the Experimental Analysis of Behavior, 60,* 17–40.

Dunn, J. C., & Kirsner, K. (1988). Discovering functionally independent mental processes: The principle of reversed association. *Psychological Review, 95,* 91–101.

Feinberg, T. E., & Mallatt, J. M. (2016). *The ancient origins of consciousness: How the brain created experience.* Cambridge, MA: MIT Press.

Ferster, C., & Skinner, B. F. (1957). *Schedules of reinforcement.* New York: Appleton-Century-Crofts.

Foster, J. K., & Jelicic, M. E. (1999). *Memory: Systems, process, or function?* Oxford: Oxford University Press.

Friston, K. (2005). A theory of cortical responses. *Philosophical Transactions of the Royal Society of London B: Biological Sciences, 360,* 815–836.

Ginsburg, S., & Jablonka, E. (2010a). The evolution of associative learning: A factor in the Cambrian explosion. *Journal of Theoretical Biology, 266,* 11–20. doi:10.1016/j.jtbi.2010.06.017.

Ginsburg, S., & Jablonka, E. (2010b). Experiencing: A Jamesian approach. *Journal of Consciousness Studies, 17,* 102–124.

Gormezano, I., & Tait, R. W. (1976). The Pavlovian analysis of instrumental conditioning. *Integrative Psychological and Behavioral Science, 11,* 37–55.

Heyes, C. (2012). Simple minds: A qualified defence of associative learning. *Philosophical Transactions of the Royal Society of London B: Biological Sciences, 367,* 2695–2703.

Kamin, L. J. (1968). Attention-like processes in classical conditioning. In M. R. Jones (Ed.), *Miami symposium on the prediction of behavior: Aversive stimulation.* Coral Gables: University of Miami Press.

Kamin, L. J. (1969). Predictability, surprise, attention, and conditioning. In B. A. Campbell & R. M. Church (Eds.), *Punishment and aversive behavior.* New York: Appleton-Century-Crofts.

Kimmel, H. (1965). Instrumental inhibitory factors in classical conditioning. In W. F. Prokasy (Ed.), *Classical conditioning: A symposium.* New York: Appleton-Century-Crofts.

Lorenzetti, F. D., Mozzachiodi, R., Baxter, D. A., & Byrne, J. H. (2006). Classical and operant conditioning differentially modify the intrinsic properties of an identified neuron. *Nature Neuroscience, 9,* 17–19.

Macphail, E. M. (1982). *Brain and intelligence in vertebrates.* Oxford: Clarendon Press.

Macphail, E. M. (1987). The comparative psychology of intelligence. *Behavioral and Brain Sciences, 10,* 645–656.

Maynard Smith, J., & Szathmary, E. (1995). *The major transitions in evolution.* Oxford: Oxford University Press.

Merker, B. (2007). Consciousness without a cerebral cortex: A challenge for neuroscience and medicine. *Behavioral and Brain Sciences, 30,* 63–81.

Moore, B. R. (2004). The evolution of learning. *Biological Reviews, 79,* 301–335.

Mozzachiodi, R., & Byrne, J. H. (2010). More than synaptic plasticity: Role of nonsynaptic plasticity in learning and memory. *Trends in Neurosciences, 33*, 17–26.

Öhman, A., & Soares, J. J. (1993). On the automatic nature of phobic fear: Conditioned electrodermal responses to masked fear-relevant stimuli. *Journal of Abnormal Psychology, 102*, 121–132.

Ostlund, S. B., & Balleine, B. W. (2007). Orbitofrontal cortex mediates outcome encoding in Pavlovian but not instrumental conditioning. *Journal of Neuroscience, 27*, 4819–4825.

Papini, M. R. (2002). Pattern and process in the evolution of learning. *Psychological Review, 109*, 186.

Pavlov, I. P. (1927). *Conditioned reflexes: An investigation of the physiological activity of the cerebral cortex.* London: Oxford University Press.

Perry, C. J., Barron, A. B., & Cheng, K. (2013). Invertebrate learning and cognition: Relating phenomena to neural substrate. *Wiley Interdisciplinary Reviews: Cognitive Science, 4*, 561–582.

Pszczolkowski, M. A., & Brown, J. J. (2005). Single experience learning of host fruit selection by lepidopteran larvae. *Physiology and Behavior, 86*, 168–175.

Rankin, C. H. (2004). Invertebrate learning: What can't a worm learn? *Current Biology, 14*, R617–R618.

Rao, R. P., & Ballard, D. H. (1999). Predictive coding in the visual cortex: A functional interpretation of some extra-classical receptive-field effects. *Nature Neuroscience, 2*, 79–87.

Razran, G. (1971). *Mind in evolution.* Boston: Houghton Mifflin.

Rescorla, R. A., & Solomon, R. L. (1967). Two-process learning theory: Relationships between Pavlovian conditioning and instrumental learning. *Psychological Review, 74*, 151–182.

Rescorla, R. A., & Wagner, A. R. (1972). A theory of Pavlovian conditioning: Variations in the effectiveness of reinforcement and nonreinforcement. In A. H. Black & W. F. Prokasy (Eds.), *Classical conditioning II: Current research and theory.* New York: Appleton-Century-Crofts.

Schacter, D. L., & Tulving, E. (1994). *Memory systems 1994.* Cambridge, MA: MIT Press.

Schultz, W. (1998). Predictive reward signal of dopamine neurons. *Journal of Neurophysiology, 80*, 1–27.

Schultz, W., Dayan, P., & Montague, P. R. (1997). A neural substrate of prediction and reward. *Science, 275*, 1593–1599.

Schultz, W., & Dickinson, A. (2000). Neuronal coding of prediction errors. *Annual Review of Neuroscience, 23*, 473–500.

Sherry, D. F., & Schacter, D. L. (1987). The evolution of multiple memory systems. *Psychological Review, 94*, 439.

Shettleworth, S. J. (1998). *Cognition, evolution, and behavior.* Oxford: Oxford University Press.

Skinner, B. F. (1935). Two types of conditioned reflex and a pseudo type. *Journal of General Psychology, 12*, 66–77.

Skinner, B. F. (1938). *The behavior of organisms: An experimental analysis.* New York: Appleton-Century-Crofts.

Smith, J. M. *See* Maynard Smith, J.

Smith, K. (1954). Conditioning as an artifact. *Psychological Review, 61*, 217–225.

Staats, A. W., Minke, K. A., Martin, C. H., & Higa, W. R. (1972). Deprivation-satiation and strength of attitude conditioning: A test of attitude-reinforcer-discriminative theory. *Journal of Personality and Social Psychology, 24,* 178–185.

Thorndike, E. (1898). Some experiments on animal intelligence. *Science, 7,* 818–824.

Treisman, A. M., & Gelade, G. (1980). A feature-integration theory of attention. *Cognitive Psychology, 12,* 97–136.

Tulving, E. (1985). How many memory systems are there? *American Psychologist, 40,* 385.

Wells, M. J. (1968). Sensitization and the evolution of associative learning. In J. Salánki (Ed.), *Neurobiology of invertebrates.* New York: Springer.

Wolf, R., & Heisenberg, M. (1991). Basic organization of operant behavior as revealed in Drosophila flight orientation. *Journal of Comparative Physiology A, 169,* 699–705.

Yonelinas, A. P. (2002). The nature of recollection and familiarity: A review of 30 years of research. *Journal of Memory and Language, 46,* 441–517.

Dialogue on Learning

Participants: Michael J. Dougher, Derek A. Hamilton, Steven C. Hayes, and Eva Jablonka

Steven C. Hayes: Maybe a good place to start is just this odd fact that it doesn't seem like it has been historically a vibrant conversation between those two wings: evolution and learning. Why is that?

Michael J. Dougher: There's an irony to this. On the behavior analytic end, it's surprising to me because Skinner based operant conditioning on evolution, that was the model. He talks about operant conditioning being an ontogenetic adaptation in the same way that evolution is at the phylogenetic level. E. O. Wilson and B. F. Skinner are having conversations between the two of them about how similar the epistemologies are and "Shouldn't we get together to do something here?" but they never did.

Some of that is because when you're doing research in laboratories with pigeons and rats in Skinner boxes, it doesn't really look like the kind of behavior that people are studying in natural environments, and so ecologists aren't particularly interested.

Eva Jablonka: I think that learning was seen as an ontogenetic kind of behavioral adaptation and evolution is dealing with the inheritance of variations that are transmitted between generations, not within a generation. So, although there is a selection principle that is common to both, these processes were seen as very different; one is phylogenetic and the other is ontogenetic. There were theories from the very beginning—such as Herbert Spencer's, for example—that tried to unify the whole picture by saying, "Look, these principles are going across the board, and when we're thinking about evolution we have to think about evolution at every level of organization, from the physical to the symbolic and cultural and so on." This was criticized for many reasons. Some of them good reasons, because sometimes the analogies were too broad and sometimes they missed the point of what each particular discipline was trying to show in a rigorous manner. There were always people, always, who were thinking about the unifying principles and trying to understand them. Baldwin, for example; Waddington; Gottlieb; Patrick Bateson. But the idea that "this is phylogeny and this is ontogeny; they are different and we have to think about them differently" was a very important source of this disconnection.

Steve: I was taught to think of evolution as relating to the organism itself and then ontogenetically that's modified, but with no sense that it could go in the other direction. Now there are articles in *Science* trying to make the argument that instincts are the echoes of ancient learning. It is catawampus. It's not just niche construction or selection, but all the way down to genetic accommodations. How do your different perspectives touch on this larger issue of the directionality between these different domains?

Eva: My own approach is very much a developmental evolution approach. Organisms don't wait for lifesaving mutations to happen; they're active, they learn, they change. And they're plastic. There is no organism that isn't; such an organism would go extinct very quickly. Organisms first adapt ontogenetically—those that have the variations that allow them to adapt better than others. So the biological infrastructure that supports this kind of adaptation will be selected if it has a heritable component. There is more than one type of heritable variation—it can be not only genetic but also epigenetic. And in social animals, also socially learned variations; and in humans, also symbolic. Genes, in this sense, are followers in evolution. This is a phrase that Mary Jane Eberhard used—genes are followers, they stabilize what has been achieved ontogenetically.

Mike: From a functional perspective, our basic unit of analysis is a relationship between the organism and that environment—and so when I think about going backward, how a functional approach to learning might go back, it doesn't go back on the organism necessarily, it goes back on the susceptibility of that organism to environmental influence.

Derek A. Hamilton: The concept of selection is universal in evolution science, but not among all learning theorists or experimental psychologists. From the perspective of our chapter, there simply has to be a capacity in the organism for that sort of selection process to take place.

Steve: Both of the chapters have dealt with taxonomic considerations to a degree, and definitional ones. What difference do they make, going forward, in terms of the kinds of research that may touch on the relationship between these two domains?

Mike: There are assumptions built into the term "associative learning": there are these associations; they are structures; they exist at a neurological level or they exist at some sort of conceptual level. If you take a more functional approach to this and you think about learning as a word that is used in an abstract way to talk about behavioral changes in response to environmental regularities, you can see that there's a sort of different emphasis there. What organisms respond to are environmental regularities, and the distinction between classical and operant conditioning that's brought up in that chapter is, from our perspective, not a difference in systems or mechanisms or underlying structures so much as it's a difference in the kind of environmental regularities that can influence behavior.

    What I would want evolutionists to tell us is how evolutionary selection prepares us, differentially, to respond to these environmental regularities. And again, I'm not sure that the neurological or biological level is where the answer is.

Eva: I think that talking just about adaptive response to environmental regularities as a definition of learning is insufficient. There are all kinds of regularities in the world to which organisms respond—for example, changes in the amount of light plants get during the year—and they respond to these regularities in a very specific and very highly adaptive way. Nevertheless, we should not call these responses "learning." So, what is special about the learning-based behavioral adaptation? I think the classical way of approaching this is to talk about something that happens to the organism—it has to have some kind of receptor, some kind of sensors that allow signals from the environment or from the body to get into the nervous system (we're talking about neural organisms) and this leads to some kind of adaptive response. We also require that some aspect of the reaction that the organism undergoes is retained within this organism so that when these kinds of conditions reoccur, this leads to a change in the response from the first time. So there is some kind of retention, there is some kind of memory. We have to assume memory; and we have to assume not only that things are retained but that they can be recalled, so there is some kind of recall process.

    I think you're right, it is very naïve to think, when we're thinking about mammals or humans, that we are going to find simple neural correlates of this behavior. This leads me to Steve's question. I think one of the things that could help us is if the kind of distinctions or categorizations that we make would allow us to think about different levels of selection, about different types of input into the organism that can be processed or dealt with in ways that lead to adaptive responses, that lead to some kind of cumulative evolutionary output. We also need to understand where it comes from.

The distinction that we are making in our chapter between limited associated learning and unlimited associated learning is useful, we hope, because we're trying to figure out what's behind different forms of learning. For example, what do you need, what do you actually need, for unlimited associated learning? This unrestricted kind of learning requires perceptual fusion and chaining between many actions and requests, as well as second-order conditioning. You need to have a dedicated memory system and to have integration systems, association systems at different levels, sensory association systems, multiassociation systems, systems that associate the motor and the sensory, and so on and so forth. Many animal groups have it, though not all. But we also found that in almost all cases, animals with limited associative learning have brains. So we asked ourselves, why do we need a brain for that?

Even limited associative learning is a very demanding task. Once you understand this, you can appreciate the complexity that's required even for something that seems a relatively simple type of learning. For a more complicated associative learning, for unlimited associative learning to evolve, a lot more is needed. It requires high-level integration, hierarchies of processing, dedicated memory systems. And you're looking at this and you think, *Wow, we're not yet at all at the level of humans* or even anything like that. We're not even at the level of creatures that can imagine and can dream and can plan. But if we can understand that level, then maybe we can understand the next stage and the next stage.

Derek: I teach a large undergraduate learning class at the University of New Mexico, and one of the things we struggle with through the first part of the class is how to define learning in a way that's scientifically useful. In the earlier part of the twentieth century, the first seventy years, the definitions of learning were quite complex, and really were populated largely by elements that had to do with what learning is *not*, rather than what it is. For example, the transient state, or just a brief adaptation of responsiveness to environmental factors, or states of the organism like fatigue, or drug states. An experience that modifies behavior in a relatively permanent way requires predictive value, and adaptive value in what is acquired. That can be said to be true of classical conditioning or instrumental conditioning—the concept of the unlimited associative learning that Eva is talking about might then expand that predictive capacity in all sorts of other kinds of complicated situations. So, for example, contiguity might be very important for very simple organisms, but we know in higher-order organisms, contiguity is neither necessary nor sufficient for learning.

More mechanistic schools of thought talk about the products of leaning, like what sorts of representations are happening in nervous systems—and that leads to a very different trajectory of research questions one might add or ask and how those might be integrated into larger questions about evolution. Appealing to representation is an underdetermined problem—you don't have enough information to predict important dimensions (e.g., timing) of what's going to happen in the future using that sort of construct.

Mike: Things like retention must be accounted for in whatever account we have. But where I get concerned is when we say because there must be retention, therefore there's memory. Then we start looking for the memory, its physiological basis and its nature as a mechanism. Soon we're starting to identify parts of the brain in terms of the functions that they serve, so you start to hear things like "hippocampal behavior" instead of configural learning or complex associations. We start to identify memory with a part of the brain, or a part of the nervous system. There is no one thing called memory; memory is all over the brain. You can get operant conditioning at the neuronal level; neurons respond to consequences. An explanation of the complexity, to be useful for me, needs to be in terms of the environmental-behavior relationships that are allowed by this complexity, not the structures that underlie them, because those tend to be conceptual and hypothetical, not really anatomical.

Eva: I agree with a lot of what you say, but still, when you're talking about relations between the environment, environmental regularities, and the organism, the organism is the interpreting system of the regularities. The regularities in the environment may be the same for a flea and for a human; they interpret them differently. So you have to understand the interpreting system of the receiver—what is it about the interpretation of this type of animal versus that type of animal that makes this animal learn in this way and the other animal learn in that way. And in order to do that, you have to understand something about the details of how this animal has learned and also about the neuroanatomical and the neuronal and the physiological infrastructure.

Steve: If you have a clear unit at the level of organism-environment interactions, and a change of that relationship that's based on contact with regularities, it may allow you to move into more phenotypic processes. Take something like taste aversion. Everything that I have seen in the literature on taste aversion tells me it's a specialized form of classical conditioning. I think most people now would look at it as an evolutionary adaptation of a functional process that altered typical temporal parameters. That adaptation was due to the particular importance of learning temporally extended relations between stimuli that are food-related stimuli for the organism given its ecological history, and illness outcomes that are also food-related for that organism. So the temperature of food is important with this organism; sight, with that organism; smell, with that organism; and this dramatic distortion of temporal parameters gradually evolves where you go from a learning based on gaps of seconds to gaps of twenty-four hours. Might we use that as an example? I would be surprised, Mike, if you would think it was a bad idea to look at the underlying neurobiology of taste aversion, given this extraordinary distortion of these temporal parameters.

Mike: I would have no objection to an attempt to understand the neurobiology of learning to the extent that that is helpful in what we're trying to do with respect to prediction and control. But that becomes the test. So when Eva says, "I want to know what these systems are and what these processes are," I agree. But to what end and how do you know that you know about them?

With respect to taste aversion, the test for me would be whether or not the description of the system, the analysis, allows me to manipulate taste aversion in the real world, to make more of it or to be able to interfere with it. For example, extinction doesn't seem to work with taste aversion—why?

Eva: One of the things that animals that have limited associative learning cannot do is learn in a nonelemental way. Distinguishing between limited and unlimited associative learning (UAL) classifies animals in a new way. We think UAL is an evolutionary marker for minimal consciousness. One of the types of cognition that we would like to understand is our own. Now, how do we go about it? It's so complicated. One of the ways is to start with something a bit simpler. And then try to see what you need in order to get something more complicated, and then even more complicated, and finally maybe our kind of cognition. We are pragmatists. We look at every level possible to get any information about cognition to try and relate them together and form a picture. UAL could have implications for understanding more complex cognition, and which animals are conscious, which is a huge philosophical and scientific question. UAL didn't fall from the sky and we can't take it for granted. It evolved. We think that it was one of the drivers of the Cambrian explosion. It opened the way to an arms race between prey and predators, as well as social relations within a species.

Mike: The way I think of awareness and consciousness is the ability of an organism to respond discriminatively to its own behavior. With humans, we can talk about that—"Hey, I just felt this" or "I just did that"—but animals can also respond discriminatively; they can be trained to do that. And that, I think, is a very interesting development in evolution. That is where I think language really plays the most important role and may be one of the drivers in all of this. Because you've got to be able—if you're living in a social community, and people are wanting to know what you're going to do, what you did do, or how you did feel—you've got to be able to answer that by discriminating yourself, your behavior, from the rest of the group.

Eva: It's very difficult to give a definition of consciousness, as you know. It's just like the problem with life—you can characterize it, but you can't define it. Consciousness involves an ability to have something we can call subjective experience. It's very difficult to define this kind of thing. There are many characterizations of the conscious state or of a conscious animal in the literature. And what we tried to come up with was a list of these characteristics and then we tried to see, well, does an animal that has unlimited associative learning actually display them? And we came to the conclusion: yes, it does—that's why we chose it as an evolutionary marker. We don't say that unlimited associative learning and consciousness are the same thing. But we say that in order to learn, in unlimited associative learning, in this kind of way, you have to be conscious.

CHAPTER 4

Symbolic Thought and Communication from a Contextual Behavioral Science Perspective

Dermot Barnes-Holmes

Yvonne Barnes-Holmes

Ciara McEnteggart

Department of Experimental Clinical and Health Psychology, Ghent University, Belgium

In the current chapter, we present a brief summary of how contextual behavioral science has approached the topic of symbolic thought and communication. We begin by considering how behavioral psychology defined and studied symbolic relations prior to the seminal research of Murray Sidman (1994) and his colleagues on stimulus equivalence in the 1970s and 1980s. Specifically, up until that time, behavioral psychology more or less assumed that symbolic thought and communication functioned in broadly similar ways for both human and nonhuman species (Skinner, 1957), and that *all* behavior involved the same behavioral processes (e.g., classical and operant conditioning). Skinner himself appeared to break with this view in the 1960s when he proposed the concept of instructional control in explaining human problem solving, but the symbolic nature of instructions remained poorly defined. Sidman's work in the 1970s helped bring some clarity to the symbolic nature of instructional control, and this was further developed and elaborated by the work of Steven C. Hayes in the 1980s under the rubric of relational frame theory (RFT). The evolution of the ability to engage in relational framing was very much understated in the seminal volume in this area (Hayes, Barnes-Holmes, & Roche, 2001), but more recent conceptual analyses have developed the account in this area. The current chapter aims to provide an overview of the behavior-analytic approach to human symbolic thought and communication, and its evolution, as viewed predominantly through the lens of RFT.

Verbal Behavior

In his 1957 book *Verbal Behavior*, Skinner defined language as any behavior by a speaker that is reinforced through the mediation of a listener, which in turn gives rise to multiple classes of verbal behavior. These classes included mands, echoics, textuals, transcription, dictation, intraverbals, tacts, extended tacts, and autoclitics. While these concepts have been used extensively in remedial education programs (e.g., Lovaas, 1981), Skinner's text did not generate a vibrant and productive program of basic research on human language per se. One likely reason for this is that many of the verbal operants he defined could be studied in the nonhuman laboratory. For example, when a rat learned to press a lever for food in an operant chamber, the lever press could be considered a type of mand, if the lever press was reinforced by an individual who was trained by a verbal community to do so (Hayes, Blackledge, & Barnes-Holmes, 2001). Given this conceptual view, there was little motivation for behavioral researchers to study language in the *human* animal. This motivation would only emerge when key differences between human and nonhuman behavior were identified.

While the influence of Skinner's operant analysis of verbal behavior remained limited, even in behavioral psychology, some behavioral psychologists, including Skinner, did pursue broader features of human language. In the early years, this was focused on so-called instructional control or rule-following. Rules were first defined by Skinner (1969) as contingency-specifying stimuli. That is, "we tend to follow rules because previous behavior in response to similar verbal stimuli has been reinforced" (p. 148). Indeed, many studies on instructional control and rule-following emerged in the literature in the 1970s and 1980s. For example, research showed that human infants only begin to show adult-like responding on schedules of reinforcement from the age of 24 months and the likely source of this developmental change could be traced to the ability to generate simple rules that could be used to regulate one's own behavior (e.g., Bentall, Lowe, & Beasty, 1985). Researchers also demonstrated that rule-governed behavior could be insensitive to changes in reinforcement contingencies (e.g., Hayes, Brownstein, Haas, & Greenway, 1986). Although many studies on rule-governed behavior emerged in the literature, a precise functional definition of instructional control remained elusive. As noted above, Skinner defined rules as contingency-specifying stimuli, but failed to define specification in functional terms. Research on derived relational responding provided this much-needed definition.

Stimulus Equivalence and Relational Frame Theory

In the early 1970s, Murray Sidman reported a behavioral effect in his research on the acquisition of basic reading skills in individuals with learning disabilities. This effect came to be known as *stimulus equivalence*, and it eventually provided the basis for a behavior-analytic account of symbolic thought and communication in humans. The basic stimulus equivalence effect involved training a series of matching responses that could easily be explained in terms of direct reinforcement contingencies, but additional *emergent* matching behaviors were frequently observed that were difficult to explain in terms of established behavioral principles. For example, a child who was taught to match a spoken word to a picture, and a spoken word to a printed word, was able to spontaneously match the printed word to the picture without any further reinforcement (see Sidman, 1994, for a book-length treatment). Other studies, using nonhuman species, including "language-trained" higher primates, repeatedly failed to demonstrate this emergent effect (e.g., Dugdale & Lowe, 2000). When stimuli became related in this way, they were said to form equivalence classes or relations. Critically, Sidman argued that the formation of such relations may provide a behavior-analytic model of symbolic or referential relations in human language. In other words, Sidman provided the first definition of semantic meaning in behavioral psychology, which suggested that it may be unique, or at least highly dominant, in humans. As we shall see subsequently, this approach to semantic relations provided the basis for a behavioral account of specification that had previously been missing in the literature on rule-governed behavior.

The extension of Sidman's seminal work to rule-governed behavior came with S. C. Hayes and L. J. Hayes's (1989) approach to stimulus equivalence as an operant class of *arbitrarily applicable relational responding* (AARR). According to this view, a history of reinforced relations among stimuli established particular patterns of overarching or generalized relational operants, referred to as relational frames (Barnes-Holmes, Barnes-Holmes, & Cullinan, 2000). For example, imagine a young child who learns to point to the family dog upon hearing the word "dog" and to say "dog" when someone else points to the dog. The child might also learn to say "Rover" when asked, "What is the dog's name?" Each of these naming or relational responses would be explicitly prompted, shaped, and reinforced initially by the verbal community. Across many such exemplars, involving other stimuli in other contexts, the operant class of coordinating stimuli in this way becomes abstracted, such that direct reinforcement for all of the individual components of naming are no longer required when a novel stimulus is encountered. So, if a child were to be shown a picture of an aardvark and the written word, and told its name, the child may later say "That's an aardvark" when presented with a relevant picture or the word, without any prompting or direct reinforcement for doing so. In other words, the generalized relational operant of coordinating pictures, spoken words, and written words is established, and directly reinforcing a subset of the relating behaviors (spoken word/picture, and spoken word/written word) "spontaneously" generates the complete set (e.g., picture/written word).

When a pattern of generalized relating is established, that class of behavior is defined as always under some form of contextual control. Contextual cues are thus seen as functioning as discriminative for different patterns of relational responding or different relational frames. The cues acquire their functions through the types of histories described above. Thus, for example, the phrase "that is a," as in "*That is a* dog," would be established across exemplars as a contextual cue for the complete pattern of relational responding (e.g., coordinating the word "dog" with actual dogs). Once the relational functions of such contextual cues are established in the behavioral repertoire of a young child, the number of stimuli that may enter into such relational response classes becomes almost infinite (Hayes, Barnes-Holmes, & Roche, 2001).

The core analytic concept of the relational frame proposed by Hayes and Hayes (1989) involved three common properties: mutual entailment; combinatorial entailment; and the transformation of stimulus functions. First, mutual entailment refers to the relation between two stimuli. For example, if you are told A *is the same as* B, you will derive that B *is the same as* A. That is, the specified A *is the same as* B relation mutually entails the (symmetrical) B *is the same as* A relation. Second, combinatorial entailment refers to the relations among three or more stimuli. For example, if you are told A *is more than* B and B *is more than* C, you will derive that A *is more than* C and C *is less than* A. That is, the A-B and B-C relations combinatorially entail the A-C and C-A relations. Third, the transformation of stimulus functions refers to the "psychological content" involved in any instance of derived relational responding. For example, if A *is less than* B, and a reinforcing function is attached to A, then B will acquire a greater reinforcing function than A, even though the function was directly attached to A and not B. This general approach to human language and cognition became known as *relational frame theory* (RFT), and facilitated a behavior-analytic account of key elements of language, such as meaning, reference, and understanding (Barnes & Holmes, 1991), which led to a book-length treatment by the turn of the millennium (Hayes, Barnes-Holmes, & Roche, 2001).

Whereas Sidman's work on equivalence relations focused on what may be considered the most basic type of symbolic relation, RFT developed and expanded the conceptual analysis in an effort to cover the full richness and complexity of human language and cognition in whole cloth. Equivalence relations were defined as just one type of symbolic relation, with numerous other relations (defined above as relational frames) also being identified and studied from the early 1990s until the present day. These patterns of relational frames (e.g., coordination, opposition, distinction, comparison, spatial frames, temporal frames, deictic relations, and hierarchical relations) have been analyzed across numerous experimental studies, and across a variety of procedures. Some research has also explored the transformation of functions (see Hughes & Barnes-Holmes, 2016a, for a recent review). In addition, empirical evidence supported the core RFT postulate that exposure to multiple exemplars during early language development is required to establish these relational frames (see Hughes & Barnes-Holmes, 2016b). The argument that relational frames may be thought of as overarching or generalized relational operants thus gained considerable traction.

Generativity and Complexity of Human Language

The seminal text on RFT also used the basic operant unit of the relational frame to provide functional-analytic accounts of specific domains of human language and cognition, and rule-governed behavior was one of these domains. According to RFT, a rule or instruction may be considered a network of relational frames typically involving coordination and temporal relations with contextual cues that transform specific behavioral functions. Take the simple instruction, for example, "If the light is green, then go." This rule involves frames of coordination between the words "light," "green," and "go," and the actual events to which they refer. In this sense, the technical definition of the frame of coordination, outlined above, provides the functional-analytic definition of "specification" that was missing from earlier accounts of rules or instructions. In addition, the words "if" and "then" serve as contextual cues for establishing a temporal relation between the green light and the act of going (i.e., first green light, then go). The relational network thus transforms the functions of the green light itself, such that it now controls the act of "going" whenever an individual who was presented with the rule observes the green light being switched on.

Additional conceptual developments generated experimental and applied analyses of verbal rules or instructions in terms of complex relational networks composed of multiple relational frames, analogical and metaphorical reasoning in terms of relating relational frames, and problem solving in terms of increasingly complex forms of contextual control over relational framing itself. To illustrate, consider the example of an analogy, *pear is to peach as cat is to dog*. In this example, there are two relations coordinated through class membership (controlled by the cue *is to*) and a coordination relation that links the two coordination relations (controlled by the cue *as*). From an RFT point of view, analogical reasoning thus involves the same psychological process involved in relational framing more generally (i.e., AARR), but applied to framing itself (see Stewart & Barnes-Holmes, 2001a).

RFT research has also focused, both conceptually and empirically, on the role of human language in perspective taking. For instance, for RFT, basic perspective taking involves three deictic relations: the interpersonal relations I-YOU, spatial relations HERE-THERE, and temporal relations NOW-THEN (Y. Barnes-Holmes, 2001). The core postulate here is that as children learn to respond in accordance with these relations, they become able to locate the self in time and space and in relation to others. Imagine a very young child who is asked "What did you have for lunch today?" while she is eating the evening meal with her family. If the child responds simply by referring to what a sibling is currently having for dinner, she might well be corrected with "No, that's what your brother is eating *now*, but what did *you* eat earlier today?" In effect, this kind of ongoing refinement of the three deictic relations enables the child to respond appropriately to questions about her own behavior in relation to others, as it occurs in specific times and specific places (McHugh, Barnes-Holmes, & Barnes-Holmes, 2004).

At this point we could continue, providing many examples of ways in which RFT has been used to provide functional accounts and approaches to various domains in psychology, including intelligence, implicit cognition, prejudice, and so on (Hughes & Barnes-Holmes, 2016b). At a more general level, however, it may be useful to consider a recent framework that has been proposed that highlights the potential that RFT has to take a simple human ability called AARR and to construct increasingly complex analyses of the ability to engage in symbolic thought and communication. Specifically, researchers have recently offered what they describe as a multidimensional, multilevel (MDML) framework for analyzing AARR. According to this framework, AARR may be conceptualized as developing in a broad sense from mutual entailment, to simple networks involved in frames, to more complex networks involved in rules and instructions, to the relating of relations and relational networks involved in analogical reasoning, and finally to relating relational networks. The framework also conceptualizes each of these levels as having multiple dimensions: *derivation, complexity, coherence,* and *flexibility* (Barnes-Holmes, Barnes-Holmes, Luciano & McEnteggart, 2017).

In simple terms, *derivation* refers to how well practiced a particular instance of AARR has become. Specifically, the first time an AARR is emitted, derivation will be high, but across repeated instances of that class, the level of derivation will fall. *Complexity* refers to the level of detail or density of a particular pattern of AARR. As a very simple example, an AARR involving mutual entailment alone is less complex than an AARR involving combinatorial entailment. *Coherence* refers to the extent to which an AARR is generally predictable based on prior histories of reinforcement. For example, the statement "A mouse is larger than an elephant" would typically be seen as lacking coherence with the relational networks that operate in the wider verbal community. Note, however, that such a statement may be seen as coherent in certain contexts (e.g., when playing a game of "everything is opposite"). *Flexibility* refers to the extent to which a given instance of AARR may be modified by current contextual variables. Imagine a young child who is asked to respond with the wrong answer to the question "Which is bigger, a mouse or an elephant?" The easier this is achieved, the more flexible the AARR.

A detailed treatment of the MDML is beyond the scope of the current chapter. The critical point to appreciate, however, is that RFT may be used to generate a conceptual framework that begins with a very simple or basic scientific unit of analysis, the mutually entailed relational response. From an RFT perspective, this unit is not synonymous with naming in a traditional analysis of symbolic meaning and communication, but it is seen to be intrinsic to it in a psychological analysis of naming as an act in context. In other words, the concept of mutual entailment strips bare the informal concept of naming, leaving nothing but the raw relational properties of the psychological or behavioral process. What the MDML adds to this conceptual analysis is a framework for considering what appear to be the key dimensions along which mutual entailment as a behavioral process may vary (e.g., mutually entailed responding may vary in terms of coherence, flexibility, complexity, and derivation). In addition, the MDML emphasizes that more complex units of analysis may evolve from mutual entailment, such as the simple relational networks involved in relational frames, more complex networks involving combinations of frames, the relating of relational frames to relational frames, and ultimately the relating of entire complex relational networks to other complex relational networks. And in each case, these different levels of AARR may vary along the four dimensions listed above, and perhaps others that remain to be identified.

When RFT is viewed through the lens of the MDML, the potential power that it may have to analyze the complexities and dynamics of human symbolic thought and communication quickly become apparent. In much the same way that mutual entailment provides a purely relational approach to understanding naming as a language process, the concepts of frames, networks, relating relations, and relating relational networks provide purely relational analyses of increasingly complex human language phenomena. As outlined above, for example, the concept of relating relations appears to be relevant to, if not synonymous with, analogical reasoning. Similarly, relating relational networks may be relevant to the telling and understanding of complex stories (Stewart & Barnes-Holmes, 2001b).

At this point, it must be acknowledged that the MDML is a relatively new development in the RFT literature, but it does coincide with another relatively new development in the theory, which aims to situate this behavioral approach to human language and cognition within the broader discipline of evolutionary science. As we will outline below, for example, considerable attention has been given recently to how the human propensity for cooperation may have been instrumental in the evolution of mutual entailment as the core or most basic unit of human symbolic thought and communication.

RFT and Evolutionary Science

As a behavioral account of human language and cognition, RFT has traditionally focused on the learning experiences that occur within the lifetime of the individual. This focus is understandable because the theory has very much been driven by a pragmatic concern with predicting and influencing human language and cognition itself in clinical, educational, and wider social settings. On balance, it has always been recognized that the ability to acquire the relational operants identified by RFT with relative ease is likely to have emerged from a particular evolutionary history, but until recently, work in this area has been limited (Hayes & Sanford, 2014; Wilson, Hayes, Biglan, & Embry, 2014).

On the one hand, it appears that language required a massive evolutionary leap, and considerable attention has been devoted to explaining the potential relationship between the many examples of nonhuman communication (e.g., mating calls, the honeybee dance, facial expressions in primates) and the richness and complexity of human language (see Hauser et al., 2014). On the other hand, if the focus is on the ability to AARR, rather than on the ill-defined concept of human language, the scientific challenge appears more manageable. Specifically, it seems wise to start with the relatively simple question, "How did the behavior of mutual entailment (under contextual control) emerge so strongly in the human species?" If we can answer this question, perhaps more complex questions about the full richness of human symbolic thought and communication may be addressed, in part, through explaining how mutual entailment facilitates combinatorial entailment, the growth of more complex relational responding, and so forth.

In what follows, we have adopted exactly this strategy. Of course, what we offer must remain wildly speculative, but the aim here is to *begin* a meaningful dialogue with experts in other domains, and in particular in evolutionary science, rather than seeking to provide some final answer to what is an incredibly complex question (i.e., How did human symbolic thought and communication evolve?). D. S. Wilson and E. O. Wilson (2007) summarized human evolution as the "three Cs": cognition, culture, and cooperation. Although all three of these are embedded within early renditions of RFT, it appears that cooperation was somewhat underplayed.

In the first book-length treatment of RFT, Hayes, Barnes-Holmes, and Roche (2001) suggest that mutual entailment in a listener could improve avoidance of predators even if entailment was not present as part of a speaking repertoire. In addition, it was suggested that this small difference could give rise to a group of listeners who were then capable of reinforcing mutually entailed responses in a speaking repertoire. This account, of course, relies on the evolution of mutual entailment as an adaptation of cognition in listeners, which then spreads to speakers and throughout the culture, thereby leading to increased social cooperation. In contrast to this account, Hayes and Sanford (2014) have recently suggested that it is more evolutionarily viable to suppose that cooperation came first. Indeed, as Hayes and Sanford point out, there is a vast amount of empirical data to support the idea that cooperation was established by multilevel selection of cooperation itself, because it offered advantages for human group competition, which occurred alongside the cultural suppression of individual selfishness.

From this perspective, cooperation that originally began with pointing and grunting, for example, provided humans with highly important behavioral skills, such as social referencing and joint attention, which have been recognized as important behavioral precursors for the psychological development of AARR in the lifetime of the individual (see Pelaez, 2009). Critically, these precursors increase the likelihood that cooperation will be reinforced, as is the case with young children. For example, if a young child says "eh" while looking at and trying to reach for a toy, the mother will reinforce this cooperation by giving the toy to the child. Hayes and Sanford (2014) conclude: "The entire exchange will build cooperation, perspective taking, and joint attention as patterns that are maintained within the group because it is a functionally useful communication exchange. If we unpack this highly likely sequence it means that in the context of high levels of cooperation, and adequate skills in joint attention, social referencing, and perspective taking, *any characteristic vocalization in the presence of a desired object would likely lead to reinforced instances of symmetry or mutual entailment*" (p. 122, emphasis in original). And once mutual entailment evolves, extended cooperation further facilitates the adaptation of the species, by allowing for more complex adaptations of this functional unit, such as combinatorial entailment.

Ultimately, increasing complexity in AARR would likely facilitate the use of symbols and the ability to problem-solve in the natural and social environment. According to this more recent RFT account of the evolution of AARR, cooperation leads to more useful forms of cognition, rather than cognition leading to more useful forms of cooperation.

Once the basic unit of AARR is established, it allows for the evolution of more complex relational operant units, such as relational networks, the relating of relations (e.g., analogy and metaphor), and the relating of entire relational networks to other relational networks (e.g., extracting common themes from different narratives). In effect, this set of relational abilities evolved into complex forms of communication and problem solving in only a few thousand years. Indeed, it could be argued that the ability to AARR is a defining characteristic of the human species, and allows us to predict and influence our environment in increasingly sophisticated and powerful ways. From this perspective, once AARR evolves, the natural environment becomes thick and rich with stimuli that are symbolic, rather than direct-acting, as they appear to be for nonhuman species. For example, symbolic stimuli can be used to form new meanings and to construct new realities detached from direct experience (e.g., fiction, poetry, metaphor). As such, the transmission of behaviors, from one individual to another and from one generation to the next, is increased dramatically. This ultimately leads to greater variation in behavior and the potential for the acquisition of new behaviors that serve to increase survival at multiple levels—individuals, groups, and species.

Summary and Conclusions

In adopting the RFT approach we have outlined here, the question is not how human language evolved, but rather how mutual entailment as the basic unit of AARR evolved in the first instance, and then how more complex units of AARR likely evolved from mutual entailment, thus allowing for the emergence of grammar and syntax, complex rule-following, analogy, metaphor, and story-telling. This approach could be seen as a gross oversimplification of the processes involved in the evolution of human language, but if we cannot explain how a simple behavior such as mutual entailment evolved, there seems little hope in explaining the evolution of human language in whole cloth. As noted above, the foregoing remains highly speculative, but is does serve to highlight potentially important areas of overlap between evolution science and contextual behavioral science. It is our hope that the current chapter helps to facilitate a fruitful dialogue in this regard.

References

Barnes, D., & Holmes, Y. (1991). Radical behaviorism, stimulus equivalence and human cognition. *Psychological Record, 41*, 19–30.

Barnes-Holmes, D., Barnes-Holmes, Y., & Cullinan, V. (2000). Relational frame theory and Skinner's verbal behavior: A possible synthesis. *Behavior Analyst, 23*, 69–84.

Barnes-Holmes, D., Barnes-Holmes, Y., Luciano, C., & McEnteggart, C. (2017). From the IRAP and REP model to a multi-dimensional multi-level framework for analyzing the dynamics of arbitrarily applicable relational responding. *Journal of Contextual Behavioral Science, 6*, 434-445. doi:org/10.1016/j.jcbs.2017.08.001

Barnes-Holmes, Y. (2001). Analysing relational frames: Studying language and cognition in young children (unpublished doctoral thesis). National University of Ireland, Maynooth.

Bentall, R. P., Lowe, C. F., & Beasty, A. (1985). The role of verbal behavior in human learning: II. Developmental differences. *Journal of the Experimental Analysis of Behavior, 43*, 165–180.

Dugdale, N., & Lowe, C. F. (2000). Testing for symmetry in the conditional discriminations of language trained chimpanzees. *Journal of the Experimental Analysis of Behavior, 73*, 5–22.

Hauser, M. D., Yang, C., Berwick, R. C., Tattersall I., Ryan, M. J., Watumull J., … Lewontin, R. C. (2014). The mystery of language evolution. *Frontiers in Psychology, 5*, 401. doi:10.3389/fpsyg.2014.00401

Hayes, S. C., Barnes-Holmes, D., & Roche, B. (Eds.). (2001). *Relational frame theory: A post-Skinnerian account of human language and cognition.* New York: Plenum.

Hayes, S. C., Blackledge, J. T., & Barnes-Holmes, D. (2001). Language and cognition: Constructing an alternative approach within the behavioral tradition. In S. C. Hayes, D. Barnes-Holmes, & B. Roche (Eds.), *Relational frame theory: A post-Skinnerian account of human language and cognition.* New York: Plenum.

Hayes, S. C., Brownstein, A. J., Haas, J. R., & Greenway, D. E. (1986). Instructions, multiple schedules, and extinction: Distinguishing rule-governed from schedule controlled behavior. *Journal of the Experimental Analysis of Behavior, 46*, 137–147.

Hayes, S. C., & Hayes, L. J. (1989). The verbal action of the listener as a basis for rule-governance. In S. C. Hayes (Ed.), *Rule-governed behavior: Cognition, contingencies, and instructional control.* New York: Plenum.

Hayes, S. C., & Sanford, B. (2014). Cooperation came first: Evolution and human cognition. *Journal of the Experimental Analysis of Behavior, 101*, 112–129.

Hughes, S., & Barnes-Holmes, D. (2016a). Relational frame theory: The basic account. In R. D. Zettle, S. C. Hayes, D. Barnes-Holmes, & A. Biglan (Eds.), *The Wiley handbook of contextual behavioral science.* West Sussex, UK: Wiley-Blackwell.

Hughes, S., & Barnes-Holmes, D. (2016b). Relational frame theory: Implications for the study of human language and cognition. In R. D. Zettle, S. C. Hayes, D. Barnes-Holmes, & A. Biglan (Eds.), *The Wiley handbook of contextual behavioral science.* West Sussex, UK: Wiley-Blackwell.

Lovaas, O. I. (1981). *Teaching developmentally disabled children: The ME book.* Austin: TX: Pro-Ed.

McHugh, L., Barnes-Holmes, D., & Barnes-Holmes, Y. (2004). A relational frame account of the development of complex cognitive phenomena: Perspective-taking, false belief understanding, and deception. *International Journal of Psychology and Psychological Therapy, 4,* 303–324.

Pelaez, M. (2009). Joint attention and social referencing in infancy as precursors of derived relational responding. In R. A. Rehfeldt, & Y. Barnes-Holmes (Eds.), *Derived relational responding: Applications for learners with autism and other developmental disabilities.* Oakland, CA: New Harbinger Publications.

Sidman, M. (1994). *Stimulus equivalence: A research story.* Boston: Authors Cooperative.

Skinner, B. F. (1957). *Verbal behavior.* New York: Appleton-Century-Crofts.

Skinner, B. F. (1969). *Contingencies of reinforcement: A theoretical analysis.* New York: Appelton-Century-Crofts.

Stewart, I., & Barnes-Holmes, D. (2001a). Understanding metaphor: A relational frame perspective. *Behavior Analyst, 24,* 191–199.

Stewart, I., & Barnes-Holmes, D. (2001b). Relations among relations: Analogies, metaphors, and stories. In S. C. Hayes, D. Barnes-Holmes, & B. Roche (Eds.), *Relational frame theory: A post-Skinnerian account of human language and cognition.* New York: Plenum.

Wilson, D. S., Hayes, S. C., Biglan, T., & Embry, D. (2014). Evolving the future: Toward a science of intentional change. *Behavioral and Brain Sciences, 34,* 1–22.

Wilson, D.S., & Wilson, E.O. (2007), Rethinking the theoretical foundation of sociobiology. *Quarterly Review of Biology, 82,* 327–348.

CHAPTER 5

Beneath Symbols: Convention as a Semiotic Phenomenon

Terrence W. Deacon

*Department of Anthropology,
University of California, Berkeley*

Introduction

Symbolic reference is a distinguishing feature of human language. In this respect, language contrasts with other species-typical vocalizations and most communicative gestures, which only provide reference iconically or indexically. Because of its arbitrary and conventional nature, symbolic reference must be acquired by learning, and lacks both the natural associations and transgenerational reproductive consequences that characterize other innately evolved communicative adaptations, like laughter and sobbing. This is why there are no innate words and why symbolic reference is so extensively reliant on social (as opposed to genetic) transmission.

Iconic and indexical forms of communication are ubiquitous in the animal world as well as in human communication. They provide reference by virtue of formal and physical features shared by the sign vehicle and that to which it refers. In contrast, it is the irrelevance of any shared properties between sign vehicles (e.g., word sounds) and what they refer to—often referred to as arbitrariness—that facilitates the capacity to combine symbolic forms into many vastly complex structures (e.g., sentences and narratives) that are able to specify highly diverse and precise communicative contents.

To say that symbolic reference is arbitrary is to say that it is determined by convention, rather than by any intrinsic sign vehicle properties. But what is entailed in the concept of convention when used in this way? The Merriam-Webster dictionary lists three related meanings that are relevant to this issue: a convention can be a usage or custom, especially in social matters; a rule of conduct or behavior; or an established technique, practice, or device.

Probably the two most common social phenomena attributed to social convention in the course of intellectual history are money and language. The claim that language is the expression of social convention is ancient. In Aristotle's work *On Interpretation* he describes a name[1] as a convention because it is not a natural feature of what it refers to. He explicitly notes that this is what makes a name a symbol. The concept of convention has also been associated with the notion of a social contract (e.g., by Rousseau and Hobbes), understood as an agreement, mostly with respect to its opposition to natural tendencies or to the so-called "state of nature" imagined to predate civilization. However, John Locke recognized that tacit conventions may arise even if no explicit agreement has been negotiated.

The concept of convention was a major focus of David Hume's analysis of many regular practices found in human societies. He argued that social conventions are necessary for establishing such social phenomena as property, agreements, laws, and so forth, in which individuals elect to all conform to certain limits to or habits of behavior out of an expectation of mutual benefit. He also explicitly critiques the notion that conventions need to be the result of explicit agreements. He illustrates this with a memorable example: of two men in a row boat, each with one oar, who need to coordinate and synchronize their rowing in order to reach any particular destination. Hume proposes a conventional theory of language regularities that need not arise from negotiated agreements when he states that "languages [are] gradually establish'd by human conventions without any explicit promise" (1739–1740 p. 490).

Twentieth-century philosophers revisited the conventional theory of language in a half-century of debates about the nature of semantics and truth, especially as it impacted the foundations of logic. Such intellectual giants of the field as Carnap and Quine battled over the coherence of conventional theories of meaning, truth, and mathematics. At mid-century, this led to an intense debate in linguistics, particularly fueled by Noam Chomsky's strident denial of the relevance of semantics (and symbolic reference in general) to grammar and syntax; a debate that still rages, and which will be addressed below.

An interesting reassessment of the logical structure of conventionality was provided by the philosopher David Lewis in his 1969 book appropriately titled *Convention*. Lewis argued that convention should be considered a solution to a coordination problem and that it need not involve explicit or implicit agreement. Beginning by recounting Hume's metaphor of the rowers and framing the problem in terms of game theory, he developed progressively more complex models of how groups of individuals might spontaneously arrive at what might otherwise appear

---

1   By "name" Aristotle appears to mean any general term, not merely a proper name.

as agreed-upon collective behaviors, with nothing resembling agreement, tacit or otherwise.

The Semiotics of Conventionality

Conventionality is not the critical determiner of symbolic reference, even though symbols must involve conventionalized sign vehicles. There can be conventionalized icons and conventionalized indices. Conventionalized icons include the stick figures on restroom doors, the skull and crossed bones on bottles containing poison, and the cigarette drawing with a superimposed cross-out slash across it. Of course, all three also are used indexically. The placement of a male or female icon on a restroom door *indicates* that it opens to a sex-specific restroom; the skull and crossed bones insignia *indicates* something about the substance in the marked container; and the crossed-out smoking cigarette *indicates* a no smoking zone. So both the formal likenesses and the factual correlations of these signs are relevant conventions. A conventional index that is minimally iconic is the white line down the middle of a two-way road. Though it might be iconically compared to a "property line" or national border, or even to the outline of a coloring book figure, it does not physically prevent drivers from crossing it. However, despite its conventional nature and dependence on traffic laws, the line itself is not symbolic and can only metaphorically be said to have a meaning or definition.

This distinction is important, because it helps to untangle a troublesome tendency to simply equate symbols with convention. As Charles Peirce clearly demonstrated, being a conventional sign vehicle and referring conventionally are not the same. Thus, he designated conventional sign vehicles "legisigns" and nonconventional sign vehicles "sinsigns" or "qualisigns," and argued that there could be iconic, indexical, and symbolic legisigns, but not symbolic sinsigns or symbolic qualisigns. In other words, symbolic reference can only be borne by conventional sign vehicles. Thus, symbols are doubly conventional in that they involve conventional sign vehicles that refer conventionally as well. To explain the development of symbolic referential capacities either in childhood language acquisition or in the case of the evolution of this capacity in humans, it is thus necessary to account for both forms of conventionality.

This means we must answer a number of related questions: If linguistic conventions are shared dispositions that were not established via explicit social agreement or social conformity, how else could they have been achieved in evolution and child development? Does this inevitably lead us to accept an innate source? And what might this mean with respect to language structure as well as other conventional forms of semiosis?

This is where insights provided by Hume's and Lewis's analyses provide an important clue. Coordination is defined with respect to achieving a common end, whether or not the agents involved know about one another's goals. This is exemplified in Lewis's framing of coordination in game theory terms. Games, as generalized in this abstract sense, are activities with explicit payoffs or goals. Even if there is no explicit communication between the agents, they still may gain information about each other's goals by observing the consequences of each other's behaviors. Thus in the case of the rowers, even if they are for some reason unable to interact in any way other than by rowing, they can still converge on a coordinated pattern that leads to reaching a specific destination.[2]

What is often overlooked when describing this sort of thought experiment is that there is still information (a.k.a., semiosis) involved. The solution is achieved semiotically, just not with explicit use of language or pointing, or other explicit means for people to share their separate intentions, but by interpreting the dispositions exhibited in behavioral responses. Indeed, this could be the case even if there were only one human agent involved and an automatic rowing machine controlling the other oar. The point is that, once we expand our analysis beyond language-like communication and even beyond intentionally produced communication to consider semiosis in the broadest sense, it becomes clear that the development of conventionality requires extensive semiotic activity. More specifically, to acquire or evolve the capacity to determine reference via convention—that is, symbolic reference—prior nonconventional communication is required at some point to establish this conventionality. To restate this hypothesis in semiotic terms: in order to develop a symbolic communication system such as a language, its conventional properties must be established using iconic and indexical means.

But I want to make a far stronger claim. I will argue that the conventionality of language is itself a reflection of these iconic and indexical relations re-emerging in the form of relations between symbols. These intersymbolic relations go by more familiar linguistic terms: grammar and syntax. My goal is to recast the concept of linguistic convention in semiotic terms in order to disentangle it from the conception of linguistic convention as mere arbitrary mapping between signifiers and signifieds. I will argue that this presumed arbitrariness in the relation between sign vehicle and referent properties is enabled by the nonarbitrary iconic and indexical structure of between-symbol relations. To put this another way, communicating with sign vehicles (e.g., words) that refer by convention alone is made possible by combining them via nonconventional iconic and indexical properties. So not only are the coordinating conditions for language achieved by

---

2 Of course, if the agents involved have conflicting goals, whether they are aware of this or not, coordination becomes far more difficult, and may result in suboptimal results for both.

social convention, but the structure that results reflects the semiotic structure of convention itself.

The Symbol Ungrounding Process

In a now famous paper published in 1990, the cognitive scientist Stevan Harnad articulated a worry that had long puzzled philosophers of language and cognitive scientists in general. He called it the "symbol grounding problem." The mystery was how arbitrary marks, such as the sounds of speech or the states of a brain, could reliably become correlated with specific referents so that symbolic communication is possible. In other words, without determining this mapping extrinsically, that is, by using symbolic communication to negotiate the establishment and sharing of these correspondences, how could these mappings ever be established in the first place? If it takes communication with symbols to establish this shared mapping convention between symbols and their referents, then we are faced with a vicious regress. To clarify a potentially confusing difference of terminology, in that context, conventionality of sign vehicle and conventionality of reference are not distinguished, and yet in that paper he speculates that symbol grounding must therefore be achieved using nonsymbolic means.

In recent lectures and forthcoming papers, Joanna Raczaszek-Leonardi and I invert this framing of the problem. We point out that the problem is actually to explain how iconic and indexical forms of communication—which are intrinsically "grounded" due to the sign vehicles sharing features with their referents—can be used to develop communication using ungrounded sign vehicles (a.k.a., words/symbols). This is, of course, the challenge faced by every human toddler. The reason that the challenge is seldom framed this way is because the infant-caretaker interactions are not generally understood in semiotic terms (either in psychology or development linguistics) and because the development of language competency is not seen as a transition from an earlier to a later, more developed semiotic process. From a semiotic perspective, however, there is a rich and complex set of social semiotic skills being acquired during the first year of life and significantly prior to the early stages of explicit language acquisition. Seen from this semiotic perspective, then, the explosive growth of language during the second and third years of life is a process in which these earlier iconic and indexical capabilities aid the child's discovery of how to use words and word combinations symbolically.

This is an *ungrounding* process to the extent that the toddler has to discover how to transfer from using intrinsically grounded sign vehicles to using ungrounded sign vehicles, *all the while maintaining referential grounding*. This can only be maintained if these iconic and indexical relations are in some way preserved in the transition to symbolic communication. Since properties that could provide referential grounding are absent from linguistic sign vehicles, grounding can only be preserved by means extrinsic to them, that is, in the relations between them. The logic of symbol ungrounding in language is depicted in figures 1 and 2.

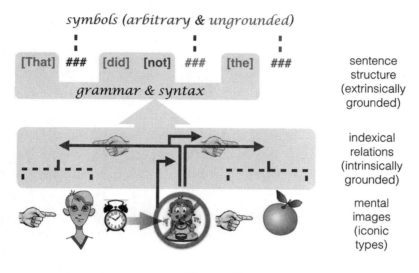

Figure 1

Figure 2

Universal Grammar from Semiotic Constraints

This raises an important question: Can the properties that linguists understand as grammatical be explained in terms of iconic and indexical properties? In this section, I will explore the possibility of explaining some of the most ubiquitous grammatical principles—so-called grammatical universals—in terms of semiotic constraints. Specifically, I will argue that the most nearly universal features of grammar are the least arbitrary aspects of language because they are constrained by the requirements for iconic and indexical reference.

Grammatical relationships do not automatically come to the fore with all forms of symbolic communication. This is because grammar is a property of symbolic reference that emerges when symbolic reference is amplified by combinatorial operations. Once it is recognized that symbolic reference is not a simple mapping relation, but emerges from a base of iconic and indexical relations transferred to symbol-symbol relations, the many contributions of these underlying semiotic constraints to the structure of language will become obvious.

Iconic and indexical relationships are constituted by sharing explicit properties with their referents. These are implicit ineluctable constraints that are inherited by features of grammar and syntax. These become re-expressed in operations involving symbol combinations, such as phrases, sentences, arguments, and narratives. These constraints emerge from below, so to speak, from the semiotic infrastructure that constitutes symbolic representation, rather than needing to be imposed from an extrinsic source of grammatical principles (e.g., innate universal grammar). Although this infrastructure is largely invisible—hidden in the details of an internalized system and largely automated during early childhood—using symbols in combination in communicative contexts necessarily exposes these constraints that determine iconic and indexical grounding.

These semiotic constraints have the most ubiquitous effect on the regularization of language structure, but in addition, there are sources of weaker, less ubiquitous constraints also contributing to cross-linguistic regularities. These include processing constraints due to neurological limitations, requirements of communication, and cognitive biases specific to our primate/hominid evolutionary heritage. Although none of these sources of constraint play a direct role in generating specific linguistic structures, their persistent influence over the course of countless thousands of years of language transmission tends to weed out language forms that are less effective at disambiguating reference, harder to acquire at an early age, demand significant cognitive effort and processing time, and are inconsistent with the distinctive ways that primate brains tend to interpret the world.

The sources of constraint on main categories include semiotic constraints, neural processing constraints, evolved sensorimotor schemas and cognitive biases, and pragmatic social communication constraints. These categories and specific constraints within each category are listed in Table 1 (modified from Deacon, 2012) in an order that roughly corresponds to their relative strength of influence on language structure. The combined effect of these multiple constraints significantly reduces the "phase space" of probable language forms (shown as a complex Venn diagram in Figure 3). Different linguistic paradigms may prioritize one or the other of these major categories of constraint to explain certain highly regular structural features of language. For example, cognitive grammars often highlight the influences of sensorimotor schemas and cognitive biases, whereas systemic functional linguistic approaches place considerable emphasis on the pragmatics of social communication. In the following discussion, however, I will focus only on some of the most ubiquitous semiotic constraints.

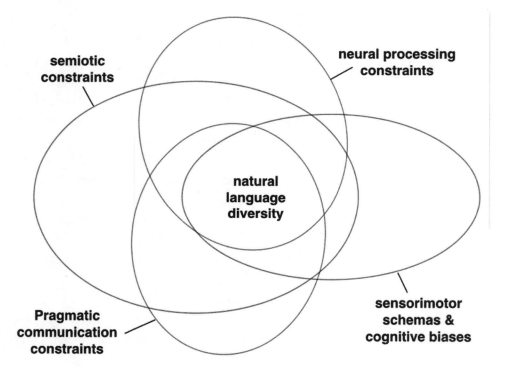

Figure 3

Perhaps the most radical implication of this analysis is that the most important and ubiquitous sources of constraints on language organization arise *neither from nature nor from nurture*. That is, they are not the result of biological evolution producing innate predispositions and they are not derived from the demands of discourse or the accidents of cultural history. Semiotic constraints are those that most directly reflect the grammatical categories, syntactic limitations, and phrasal organization of language. They are in a real sense *a priori* constraints, which precede all others. Consequently, they are most often confused with innate influences.

Table 1

A. Semiotic constraints

    1. Recursive affordance (only symbols can provide nondestructive [opaque] recursion across logical types)

    2. Predication structure (symbols must be bound to indices in order to refer)

    3. Transitivity and embedding constraints (indexicality depends on immediate correlation and contiguity, and is transitive)

    4. Quantification (symbolized indices need respecification)

    5. Semiotic constraints can be discovered pragmatically and "guessed" prior to language feedback (because of analogies to nonlinguistic iconic and indexical experiences)

B. Neural processing constraints

    6. Chunking-branching architecture (mnemonic constraint)

    7. Algorithmic regularization (procedural automatization)

    8. Neural substrates will vary on the basis of processing logic, not linguistic categories (there should be language-specific localization differences)

C. Evolved sensorimotor schemas and cognitive biases

    9. Standard schema/frame units (via cognitive borrowing)

    10. Vocal takeover (an optimal medium for mimicry)

D. Pragmatic communication constraints

    11. Pragmatic constraints (communication roles and discourse functions)

    12. Culture-specific expectations/prohibitions (e.g., distinctive conventions of indication, ways of marking discourse perspective, prohibitions against certain kinds of expressions, etc.)

Recursive Affordance

In a recent and now well-known theoretical review of the language origins problem (Hauser, Chomsky, & Fitch, 2002), Noam Chomsky appeared to retreat from a number of earlier claims about the innate "faculty" for language. In his new minimalist program, he instead focuses on the ubiquity of the hierarchic combinatorial structure of language and the recursive application of an operation described as "merge." This shift in focus doubles down on his long-term insistence that what makes the human mind unique is an innate capacity to handle recursive relationships. Most languages do indeed make extensive use of recursive combinatorial operations that are not found in nonhuman communication. Like many related claims for an innate grammatical faculty, however, this one also follows from a reductionistic conception of symbolic reference. If instead we recognize that only human communication is symbolic, whereas nonhuman communication is limited to iconic and indexical communication, another possible explanation for this uniquely human cognitive difference becomes available: recursion is only possible symbolically.

Because the sign vehicles used for symbolic communication (e.g., words) require no intrinsic properties linking them to their referents, they can refer to one another or to combinations of other symbols without equivocation. This allows substitutions that refer across logical types (e.g., part for whole, member for class, word for phrase) and thus across hierarchic levels in linguistic communications. Neither icons nor indices can refer across logical types because of the involvement of sign vehicle properties (e.g., similarity of form, correlation in space or time) in determining reference. But because of the independence of sign vehicle properties from the objects of reference, symbols can represent other symbolic relationships, including nearly unlimited levels of combinations of symbols (such as phrases, whole sentences, and even narratives). Recursion is not therefore an operation that must be "added" to human cognition over and above symbolic capabilities. It is a combinatorial possibility that comes for free, so to speak, as soon as symbolic reference is available. So the absence of recursion in animal communication is no more of a mystery than the absence of symbolic communication. It is simply due to their lack of symbolic abilities.

Though recursion is made available with symbolic communication, it need not be taken advantage of. So its paucity in child language and pidgins, as well as its absence in some languages (e.g., Everett, 2005) is not evidence against its universal *availability* in language. Recursion is an important means for optimizing linguistic communication because it provides a way to condense symbol strings. For example, repeated recursive operations make it possible to use a single word (e.g., pronoun) or phrase (e.g., anaphor) to refer to an extensive corpus of prior discourse. This not only optimizes communicative effort, it also reduces working memory load. Nevertheless, recursion also creates new "record-keeping" demands that help to avoid the confusions made possible by this condensation. This requires incorporating iconic and indexical constraints into the ways symbols can be combined. These infrasymbolic constraints on the relationships *between* words constitute the core features of grammar and syntax.

Predication Structure

Another nearly ubiquitous semiotic constraint is reflected in the combinatorial chunking that constitutes phrase and sentence structures. Combinatorial units such as complex words, clauses, and sentences are composed of elements that necessarily complement one another's semiotic functions. In other words, what can be "merged" in a way that constitutes a recursively higher order combinatorial unit is highly constrained. Such a functional unit must include at least two semiotically distinct components, one operating on the other. For example, all languages require at least a dyadic sentential structure, i.e., something like a subject-predicate or a topic-comment sentential form. Although holophrastic utterances, commands, and expletives are not uncommon, they typically are embedded in a pragmatic context in which what they refer to is made salient by immediate embedding in a semiotic context that fixes the reference; typically some salient feature of the immediate physical or social context. It has been suggested that this ubiquitous structure might reflect an action-object, agent-patient, or what-where dichotomy. But the ease with which these cognitive categories can be interchanged in their grammatical roles indicates that there is a more basic common constraint behind all of them.

Since long before efforts to formalize logical inference, scholars have recognized that isolated terms express a sense but lack specific reference unless they are embedded in a combinatorial construction roughly corresponding to a proposition. The assignment of a specific reference to an expression or formula in order to make an assertion about something is called predication. In symbolic logic, for example, a well-formed (i.e., referring) expression requires both a function and an argument (i.e., that to which the function is applied). First-order predicate logic is often considered the semantic skeleton for propositional structure in language, though its primary form is seldom explicitly exhibited in natural language. It is characterized by a "predicate (argument)" structure of the form $F(x)$, where F is a function and x is a variable or "argument" operated on by that function. Such an expression is the basic atomic unit of predicate logic. Such an expression may refer to an event, state, or relationship, and there can be one-, two-, three-, and zero-place predicates determined by how many arguments they take. So, for example, the function "lives" typically is a one-place predicate, "likes" is a two-place predicate, and "gives" is a three-place predicate.

This suggests the following hypothesis: Predicate (argument) structure expresses the dependency of symbolic reference on indexical reference, as in Symbol (index). Evidence for this semiotic dependency is implicit in the way that deictic procedures (e.g., pointing and other indicative gestures) are used to help fix the reference of an ambiguous term or description, and can even be substituted for the subjects and arguments of a sentence. Thus, for example, uttering the word "smooth" in a random context only brings attention to an abstract property, but when uttered while running one's hand along a table top or pointing to the waveless surface of a lake, reference is thereby established. It can also refer even if uttered in isolation from any overt index in a social context where the speaker and listener have their joint attention focused on the same flawless action. In this case, as with holophrastic utterances in general, the symbolic reference is established by implicit indication presupposed in the pragmatics of the communicative interaction. Indeed, where explicit indexing is not provided, it is assumed that the most salient agreeing aspect to the immediate context is to be indicated. In general, then, any symbolic expression must be immediately linked to an indexical operation in order to refer. Without such a link there is sense but no reference.

This is a universal semiotic constraint (though not a universal rule) that is made explicit in logic and is implicit in the necessary relational structure of sentences and propositions. It is a constraint that must be obeyed in order to achieve the establishment of joint reference, which is critical to communication. Where this immediate link is missing, reference is ambiguous, and where this constraint is violated (e.g., by combinations that scramble this contiguity between symbolic and indexical operations; so-called word-salad), reference typically fails.

This constraint derives from the unmasking of indexical constraints implicit in the interpretation of symbolic reference. Because symbolic reference is indirect and "virtual," by itself it can determine only ungrounded referential possibility. The subject, topic, or argument (= variable) performs a locative function by symbolizing an indexical relationship—a pointing to something else linked to it in some actual physical capacity (e.g., contiguous pragmatic or textual context). This reference determination cannot be left only in symbolic form, because isolated symbols (e.g., words and morphemes) only refer reciprocally to their "position" in the system or network of other symbols.

The importance of immediate contiguity in this relationship reflects the principal defining constraint determining indexical reference. Indexical reference must be mediated by physical correlation, contiguity, containment, causality, and so on, with its object in some way. Indexicality fails without this immediacy. There are, of course, many ways that this immediacy can be achieved, but without it, nothing is indicated. These constraints on indexicality are inherited by the grammatical categories and syntactic organization of sentences, propositions, and logical formulas.

To state this hypothesis in semiotic terms: A symbol must be contiguous with the index that grounds its reference (either to the world or to the immediate agreeing textual context, which is otherwise grounded), or else its reference fails. Contiguity thus has a doubly indexical role to play: contiguity (textually or pragmatically) of an index with the symbolizing sign vehicle as well as with some feature in the immediate context bridges between the symbol and that feature. This is an expression of one further feature of indexicality: transitivity of reference.

Simply stated, a pointer pointing to another pointer pointing to some object effectively enables the first pointer to also point to that object. This property is commonly exploited outside of language. Thus, the uneven wear on automobile tires indicates that the tires have not been oriented at a precise right angle to the pavement, which may indicate that they are misaligned, which may in turn indicate that the owner is not particularly attentive to the condition of the vehicle. Similarly, the indexical grounding of content words in a sentence can also be indirect, but only so long as no new symbolically functioning word is introduced to break this linear contiguity.

Of course, every word or morpheme in a sentence functions symbolically, and a word or phrase may take on a higher-order symbolic or indexical role in its combinatorial relationships to other language units at the same level. This flexibility provides a diversity of symbolized indexical relations. So, for example, arguments can be replaced by pronouns, and pronouns can point to other predicates and arguments, or they can point outside the discourse, or if a language employs gender marking of nouns a gender-specified pronoun can refer to the next most contiguous noun with agreeing gender expressed in the prior interaction, even if separated by many nonagreeing nouns and noun phrases. A sentence that lacks inferrable indexical grounding of even one component symbolic element will be judged ungrammatical for this reason. However, the basis for this judgment by nonlinguists is not determined with respect to either explicit rules or constraints. It is determined by the fact that the sentence does not have an unambiguous reference.

The exception that proves the rule, so to speak, is exemplified by highly inflected and/or agglutinated languages where indexical marking is incorporated directly into word morphology. In comparison with English, which maintains the indexical grounding of most of its symbolic functions by strict word-order constraints, these languages tend to have relatively free word order. This leads to a prediction: the more completely that indexical functions are incorporated into word morphology, the less restrictive the syntax, and vice versa.

Quantification and Transitivity

Related to this indexical function is the role of quantification in natural language and symbolic logic. In language, only nouns and the arguments of a verb require quantifiers. In logic, a well-formed expression requires more than just a function and its argument. Unambiguous predication requires "quantifying" the argument (unless it is a proper name). This latter requirement and exception are telling. In English, quantifiers include such terms as "a," "the," "some," "this," "these," and "all." These terms indicate the numerosity of what is being referred to, even if just in relative terms (such as "some"). Proper names are the exception because they refer to single individuals, whether an individual person or a named place, like a city or country. Reference in that case is unambiguous. It can also be unambiguous in the case of so-called mass terms like "water" or abstract properties such as "justice," since they have no clear individuality. This basic structural constraint is again due to the complex infrastructure behind symbolic reference.

Words like "a," "the," "some," "many," "most," "all," and so on, symbolize the virtual result of various forms of iterated indications or virtual ostentions (pointings). They are effectively virtual pointings that take advantage of transitive correlation with other indexical relationships, such as proximity information ("this," "that") or possession information ("his," "your") to differentiate indexicality.

Analogous to the case of implicit presupposed indexicality in holophrastic utterances, there are also contextual conditions where explicit quantification in language may be unnecessary. This is most obvious in cases where the possibility of specifying individuals is inappropriate (as in some mass nouns; e.g., "a water," "all waters," "few waters"). Pronominal reference does not require quantification because it is supplied by the text that it indicates (transitivity of indication). But when general terms are substituted for pronouns or other words serving overt indexical functions (e.g., "this" or "that") they inevitably require the addition of quantification. There are also, of course, many other exceptions to the need for quantification. Proper names and numbers do not require quantification when they are used to refer to a type as a singular class, because indicating would again be redundant.

Consequences of a Semiotic Reframing of Language

The long-unquestioned assumption that symbolic reference lacks intrinsic structure has tricked linguists into postulating ad hoc rule systems and algorithms to explain the structural constraints of language. Failure to pay attention to the iconic and indexical underpinnings of symbolic reference has additionally exaggerated the complexity of the language acquisition problem. This myopic avoidance of semiotic analysis has led to the doctrine of an innate language faculty that includes some modicum of language-specific knowledge, and this seeming logical necessity has supported an almost religious adherence to this assumption despite the biological implausibility of its evolution and the lack of neurological support for any corresponding brain structures or functions.

Unfortunately, contemporary semiotic theory has not been of much assistance, primarily because it has remained a predominantly structural theory tied to a static taxonomic understanding of semiotic relationships. However, when semiosis is understood as a process of interpretive differentiation in which different modes of reference are understood as dynamically and hierarchically constituent of one another, these many conundrums dissolve and these once apparently independent aspects of the language mystery turn out to have a common foundation.

So approaching language acquisition semiotically provides a functional account that can unify a wide range of grammatical and syntactic relationships. It also suggests that our naïve intuition about these linguistic regularities may be more accurate than the formal rule-governed approach would suggest. A naïve speaker seldom comments that an ungrammatical sentence breaks a rule, and is generally hard-pressed to articulate such a rule. Rather, the usual comment is that it just sounds wrong or that it does not make sense said that way.

In the case of ungrammatical sentences, naïve speakers know there is something wrong even if they cannot articulate it except to say that they are awkward or difficult to interpret and require some guesswork to make sense of them. Moreover, in everyday conversational speech, the so-called rules of grammar and syntax are only very loosely adhered to. This is usually because common interests and joint attention as well as culturally regularized interaction frames provide much of the indexical grounding, and so in such circumstances adherence to these strictures tends to be preferentially ignored. Not surprisingly, it was with the widespread increase in literacy that scholarly attention began to be focused on grammar and syntax, and with education in reading and writing, these "rules" began to get formalized. With the written word, shared immediate context, common pragmatic interests, and implicit presuppositions are minimally, if at all, available to provide indexical disambiguation, and so language-internal maintenance of these constraints becomes more critical.

Richness of the Stimulus

Finally, this semiotic functional analysis also provides an alternative understanding of the so-called poverty of the stimulus problem that is often invoked to argue that knowledge of grammar must be largely innate. Consistent with the fact that naïve speakers are generally unable to articulate the "rules" that describe their understanding of what is and is not a well-formed sentence, young children learning their first language are seldom corrected for grammatical errors (in contrast to regular correction of pronunciation).

Moreover, children do not explore random combinatorial options in their speech, testing to find the ones that are approved by others. They make remarkably prescient guesses. It has been assumed, therefore, that they must have some implicit understanding of these rules already available. In fact, children do have an extensive and ubiquitous source of information for learning to produce and interpret these basic semiotic constraints on predication, but it is not in the form of innate knowledge of grammar. It is in the form of knowledge about the intrinsic constraints of iconic and indexical reference that are discovered and internalized from social interactions prior to and during infancy and early childhood. We humans come into the world with attentional biases and behavioral tendencies that facilitate this learning.

First of all, discerning indexicality is a capacity that is basic to all cognition, animal and human. It requires no special training to become adept at the use of correlation, contiguity, and so on, in order to make predictions and thus to understand indexical relationships. This is essential to all forms of learning.

Evolved predispositions to point to or indicate desired objects or to engage joint attention have long been recognized as universally shared human predispositions that are poorly developed in other species. This universal human indexical predisposition provides the ideal scaffold to support what must be negotiated and must be progressively internalized within language structure. The early experience of communicating with the aid of pointing also provides additional background training in understanding the necessary relationship between symbols and indices.

Second, although there is little if any correction of the grammar and syntax in children's early speech, there is extensive pragmatic information about success or failure to refer or to interpret reference. This is in the form of pragmatic feedback concerning the communication of unambiguous reference; and this source of information attends almost every use of words. So I would argue that children do not "know" grammar innately, nor do they learn rules of grammar, and yet they nevertheless quickly "discover" the semiotic constraints from which grammars derive.

Although it is necessary to learn how a given language implements these constraints, the process is not inductive. It is not necessary for a child to derive general rules from many instances. Young children make good guesses about sentence structure—as though they already know "rules" of grammar—by tapping into more natural analogies to the nonlinguistic constraints and biases of iconicity and indexicality, and by getting pragmatic feedback about confused or ambiguous reference.

Universality?

Semiotic constraints should be agent-independent, species-independent, language-independent, and discourse-independent. They have been mistakenly assumed to be either innate structures or else derived from cognitive schemas or determined by sensorimotor biases and/or social communicative pragmatics. Though they are prior to language experience, and some are prerequisites to successful symbolic communication, they are neither innate nor socially derived.

They are emergent from constraints that are implicit in the semiotic infrastructure of symbolic reference and interpretive processes. They are in this way analogous to mathematical universals (e.g., prime numbers) that are "discovered" (not invented) as mathematical representation systems become more powerful. Though each form of symbol manipulation in mathematics has been an invention and thus a convention of culture, we are not free to choose just any form if we want to maintain consistency of quantitative representation. Likewise, as languages become more complex and expressively powerful, they also become more constrained; and as literary forms have become removed from the pragmatic contexts of day-to-day spoken communication, the loss of extralinguistic indexicality has demanded more rigorous adherence to semiotic constraints of grammar and syntax to avoid referential ambiguity and equivocation. It should not be surprising, then, that the rise of widespread literacy leads to official efforts to establish norms of "proper" grammar and syntax.

Semiotic constraints are the most ubiquitous influences on language structure, and indeed they are even more universal than advocates of mentalese could have imagined—because they are not human-specific. They are universal in the sense that the constraints of mathematics are universal. They would even be relevant to the evolution of symbolic communication elsewhere in the universe. But they are not like exceptionless "rules." Different languages, everyday spoken interactions, and artistic forms of expression can diverge from these constraints to varying extents, but at the cost of ambiguity and confusion of reference. In general, these constraints will probably be the most consistent regularities across the world's languages because means to minimize this divergence will be favored by the social evolution-like processes of language transmission from generation to generation.

Of course, reflecting on the larger list of factors contributing to the properties most widely shared across languages (cf. Table 1), we must acknowledge the contributions of both human-specific neurological constraints and historically contingent social constraints. They, too, contribute to the many nearly universal regularities that characterize the world's languages. And although many do indeed reflect innate influences that may have evolved specifically due to their contributions to easing language acquisition and performance, none determine language organization in a generative sense. Rather, along with the ubiquitous semiotic constraints discussed in this chapter, they add to the collective influences of the whole set.

References

Aristotle (ca. 350 BCE). *On interpretation* (Translated by E. M. Edghill). http://classics.mit.edu//Aristotle/interpretation.html

Convention. (n.d.). In *Merriam-Webster's online dictionary* (11th ed.). Retrieved from http://www.m-w.com/dictionary/convention

Deacon, T. (2012). Beyond the symbolic species. In T. Schilhab, F. Stjernfeldt, & T. Deacon (Eds.), *The symbolic species evolved*. New York: Springer.

Everett, D. L. (2005). Cultural constraints on grammar and cognition in Pirahã:Another look at the design features of human language. *Current Anthropology, 46,* 621–646.

Harnad, S. (1990). The symbol grounding problem. *Physica D, 42,* 335–346.

Hauser, M., Chomsky, N., & Fitch, W. T. (2002). The faculty of language: What is it, who has it, and how did it evolve? *Science, 298,* 1569–1579.

Hume, D. (1739–1740). *A treatise of human nature: Being an attempt to introduce the experimental method of reasoning into moral subjects.* London.

Lewis, D. (1969). *Convention: A philosophical study.* Princeton: Princeton University Press.

Peirce, C. S. (1931–1935, 1958). *The collected papers of Charles Sanders Peirce* (vols. 1–6, ed. Charles Hartshorne & Paul Weiss; vols. 7–8, ed. Arthur W. Burks). Cambridge, MA: Harvard University Press.

Raczaszek-Leonardi, J., & Deacon, T. (in press). *The symbol ungrounding problem.*

Dialogue on Symbolic Thought and Communication

Participants: **Dermot Barnes-Holmes, Yvonne Barnes-Holmes, Terrence W. Deacon, and Steven C. Hayes**

Steven C. Hayes: Maybe we could start at 35,000 feet and just ask: What do we expect of a theory of symbolic thought and communication?

Terrence W. Deacon: What I focus on in my chapter—though not necessarily in my work in general—is the between-symbol relationships, the things that are called grammar and syntax. And yet, my book *The Symbolic Species*—the title gives this away—is all about what's unusual about symbols. And, of course, I don't talk about this mutual entailment relationship. Recently, people have begun to look at stimulus equivalence in fMRI studies, and it's looking as though it depends on prefrontal cortex activity. Interestingly, the argument I make in *The Symbolic Species* suggests that there has been a significantly increased role of the prefrontal cortex in humans, both in general brain function and specifically in language.

An aim of my work was to show that the prefrontal cortex has a kind of map that has to do with what I call orienting space. And the way that I worked this out was to look at prefrontal connections with an area in the midbrain called the superior colliculus; specifically the deep superior colliculus, which is involved in orienting attention in personal space. I was very interested in the fact that the prefrontal cortex doesn't have a sensory map or a motor map, whereas every place else in the cerebral cortex is map-like. To find that there is an orienting map in the prefrontal cortex that apparently plays a role in controlling and regulating the possible orientations you could take toward a thing helped me understand why the prefrontal cortex is so much involved in combinatorial assessments of various kinds. I think that's why it's involved in this stimulus equivalence effect. So I think the stimulus equivalence effect—or mutual entailment—plays a significant role in our facility for language. I don't think it's necessarily what you might call the smoking gun that made language possible; but I think that biasing the human tendency toward stimulus equivalence has made language easier, and I think that human prefrontal enlargement is the result of selection for increased language facility.

Dermot Barnes-Holmes: The study of stimulus equivalence, as we indicated in our chapter, went back to the early work of Murray Sidman, who I think coined the phrase and has been the source of many, many studies. Years ago, David Dickens showed that when people were demonstrating equivalence, areas associated with language were more active during the equivalence responding than during a control task. There is considerable evidence, mounting evidence, that the areas of the cortex associated with symbolic relations in natural language are also involved when human beings are demonstrating equivalence relations in a scanner, or when EEGs are being recorded.

One of Steve's early studies, with Gina Lipkins in 1993, showed the emergence of mutual entailment before you see the emergence of combinatorial entailment in the developmental trajectory of relational framing, or equivalence relations. Symmetry or mutual entailment appears to emerge, if only by a matter of months, before you see combinatorial entailment. In our early work on RFT, we just defined a frame as mutual entailment, combinatorial entailment, and transformation of function, and I think we saw the frame as the fundamental unit. But now I think that we're seeing mutual entailment as the fundamental unit. On balance, mutual entailment gets you some way toward very basic symbolic relations and not much more; combinatorial entailment, increasingly complex relational networks, and so on, as outlined in our chapter, is where human language and cognition really take off.

Steve: What do we expect of our theories? The idea is that stimulus equivalence is an example of relating and we're evolutionary prepared to do it, but it still requires training within the history of the individual—and so some of these accounts suggest how that training should go. In that approach, if you can't remediate it then there's something wrong with the theory. Like, Terry, if we could take the things you're pointing to and build a training program, would it bother you if it didn't lead to the outcomes that you were hoping for, say with developmentally delayed children?

Terry: I think there is no question that in autistic children there are some neurological issues that we don't fully understand. And, just as with trying to work with adult aphasia patients after a stroke, there are some things you just cannot bypass. You can't repair it, though you may be able to shift to secondary functional systems that can carry some of the load, but usually you can't completely compensate. And my guess is that's the case with autism as well.

One other thing that I'd be interested in focusing on in this respect has to do with recent work that has come out of Berkeley, just this last year from the Gallant Lab. It's a paper published in *Nature* by Huth and colleagues in which they played long segments of stories to subjects during fMRI scanning, and used a complex principle complements analysis to attempt to identify what areas of the cerebral cortex were preferentially active and thus associated with particular semantic categories. This resulted in an almost unbelievable map of the cerebral cortex overlayed with a mosaic of words, in which almost every area of the cerebral cortex is covered, showing that nearly the whole cortex plays a role in making semantic associations. Very few areas of the cerebral cortex were not involved. And that includes areas that we only consider sensory or motor, for example—not "language areas" per se. This widespread use of diverse cortical areas involved in the semantic analysis of language also suggests that there has been a really remarkable recruitment of the entire cerebral cortex in language during evolution; and that when we focus on so-called "language areas," we're looking at only a small part of the picture. This suggests a very different view of language than we have traditionally been taught.

Dermot: Going back to the question about the extent to which a theory, or account, should lead to some practical gain, the very units of analysis that we propose in relational frame theory are built around that absolutely essential philosophical need for prediction and influence, remediation, improvement, and so on. That's not to say that you don't recognize there may be structural deficits or structural issues that cannot be resolved. But you start from the assumption that you have to come up with units of scientific analysis that specify the very variables that you need to manipulate, at least in principle, to remediate deficits or control for excesses, or to construct behaviors that don't yet exist.

With mutual entailment, you want to be able to show that multiple exemplar training with young children matters—teach them that they can relate things as being the same, as being the opposite, as being different, as being above and below, to the left of and to the right of, and do it across various settings and contexts, and so forth, and do it to a level of fluidity, or flexibility, so you can get children to do it one way, then bring in a cue that says do it the other way and they will do it the other way for you, and so on. And then build up to more complex relations under contextual control. So your very units of analysis must specify what you need to do: the variables you need to target to produce the behaviors you're defining as language. But you want the remediation, you want the behaviors you're targeting with your theory when you treat them with your theory, to show improvements then in language and cognition. If you don't get that, relational frame theory is not a theory of language and cognition.

Steve: And this links to something that's brought into high relief throughout this volume—if you read all the chapters, you'll see it—because here you have two approaches that are specifically *not* mentalistic. They're evolutionarily oriented in the broadest sense, both within the lifetime of individuals and across lifetimes. And yet they haven't been really speaking to each other. This sort of archipelago nature of the unvisited islands even within the same basic philosophical approach is remarkable. And it is notable that there's really not a robust applied evolution science. There are a few efforts, a few, but they're not robust. Why is that?

Terry: Part of it has to do with this disciplinary siloing that's been going on. Very few people in psychology follow the recent developments in Evo-Devo, linking development and evolutionary theory. There really is so much of a difference between the simple-minded neo-Darwinian perspectives and the current perspective on evolutionary theory that's driving a lot of interesting research that hasn't made it out of the discipline. And I would say the same thing is probably true in the reverse.

Another feature is that relevant to applied research, you really need to have an understanding of the neural substrates involved. All of my work on the evolution of the brain has involved comparing brains of different species and focusing on the developmental processes that produce their differences in structure. Only then can we ask the question, "What makes humans different? What developmental processes might be involved?" That's indirect evidence, to be sure, very indirect, and as a result it's hard to apply, particularly with respect to the nervous system. It's hard to assume, "Okay, now I can use that for therapeutic purposes." My focus has not been therapeutic. As an evolutionary biologist, I aim to understand these process themselves, to understand how brain development has produced brain differences in different species, and how it might help to explain the distinct features of human mentality. So I think that that kind of—I wouldn't say necessary, but hard-to-avoid—island-like effect that you refer to is difficult to overcome. And that's why conversations like this should not be just one-off. They have to continue and we have to find ways—I suspect we have to force them—to regularly take place.

The therapeutic relevance of my work involves the acquisition of grammar and syntax. It effectively resolves a long-standing mystery that Noam Chomsky called the "poverty of the stimulus." The poverty of the stimulus concerns the fact that children don't get corrected for their use or misuse of the rules presumed to comprise grammar and syntax. I argue that this way of conceiving the problem is confused. In fact, children do get extensive feedback, but it's not about rules of grammar and syntax. It's about failure or ambiguity of reference. And that is what the grounding is all about, and what is learned in prelinguistic communication, and what language must preserve despite the lack of clues provided by the arbitrary symbols. Reference fails because of indexical problems, and in language, this is either provided by extralinguistic context or the iconic and indexical aspects of grammar and syntax.

Dermot: This bears on the literature on exemplar training, or on attempts to use exemplar training to generate derived relations with nonhumans. And again, the results were largely negative, it just doesn't seem to work with nonhuman species. This sits more readily, I think, with the idea that even mutual entailment, the simplest unit, is built, or very much embedded, in learning and in cooperative acts.

Steve: There's a small number of studies looking at exemplar training with very young children—in fact, your team is involved in some of that, and my takeaway from that seemed to be that if you didn't provide that kind of training to children, they weren't going to spontaneously do it. Do you think that's been settled? That basically, you have to train symmetry as an operant in order to be able to get it and see it that way?

Dermot: I think the evidence indicates that you need that history of exemplar training for particular types of frames to emerge.

Steve: Some of the things that are in Terry's chapter, it seems to me, actually kind of involve the things that need to be done to do that. Take things like the traditional Skinnerian operants; they include fair amounts of these kinds of indexical relations that are directly trained. Same thing, for example, if you take the recent evidence with Mark Dixon's protocol, where he starts training these higher-order derived relations, but only after he trains a robust ability to respond more in these nonarbitrary ways that sort of form a foundation for it. In other words, I think this complementary nature of it might actually be farther along than it looks.

Yvonne Barnes-Holmes: Good contextual control accounts for a lot of our success in this training and I think connects to what Terry is saying. We have children, for example, typical and atypical, particularly atypical, who can do mutual entailment in one context and clearly cannot in another. There's not sufficient behavioral momentum in the system for the mutual entailment to generalize from one context to another. I think what Terry's work does is it alludes to context, and there are many contextual features that we have now started to focus on when training relational repertoires with young children.

Steve: It seems possible that behavioral community may almost be coming into a partnership with some of the things that Terry and his colleagues are pointing to as kind of an experimental and extended arm.

Terry: One of my earliest studies as a graduate student was to reconsider the Piagettian sensorimotor tasks in terms of prefrontal cortex maturation. I suspected that what we were seeing in the stages of sensorimotor "learning" was development of the capacity for transfer learning—the capacity to use what is learned in one context and shift that knowledge to a new context that shares relational features with the first context, but with different stimulus features. Stimulus equivalence, or whatever you want to call it, involves relational features, not just stimulus features. Transfer learning turns out to be poorly developed across species—it is even poorly developed in apes such as chimpanzees and gorillas.

Children become good at transfer learning between two and three years of age, similar to when they readily exhibit stimulus equivalence. Older children solve these kinds of tasks without difficulty. So I think that one of the factors that we have to consider, when we talk about development, is that the changes we observe are not just due to experience; it takes both experience and having a nervous system capable of taking advantage of that experience. This seems to be a crucial factor that is often overlooked.

Dermot: I think that we have been talking past each other in the history of the discipline, particularly when trying to deal with language, in a way that wasn't necessarily good for any of us. But I think what we're seeing here is different. I have Terry's chapter in front of me now, and I can see my notes all over the manuscript with things like "Yes, yes, RFT-consistent; yes, yes, makes sense; oh, that's very useful; oh, I didn't know that; I will use that in my lectures on RFT." I don't think there's a single big "No" in there that I've written where I've found I fundamentally disagree.

Steve: Well I think people will see that this is largely true in the whole volume. You're going to see a remarkable amount of overlap. And not just "me too," but an opportunity, despite differences, to build and support each other.

Terry: And neither nature nor nurture gets at the whole issue. There is a whole third realm that we've ignored, what I call semiotic constraint. We've ignored it because we've thought that if symbols are arbitrary, grammar either had to be built in, given in our brains, or that we had to create it by some conventional process which we have not yet analyzed semiotically either. So my interest in this chapter and project is to get people to say, "Oh, this is an unexplored territory."

Steve: Well, let's see where the community takes these opportunities.

CHAPTER 6

The Evolutionary Basis of Risky Adolescent Behavior

Bruce J. Ellis

University of Utah

B ehaviors such as aggression, crime, promiscuity, reckless driving, and drug use are often called risky because they are likely to harm the individuals who engage in them, others around them, or society as a whole. Adolescents are more likely to engage in these behaviors than people at any other stage of the life cycle (Institute of Medicine [IOM] & National Research Council [NRC], 2011; Steinberg, 2008). Thus, the legal system, policy makers, and scientists have focused an enormous amount of attention on risky adolescent behavior as a problem in need of a solution.

Given the problems caused by risky adolescent behaviors, it is tempting to regard them as maladaptive. Indeed, the prevailing conceptual framework for thinking about these behaviors considers them to be negative or disturbed developmental outcomes arising from stressful life experiences (together with personal or biological vulnerabilities). According to this framework, children raised in supportive and well-resourced environments (e.g., who live in communities with social networks and resources for young people; who have strong ties to schools and teachers; who benefit from nurturing and supportive parenting that includes clear and consistent discipline; who are exposed to prosocial peers) tend to develop normally and exhibit healthy behavior and values. In contrast, children raised in high-stress environments (e.g., who experience poverty, discrimination, low neighborhood attachment, and community disorganization; who feel disconnected from teachers and schools; who experience high levels of family conflict and negative relationships with parents; who are exposed to delinquent peers) often develop abnormally and exhibit problem behaviors that are destructive to themselves and others. Different developmental outcomes are regarded as "adaptive versus maladaptive" depending on the extent to which they promote versus threaten young people's health, development, and safety. I refer to this set of guiding assumptions as the *developmental psychopathology model* of risky adolescent behavior (Ellis et al., 2012).

Although the validity of the developmental psychopathology model seems self-evident, one purpose of this chapter is to show that it is at once inadequate and incomplete. To understand why, consider the basic definition of risk as "the possibility of suffering harm or loss" (YourDictionary.com). This definition—the backbone of the "risk factor" approach to psychiatric and biomedical disorder— only captures the downside of risk without considering why people take risks. Risky behaviors are not maladaptive if the expected benefits outweigh the expected costs. People take *calculated* risks all the time, at all stages of the life cycle. We cannot legitimately regard risky behaviors as maladaptive based only on their costs. Yet a balanced cost-benefit analysis is seldom performed in the developmental psychopathology literature, which is dominated by pathologizing views of risk.

Then there is the question of who benefits and in what metric. As much as we might endorse the principle of "Do unto others as you would have others do unto you," people often act for the benefit of themselves or a circumscribed group at the expense of, or at least without considering the welfare of, other individuals and groups. Further, risky adolescent behaviors can result in net harm in terms of a person's own phenomenology and well-being (e.g., producing miserable feelings or a shortened life), the welfare of others around them, or the society as a whole, but still be *adaptive* in an evolutionary sense. Consider, for example, high-risk behaviors that expose adolescents to danger and/or inflict harm on others but increase dominance in social hierarchies and leverage access to mates (e.g., Gallup, O'Brien, & Wilson, 2011; Palmer & Tilley, 1995; Sylwester & Pawłowski, 2011).

These examples should make it clear that "risky" is not the same as "maladaptive." The problems associated with risky adolescent behaviors are real, and there is a strong need to reduce them, but regarding them as dysfunctional is merely a label, not a solution. Rather, from an evolutionary perspective, viable solutions involve understanding the functions of risk taking in the contexts of adolescents' lives.

Although taking into account both costs and benefits of risky behavior is a crucial first step, it is important to go beyond economic cost-benefit models and understand the concept of adaptive behavior from an *evolutionary* perspective. Traditional economic models make assumptions about utility maximization that are false for people of all ages and especially for adolescents (Beinhocker, 2006; Hodgson & Knudsen, 2010). We are a biological species, albeit one with remarkable capacities for psychological and cultural change. We have a long evolutionary history, and many other primates, mammals, and vertebrates share our adaptations. Adolescence, that developmental period when organisms sexually mature and attempt to break into the breeding pool, is a period of heightened risk in *many* species (Weisfeld, 1999), and it behooves us to understand why in terms of understanding the causes and consequences of risk taking. It is not an accident that risky adolescent behaviors are all about the standard metrics of evolutionary success—survival and sex—and attaining the resources, relationships, and social status that ensure these outcomes.

I will call the study of risky adolescent behavior from an evolutionary perspective the *evolutionary model*, in contrast to the developmental psychopathology model. The two models are not mutually exclusive, and they both share the same practical goal of reducing problem behaviors for the long-term benefit of individuals and society (regardless of the evolutionary adaptiveness of the behavior). The evolutionary model, however, can help us achieve that goal through increased understanding of the adaptive logic and motivation that underlie so many risky adolescent behaviors (Ellis et al., 2012). Even when risky behavior *is* genuinely pathological (i.e., harmful from both an evolutionary and a developmental psychopathology perspective), a detailed understanding of adaptations in the context of past and present environments is often needed to understand the nature of the pathology.

Both the developmental psychopathology model and the evolutionary model have a compelling internal logic that makes them appear self-evident. It seems obvious that high-risk behaviors must be dysfunctional. It seems equally obvious that there must be something in it for the kids who engage in these behaviors. The developmental psychopathology model has strongly influenced thinking about adolescent development over the past half-century; it is truly the dominant model. The evolutionary model has roots that extend into the past, but it has only started to mature during the last two decades and is still relatively unknown by developmental psychologists. Drawing on Ellis et al. (2012), this chapter attempts to pull those pieces together and apply them to risky adolescent behavior.

The evolutionary model contrasts with the developmental psychopathology model, which emphasizes that exposure to environmental adversity places children and adolescents at elevated risk for developing cognitive, social, emotional, and health problems (e.g., Shonkoff, Boyce, & McEwen, 2009). Although evolutionary models also recognize the strong effects of environmental stress, the developmental psychopathology and evolutionary models conceptualize environmental stress and adversity, as well as environmental resources and support, in a different manner. According to the developmental psychopathology model, positive or supportive environments, by definition, promote "good" developmental outcomes (as defined by dominant Western values; e.g., health, happiness, secure attachment, high self-esteem, emotion regulation, educational and professional success, stable marriage), whereas negative or stressful environments, by definition, induce "bad" developmental outcomes (as defined by that same value system; e.g., poor health, insecure attachment, substance abuse, conduct problems, depression, school failure, teenage pregnancy).

In contrast, from an evolutionary perspective, environments that are positive in character disproportionately afford resources and support that enhance fitness, whereas environments that are negative in character disproportionately embody stressors and adversities that undermine fitness. The evolutionary model posits that natural selection shaped our neurobiological mechanisms to detect and respond to the different ratios of costs and benefits afforded by positive versus negative environments. Most importantly, these responses are not arbitrary but instead function to adaptively calibrate developmental and behavioral strategies to match those environments (e.g., Belsky, Steinberg, & Draper, 1991; Del Giudice, Ellis, & Shirtcliff, 2011). This view of development challenges the prevailing psychopathology analysis of dysfunctional outcomes within settings of adversity. In particular, an evolutionary perspective contends that both stressful and supportive environments have been part of human experience throughout our history, and that developmental systems shaped by natural selection respond adaptively to both kinds of contexts (Ellis, Boyce, Belsky, Bakermans-Kranenburg, & Van IJzendoorn, 2011; Ellis et al., 2012). Thus, stressful environments do not so much *disturb* development as *direct or regulate* it toward strategies that are *adaptive* under stressful conditions (or at least were adaptive during our evolutionary history).[1]

It is important to note that optimal adaptation (in the evolutionary sense) to challenging environments is not without real consequences and costs. Harsh environments often harm or kill people, and the fact that children and adolescents developmentally adapt to such rearing conditions (reviewed in Ellis, Figueredo, Brumbach, & Schlomer, 2009; Pollak, 2008) does not imply that such conditions either promote child well-being or should be accepted as unmodifiable facts of life (i.e., David Hume's "naturalistic fallacy"). Developmental adaptations to high-stress environments enable individuals to make the best of a bad situation (i.e., to mitigate the inevitable fitness costs), even though "the best" may still constitute a high-risk strategy that jeopardizes the person's health and survival (e.g., Mulvihill, 2005; Shonkoff et al., 2009) and may still be harmful to the long-term welfare of the society as a whole. Further, there are genuinely novel environments, such as Romanian or Ukrainian orphanages (Dobrova-Krol, Van IJzendoorn, Bakerman-Kranenburg, & Juffer, 2010), that are beyond the normative range of

---

1   This model is not intended to be all-encompassing, as there are genuinely pathological conditions (e.g., genetic abnormalities, neurotoxins, head injury) that interfere with the ability of individuals to use adaptive strategies in a variety of contexts, particularly under stress.

conditions encountered over human evolution. Selection simply could not have shaped children's brains and bodies to respond adaptively to collective rearing by paid, custodial, non-kin caregivers (Hrdy, 1999). Exposures to such challenging yet (evolutionarily) unprecedented conditions can be expected to induce pathological development, not evolutionarily adaptive strategies (as discussed below).

The developmental psychopathology model has led researchers to focus on the deleterious effects of adverse environments, such as the impact of familial and ecological stressors on mental health outcomes (e.g., adolescent onset of psychopathology). Dysfunctional behavior in adolescence is seen as the natural consequence of exposure to harsh, unpredictable, or uncontrollable socio-ecological contexts. The developmental psychopathology model thus places undue emphasis on expected *costs* and largely ignores expected *benefits* of risk taking, making it difficult to explain adolescent motives for risky behavior. As discussed by Ellis and colleagues (2012), this bias has led the field to neglect a critically important question: What's in it for the adolescent?

Because different risk-taking strategies are potentially adaptive or maladaptive, depending on context, we can expect natural selection to favor risk-taking strategies that are contingent on reliable and valid environmental cues. Central to this perspective is the concept of conditional adaptations: "evolved mechanisms that detect and respond to specific features of childhood environments, features that have proven reliable over evolutionary time in predicting the nature of the social and physical world into which children will mature, and entrain developmental pathways that reliably matched those features during a species' natural selective history" (Boyce & Ellis, 2005, p. 290; for a comprehensive treatment of conditional adaptation, see West-Eberhard, 2003). Conditional adaptations underpin development of contingent survival and reproductive strategies and thus enable individuals to function competently in a variety of different environments. Viewed from within this framework, the adolescent who responds to a dangerous environment by developing insecure attachments, adopting an opportunistic interpersonal orientation, engaging in a range of externalizing behaviors, and sustaining an early sexual debut is no less functional than the adolescent who responds to a well-resourced and supportive social environment by developing the opposing characteristics and orientations (see Belsky et al., 1991; Ellis et al., 2011).

In summary, the developmental psychopathology model is limited in its ability to explain patterns of risky adolescent behavior because it does not explicitly model evolutionary constraints—how natural selection shaped the adolescent brain to respond to environmental opportunities and challenges—and does not adequately address why adolescents engage in risk-taking behaviors in the first place. Explaining high-risk behaviors as adaptive in an evolutionary sense does not justify high-risk behavior in a normative sense; however, by providing a unique vantage point on the functions of risky adolescent behavior, the evolutionary model can lead to practical solutions that have not been forthcoming from the developmental psychopathology perspective (Ellis et al., 2012).

The Adolescent Transition Is an Inflection Point in the Development of Socio-Competitive Competencies and the Determination of Social and Reproductive Trajectories

Spanning the years from the onset of puberty until the onset of adulthood, adolescence is fundamentally a transition from the prereproductive to the reproductive phase of the life span. The developing person reallocates energy and resources toward transforming into a reproductively competent individual. From an evolutionary perspective, a major function of adolescence is to attain reproductive status—to develop the physical and social competencies needed to gain access to a new and highly contested biological resource: sex and, ultimately, reproduction. Both sexual promiscuity and the intensity of sexual competition peak during adolescence and early adulthood (Weisfeld, 1999; Weisfeld & Coleman, 2005), when most people have not yet found a stable partner and the mating market is maximally open. This time of heightened promiscuity and competition may help young people determine their own status and attractiveness, refine their mate preferences, and practice mate attraction strategies (Weisfeld & Coleman, 2005). These processes are central to the establishment of identity in adolescence. Most critically, the adolescent transition is an inflection point (i.e., a sensitive period for change) in developmental trajectories of status, resource control, mating success, and other fitness-relevant outcomes (Ellis et al., 2012).

To achieve success at the critical adolescent transition, natural selection has favored a coordinated suite of rapid, punctuated changes—puberty—across multiple developmental domains. Driven by maturational changes in secretion of growth hormones, adrenal androgens, and gonadal steroids, pubertal development includes maturation of primary and secondary sexual characteristics, rapid changes in metabolism and physical growth, activation of new drives and motivations, and a wide array of social, behavioral, and affective changes (Table 1). These puberty-specific processes function to build reproductive capacity and increase socio-competitive competencies in boys and girls. Thus, increases in height, weight, and muscularity; more prominent jaws and cheekbones; emergence of body and facial hair; greater cardiovascular capacity and upper-body and grip strength; and broader shoulders make the male body more hardy, formidable, and sexually attractive to females. Breast development, fuller lips, widening of the hips, fat accumulation, and attainment of adult height and weight signal fertility and make the female body more sexually attractive to males. Changes in metabolic rates, food consumption, and sleep patterns support this physical metamorphosis. The adolescent phase shift also increases nighttime activity (when most sexual and romantic behavior occurs). Heightened sexual desire increases motivation to pursue, attract, and maintain mating relationships. Increased sensation-seeking and emotional responsivity promote novelty-seeking and exploration and may increase pursuit of socially mediated rewards. Higher levels of aggression and social dominance both facilitate and reflect the higher-stakes competition that is occurring in adolescence over sex, status, and social alliances. Delinquent and risky behaviors (e.g., crime, rule-breaking, fighting, risky driving, drinking games) often have signaling functions that enhance reputations for bravery and toughness and can leverage position in dominance hierarchies, especially for males. Distancing of parent-child relationships increases autonomy and reorients the adolescent toward peer relationships and the mating arena. Increasing levels of anxiety and depression in girls may reflect heightened sensitivity to negative social evaluations (discussed below). Although any given puberty-specific change listed in Table 1 may be modest in size, taken together, the pubertal transformation is dramatic.

Table 1. Puberty-Specific Morphological and Biobehavioral Changes (Independent of Age)

Adopted from Ellis et al. (2012); citations to supporting empirical work are reported therein.

Puberty-Specific Change
1. Sexual development. Maturation of primary and secondary sexual characteristics. Growth spurt in height and weight. Each stage of pubertal development moves the adolescent toward greater physical reproductive capacity.
2. Sleep. Circadian shift in sleep timing preference, with later onset of sleep and morning rise times, occurs in mid-puberty. Increased sleepiness, which may indicate increased need for sleep, is linked to more advanced pubertal development.
3. Appetite and eating. Total caloric intake increases over the stages of pubertal development, with approximately a 50-percent increase from prepuberty to late-puberty. Sharpest increases occur from pre- to mid-puberty in girls and mid- to late-puberty in boys, corresponding to the periods of most rapid growth in females and males, respectively.

4. Sexual motivation. Each stage of pubertal development increases the probability of being romantically involved (e.g., dating), being sexually active, sexually harassing members of the other sex, and being "in love." Effects generally apply to both boys and girls.

5. Sensation-seeking (wanting or liking high-sensation, high-arousal experiences). Boys and girls with more advanced pubertal development display higher levels of sensation-seeking and greater drug use.

6. Emotional reactivity. Boys and girls with more advanced pubertal development (pre- to early versus mid- to late) display greater reactivity of neurobehavioral systems involved in emotional information processing.

7. Aggression/delinquency. Progression through each Tanner stage is associated with increasing levels of aggression and delinquency in both boys and girls.

8. Social dominance. During pubertal maturation, higher levels of testosterone are associated with greater social dominance or potency in boys. This relation appears to be strongest in boys who affiliate with nondeviant peers.

9. Parent-child conflict. Parent-child conflict/distance increases and parent-child warmth decreases over the course of pubertal maturation. Some research suggests a curvilinear relation, with conflict/distance peaking at mid-puberty. Effects generally apply to both boys and girls.

10. Depression and anxiety. More advanced pubertal maturation, as well as underlying changes in pubertal hormone levels, are associated with more symptoms of depression and anxiety and greater stress perception in girls.

Puberty-specific neuromaturational changes, together with age- and experience-dependent changes in the adolescent brain, make human adolescence a period of major and dynamic synaptic reorganization, ranging from neurogenesis to programmed cell death, elaboration and pruning of dendrites and synapses, myelination, and sexual differentiation (Blakemore & Mills, 2014; Piekarski et al., 2016). It has been hypothesized that this remodeling and refinement of behavioral circuits opens the brain to environmental input and thus creates a sensitive period for learning and developmental change (Blakemore & Mills, 2014; Piekarski et al., 2016). Adolescence may thus constitute a window of vulnerability and opportunity—an inflection point where experiences can disproportionately influence developmental trajectories.

Consistent with the sensitive period hypothesis, there is a dramatic increase during adolescence in death and disability (e.g., US morbidity and mortality rates) related to depression, eating disorders, alcohol and other substance use, accidents, suicide, homicide, reckless behavior, violence, and risky sexual behavior (IOM & NRC, 2011; Steinberg, 2008). Further, in addition to directly measureable morbidity and mortality, adolescence is also a key time in the development of many unhealthy behaviors and habits (e.g., smoking cigarettes, substance abuse, eating disorders) that will have an enormous negative impact on long-term health across the life span. At the same time, however, adolescence is a key period of opportunity to impact developmental trajectories in positive directions. It is a time when youth develop healthy habits, interests, skills, and inclinations, and align their motivations and inspirations toward positive goals. At a psychological level, these changes and the reorganization that occur in adolescence are often consolidated through identity formation.

From an evolutionary perspective, the adolescent inflection and associated identity formation processes are critical because they regulate development of alternative reproductive strategies (see Ellis et al., 2012). Maturational experiences in adolescence interact with social context to shape long-term social and reproductive trajectories. Among males, early-maturing boys tend to be taller and stronger than their same-age peers and often attain high status within the peer group (reviewed in Weisfeld, 1999). Jones (1957) found that early-maturing boys were more socially poised and less anxious in adolescence. In longitudinal analyses, these boys, although only achieving about the same final height as their later-maturing peers, remained more self-assured in adulthood, scored higher on personality characteristics associated with dominance, and were more likely to attain executive positions in their careers. In a more recent longitudinal study, height attained in adolescence, rather than final adult height, positively predicted income in adult males (Persico, Postlewaite, & Silverman, 2004), again suggesting long-term consequences of "stature" in adolescence. Finally, early-maturing boys (but not early-maturing girls) display a more unrestricted sociosexual orientation (i.e., greater willingness to engage in casual sex) and have a higher number of lifetime sexual partners in young adulthood than do later-maturing boys (Ostovich & Sabini, 2005; see also Ellis, 2004). Interestingly, pubertal status is clearly linked to levels of aggressive/delinquent behavior in pubescent boys, but timing of puberty does not feed forward to predict aggressive/delinquent behavior in young men (Najman et al., 2009). It may be that status obtained in adolescence is long lasting and obviates the need for elevated externalizing behaviors in adulthood.

Existing research has also documented the long-term sequelae of early pubertal development in girls. Women who experienced early pubertal development, compared with their later-maturing peers, tend to have higher levels of serum estradiol and lower sex hormone-binding globulin concentrations that persist through twenty to thirty years of age; have shorter periods of adolescent subfertility (the time between menarche and attainment of fertile menstrual cycles);

experience earlier ages of first sexual intercourse, first pregnancy, and first child-birth; display more negative implicit evaluations of men in early adulthood; attain lower educational outcomes and occupational status and engage in more aggressive/delinquent behavior as young adults; and are heavier, carry more body fat, and display higher allostatic loads (cumulative biological "wear and tear") in adolescence and early adulthood (reviewed in Ellis et al., 2012). These effects can be conceptualized as part of a developmental continuum in which early environmental conditions (e.g., scarcity or unpredictability of resources, conflictual family relationships, lack of parental warmth and support) predict earlier pubertal maturation in girls (Belsky, Steinberg, & Draper, 1991; Ellis, 2004; Mendle, Ryan, & McKone, 2016), which in turn regulates important dimensions of social and reproductive development (see especially Baams, Dubas, Overbeek, & Van Aken, 2015; Belsky, Steinberg, Houts, Halpern-Felsher, 2010; James, Ellis, Schlomer, & Garber, 2012).

Whether measured in terms of developmental psychopathology or reproductive fitness, much is at stake in the adolescent transition. For this reason, I hypothesize that natural selection favored especially strong emotional and behavioral responses to social successes and failures at this juncture. This hypothesis concurs with (1) animal data showing heightened vulnerability to stress in adolescence, particularly among females, as a result of increased glucocorticoid receptor expression in the cortex and stronger and more prolonged corticosterone responses following acute stress (Andersen & Teicher, 2008); and (2) human data demonstrating increasing reactivity of stress-sensitive neuroendocrine systems over the transition to adolescence (Gunnar, Wewerka, Frenn, Long, & Griggs, 2009; Stroud et al., 2009).

Among individuals whose current condition or circumstances are predictive of future reproductive failure (e.g., unemployed, unmarried, marginalized young men with few resources or prospects), low-risk strategies that minimize variance in outcomes have limited utility. In contrast, high-risk activities (e.g., confrontational and dangerous competition with other males, gang membership, criminal activities), which by definition increase variance in outcomes, become more tolerable—even appealing—because success at these activities can yield otherwise unobtainable fitness benefits for disenfranchised individuals (Wilson & Daly, 1985). In this sense, risky behavior may strategically increase access to status, resources, and mating—the pillars of reproductive success. Extensive data supports this theorizing, uniformly demonstrating markedly elevated rates of violence among young, poor, marginalized males (e.g., Archer, 2009)—a group that may largely account for the dramatic rise in serious violence and delinquency in adolescence. Further, peer aggression and risk-taking behaviors among adolescents are reliably associated with greater mating opportunities (Gallup et al., 2011; Palmer & Tilley, 1995; Sylwester & Pawłowski, 2011).

The other side of the coin is sharply elevated rates of depression and anxiety in adolescent girls. A critically important resource at stake for adolescent girls is social support. Humans are a cooperatively breeding species (Hrdy, 2009), and women in traditional societies depend on an extended social network, including both mates and female allies, to help raise very energetically expensive offspring, obtain and share resources, and provide protection. Over evolutionary history, adolescence may have been a critical time for girls to develop their social base and relationship skills. For this reason, I hypothesize that selection favored heightened sensitivity to threats and opportunities regarding formation of social relationships in adolescent girls. Consistent with this hypothesis, adolescent girls are more concerned than adolescent boys about being negatively evaluated, have a greater need for social approval, are more empathic, and are more reactive to interpersonal conflict and peer stress (reviewed in Andrews & Thomson, 2009; Rose & Rudolph, 2006). This heightened sensitivity to social evaluation and conflict may interact with perceived threats to social relationships to produce elevated levels of anxiety and depression in pubescent girls. Social inclusion and acceptance are critical in this context, as early pubertal development only predicts increasing levels of depressive symptoms among adolescent girls who are low in popularity or have problematic peer relationships (Conley & Rudolph, 2009; Teunissen et al., 2011). This finding converges with evolutionary models that conceptualize depression as an adaptation to social exclusion or other complex social problems and whose function is to minimize social risk under the circumstances (Allen & Badcock, 2003) and promote sustained analysis of the contexts that triggered the depressive episode, including generating and evaluating potential solutions (Andrews & Thomson, 2009).

Finally, the adolescent transition can be put in a broader context by considering the developmental changes that take place several years earlier, in the passage from early to middle childhood (the *juvenile transition*; Del Giudice, Angeleri, & Manera, 2009). The juvenile transition is marked by the endocrine event of adrenarche (the start of androgen secretion by the adrenal gland) and the emergence or intensification of important sex differences in behavior and cognition. Moreover, some processes that culminate in adolescence (e.g., developmental changes in aggression levels) actually begin at this earlier stage. In an evolutionary perspective, middle childhood may promote social competition before reproductive maturity, as the status and social resources acquired during this stage can increase the chances of succeeding at later stages. Further, the social feedback received during middle childhood can allow for adaptive recalibration of competitive strategies before navigating the more consequential social arena of adolescence (Del Giudice et al., 2009).

Summary and Conclusion

The issue of risky adolescent behavior is remarkably complex, and any intervention aimed at reducing it (or ameliorating its consequences) faces formidable obstacles and complications. For this very reason, we believe that risk taking in adolescence provides an excellent testing ground for evaluating the potential of an evolutionary approach to human development. The evolutionary model supplements, extends, and amends the standard approach in two interconnected ways. First, the evolutionary model delivers a deep and sophisticated theoretical foundation for understanding the meaning and manifestations of risky behavior. Second, equally important, it can inform prevention and treatment programs by highlighting key variables, indicating ways to maximize program effectiveness, revealing potential pitfalls and tradeoffs, increasing the realism of the intervention goals, and in some cases suggesting truly novel approaches and solutions (Ellis et al., 2012).

Central to an evolutionary analysis of adolescence are adaptationist hypotheses about function—why some features of adolescence have been maintained by natural selection instead of others. A guiding assumption of the current chapter is that understanding the functions of adolescence is essential to explaining why adolescents engage in risky behavior, and that successful prevention-intervention depends on working with instead of against adolescent goals and motivations (Ellis et al., 2012; Ellis, Volk, Gonzalez, & Embry, 2015). Like the theory of stage-environment fit (Eccles et al., 1993), the evolutionary model emphasizes the importance of match between adolescents' needs and opportunities.

From an evolutionary perspective, a major function of adolescence is to attain reproductive status—to develop the physical and social competencies needed to gain access to a new and highly contested biological resource: sex and, ultimately, reproduction. Puberty-specific developmental changes function to build reproductive capacity and increase socio-competitive competencies at this critical juncture. Much is at stake at the adolescent transition; it is an inflection point (i.e., a sensitive period for change) in the development of status, resource control, mating success, and other fitness-relevant outcomes. Consistent with the sensitive period hypothesis, there is a dramatic increase during adolescence in death and disability related to depression, eating disorders, alcohol and other substance use, accidents, suicide, homicide, reckless behavior, violence, and risky sexual behavior. I hypothesize that natural selection favored especially strong emotional and behavioral responses to social successes and failures during the adolescent transition—an important window of vulnerability and opportunity for setting long-term social and reproductive trajectories. Identity formation is central to the reorganization and change that occurs in adolescence as individuals discover and move into new social and reproductive niches.

The evolutionary model has myriad implications for designing prevention-intervention programs for adolescents (Ellis et al., 2012, 2015). It suggests practical solutions and new directions for research that have not been forthcoming from a developmental psychopathology perspective. The evolutionary model moves beyond psychopathology to consider "What's in it for the kids?" and, accordingly, models how natural selection shaped the adolescent brain to respond to environmental opportunities and challenges encountered during the adolescent transition. In presenting this perspective, it is my hope that new knowledge concerning the causes of risky adolescent behavior will be uncovered, and that developmentally appropriate programs and niches can be fostered that work with adolescent goals and motivations to more effectively address the problems associated with risky behavior in the second decade of life.

References

Allen, N. B., & Badcock, P. B. T. (2003). The social risk hypothesis of depressed mood: Evolutionary, psychosocial, and neurobiological perspectives. *Psychological Bulletin, 129,* 887–913.

Andersen, S. L., & Teicher, M. H. (2008). Stress, sensitive periods and maturational events in adolescent depression. *Trends in Neurosciences, 31,* 183–191.

Andrews, P. W., & Thomson, J. A., Jr. (2009). The bright side of being blue: Depression as an adaptation for analyzing complex problems. *Psychological Review, 116,* 620–654.

Archer, J. (2009). Does sexual selection explain human sex differences in aggression? *Behavioral and Brain Sciences, 32,* 249–311.

Baams, L., Dubas, J. S., Overbeek, G., & Van Aken, M. A. (2015). Transitions in body and behavior: A meta-analytic study on the relationship between pubertal development and adolescent sexual behavior. *Journal of Adolescent Health, 56,* 586–598.

Beinhocker, E. D. (2006). *The origin of wealth: Evolution, complexity, and the radical remaking of economics.* Boston: Harvard Business School Press.

Belsky, J., Steinberg, L., & Draper, P. (1991). Childhood experience, interpersonal development, and reproductive strategy: An evolutionary theory of socialization. *Child Development, 62,* 647–670.

Belsky, J., Steinberg, L., Houts, R. M., & Halpern-Felsher, B. L. (2010). The development of reproductive strategy in females: Early maternal harshness- earlier menarche- increased sexual risk taking. *Developmental Psychology, 46,* 120–128.

Blakemore, S. J., & Mills, K. L. (2014). Is adolescence a sensitive period for sociocultural processing? *Annual Review of Psychology, 65,* 187–207.

Boyce, W. T., & Ellis, B. J. (2005). Biological sensitivity to context: I. An evolutionary–developmental theory of the origins and functions of stress reactivity. *Development and Psychopathology, 17,* 271–301.

Conley, C. S., & Rudolph, K. D. (2009). The emerging sex difference in adolescent depression: Interacting contributions of puberty and peer stress. *Development and Psychopathology, 21*, 593–620.

Del Giudice, M., Angeleri, R., & Manera, V. (2009). The juvenile transition: A developmental switch point in human life history. *Developmental Review, 29*, 1–31.

Del Giudice, M., Ellis, B. J., & Shirtcliff, E. A. (2011). The adaptive calibration model of stress responsivity. *Neuroscience and Biobehavioral Reviews, 35*, 1562–1592.

Dobrova-Krol, N. A., Van IJzendoorn, M. H., Bakermans-Kranenburg, M. J., & Juffer, F. (2010). Effects of perinatal HIV infection and early institutional rearing on physical and cognitive development of children in Ukraine. *Child Development, 81*, 237–251.

Eccles, J. S., Midgley, C., Wigfield, A., Buchanan, C. M., Reuman, D., Flanagan, C., & Mac Iver, D. (1993). Development during adolescence: The impact of stage-environment fit on young adolescents' experiences in schools and in families. *American Psychologist, 48*, 90–101.

Ellis, B. J. (2004). Timing of pubertal maturation in girls: An integrated life history approach. *Psychological Bulletin, 130*, 920–958.

Ellis, B. J., Figueredo, A. J., Brumbach, B. H., & Schlomer, G. L. (2009). Fundamental dimensions of environmental risk: The impact of harsh versus unpredictable environments on the evolution and development of life history strategies. *Human Nature, 20*, 204–268.

Ellis, B. J., Boyce, W. T., Belsky, J., Bakermans-Kranenburg, M. J., & Van IJzendoorn, M. H. (2011). Differential susceptibility to the environment: An evolutionary-neurodevelopmental theory. *Development and Psychopathology, 23*, 7–28.

Ellis, B. J., Del Giudice, M., Dishion, T. J., Figueredo, A. J., Gray, P., Griskevicius, V., … Wilson, D.S. (2012). The evolutionary basis of risky adolescent behavior: Implications for science, policy, and practice. *Developmental Psychology, 48*, 598–623.

Ellis, B. J., Volk, A. A., Gonzalez, J. M., & Embry, D. D. (2015). The meaningful roles intervention: An evolutionary approach to reducing bullying and increasing prosocial behavior. *Journal of Research on Adolescence, 26*, 622-637. doi:10.1111/jora.12243

Gallup, A. C., O'Brien, D. T., & Wilson, D. S. (2011). Intrasexual peer aggression and dating behavior during adolescence: An evolutionary perspective. *Aggressive Behavior, 37*, 258–267.

Gunnar, M. R., Wewerka, S., Frenn, K., Long, J. D., & Griggs, C. (2009). Developmental changes in hypothalamus-pituitary-adrenal activity over the transition to adolescence: Normative changes and associations with puberty. *Development and Psychopathology, 21*, 69–85.

Hodgson, G. M., & Knudsen, T. (2010). *Darwin's conjecture: The search for general principles of social and economic evolution.* Chicago: University of Chicago Press.

Hrdy, S. B. (1999). *Mother nature: A history of mothers, infants and natural selection.* New York: Pantheon.

Hrdy, S. B. (2009). *Mothers and other: The evolutionary origins of mutual understanding.* Cambridge, MA: Harvard University Press.

Institute of Medicine & National Research Council. (2011). *The science of adolescent risk-taking: Workshop summary.* Washington, DC: National Academies Press.

James, J., Ellis, B. J., Schlomer, G. L., & Garber, J. (2012). Sex-specific pathways to early puberty, sexual debut, and sexual risk taking: Tests of an integrated evolutionary-developmental model. *Developmental Psychology, 48,* 687–702.

Jones, M. C. (1957). The later careers of boys who were early or late maturing. *Child Development, 28,* 113–128.

Mendle, J., Ryan, R. M., and McKone, K. M. (2016). Early childhood maltreatment and pubertal development: Replication in a population-based sample. *Journal of Research on Adolescence, 26,* 595–602.

Mulvihill, D. (2005). The health impact of childhood trauma: An interdisciplinary review, 1997–2003. *Issues in Comprehensive Pediatric Nursing, 28,* 115–136.

Najman, J. M., Hayatbakhsh, M. R., McGee, T. R., Bor, W., O'Callaghan, M. J., & Williams, G. M. (2009). The impact of puberty on aggression/delinquency: Adolescence to young adulthood. *Australian and New Zealand Journal of Criminology, 42,* 369–386.

Ostovich, J. M., & Sabini, J. (2005). Timing of puberty and sexuality in men and women. *Archives of Sexual Behavior, 34,* 197–206.

Palmer, C. T., & Tilley, C. F. (1995). Sexual access to females as a motivation for joining gangs: An evolutionary approach. *Journal of Sex Research, 32,* 213–217.

Persico, N., Postlewaite, A., & Silverman, D. (2004). The effect of adolescent experience on labor market outcomes: The case of height. *Journal of Political Economy, 112,* 1019–1053.

Piekarski, D. J., Johnson, C. M., Boivin, J. R., Thomas, A. W., Lin, W. C., Delevich, K., ... Wilbrecht, L. (2016). Does puberty mark a transition in sensitive periods for plasticity in the associative neocortex? *Brain Research, 1654,* 123-144. doi.org/10.1016/j.brainres.2016.08.042i

Pollak, S. D. (2008). Mechanisms linking early experience and the emergence of emotions: Illustrations from the study of maltreated children. *Current Directions in Psychological Science, 17,* 370–375.

Rose, A. J., & Rudolph, K. D. (2006). A review of sex differences in peer relationship processes: Potential trade-offs for the emotional and behavioral development of girls and boys. *Psychological Bulletin, 132,* 98–131.

Shonkoff, J. P., Boyce, W. T., & McEwen, B. S. (2009). Neuroscience, molecular biology, and the childhood roots of health disparities: Building a new framework for health promotion and disease prevention. *Journal of the American Medical Association, 301,* 2252–2259.

Steinberg, L. (2008). A social neuroscience perspective on adolescent risk-taking. *Developmental Review, 28,* 78–106. doi:10.1016/j.dr.2007.0 8.002

Stroud, L. R., Foster, E., Papandonatos, G. D., Handwerger, K., Granger, D. A., Kivlighan, K. T., & Niaura, R. (2009). Stress response and the adolescent transition: Performance versus peer rejection stressors. *Development and Psychopathology, 21,* 47–68.

Sylwester, K., & Pawłowski, B. (2011). Daring to be darling: Attractiveness of risk takers as partners in long- and short-term sexual relationships. *Sex Roles, 64,* 695–706.

Teunissen, H. A., Adelman, C. B., Prinstein, M. J., Spijkerman, R., Poelen, E. A., Engels, R. C., & Scholte, R. H. (2011). The interaction between pubertal timing and peer popularity for boys and girls: An integration of biological and interpersonal perspectives on adolescent depression. *Journal of Abnormal Child Psychology, 39,* 413–423.

Weisfeld, G. E. (1999). *Evolutionary principles of human adolescence*. New York: Basic Books.

Weisfeld, G. E., & Coleman, D. K. (2005). Further observations on adolescence. In R. L. Burgess & K. MacDonald (Eds.), *Evolutionary perspectives on human development* (2nd ed.). Thousand Oaks, CA: Sage.

West-Eberhard, M. J. (2003). *Developmental plasticity and evolution*. New York: Oxford University Press.

Wilson, M., & Daly, M. (1985). Competitiveness, risk taking, and violence: The young male syndrome. *Ethology and Sociobiology, 6*, 59–73.

CHAPTER 7

Shaping DNA (Discoverer, Noticer, and Advisor): A Contextual Behavioral Science Approach to Youth Intervention

Joseph Ciarrochi

Institute of Positive Psychology and Education, Australian Catholic University

Louise L. Hayes

University of Melbourne and Origen, National Centre of Excellence in Youth Mental Health

We work hard to immunize every child against physical illness, but we, as a society, have not done enough to immunize children against psychological illness. Fourteen percent of young people have experienced an anxiety disorder (Wittchen, Nelson, & Lachner, 1998), 20% have experienced at least one traumatic event, 3% have developed PTSD (Perkonigg, Kessler, Storz, & Wittchen, 2000), 14% engage in some form of self-harm (Martin, Swannell, Hazell, Harrison, & Taylor, 2010), and 17% have seriously considered attempting suicide in the previous twelve months (Kann et al., 2014). There is no evidence that these rates of psychological struggle are decreasing.

What is going wrong here? Why is developmental science making so little progress? We suggest that our failure to communicate contributes to the problem. Youth research exists on isolated islands. Some researchers focus on pathology, others on developmental stages, or social and family factors, or brain functioning, or individual cognitions, or emotion regulation strategies. We have clearly learned a lot about youth, but we do not work together to synthesize knowledge in a way that allows it to be easily disseminated to schools, families, and communities. The science gets lost on individual islands.

In this chapter, we argue that a contextual behavioral science (CBS) approach can unify the different fields of research in the service of facilitating positive youth development. We make this argument in three sections. First, we show what a CBS model can look like. We provide a concrete example of a comprehensive CBS model called DNA-V, which steps across the islands of knowledge and can be applied to helping young people across contexts. This example shows how the streams of science can come together—environment of adaptation, attachment, development, selection by consequences, learning, cognitions, emotion, and culture. Second, we take one aspect of CBS, a young person's hidden and symbolic life, and examine this in detail to show how the principles are applied to language and cognition. Finally, we draw together how a CBS model makes contextual assumptions that can be used to understand adaptation in context. Although evolutionary science (ES) and principles of adaptation, variation, selection, and retention are integral to the DNA-V adolescent model, we avoid duplication by leaving detailed evolutionary science discussion to the partner chapter by Bruce Ellis (chapter 6).

DNA-V: A CBS Approach to Improving the Lives of Young People

DNA-V is the name for a model for youth development that is an applied version of contextual behavioral science and aims to address all streams of adaptation. DNA stands for three classes of behavior that we label "Discoverer (D)," "Noticer (N)," and "Advisor (A)," and that are optimally used in the service of vitality and valued action (V). DNA-V is one of the few models that incorporates all four CBS streams of knowledge: evolutionary science (D. S. Wilson, Hayes, Biglan, & Embry, 2014), behavioral principles (Skinner, 1969), relational frame theory (Hayes, Barnes-Holmes, & Roche, 2001), and acceptance and commitment therapy (S. C. Hayes, Strosahl, & Wilson, 2012). However, unlike acceptance and commitment therapy, DNA-V is not based on treatment for psychological distress. Rather, we conceptualize it as a bottom-up model that considers how humans grow from birth into adulthood. In this chapter, we focus on adolescents specifically.

DNA-V shines a spotlight on adolescent development in the context of evolution and life history. We do not follow the clinical approach of comparing "disordered" adolescents to "normal" adolescents and trying to make the disordered ones look more normal (Ciarrochi & Bailey, 2008). Instead, we view all adolescent behavior as an adaptation to adolescents' contexts. Thus, we look at how adolescents can grow strong and flexible, and how principles of variation, selection, and retention contribute across the streams, from biology to learning to cultural transmission. We never lose sight of the context in which young people grow. We look at their social world and attachment bonds. We use operant principles to examine how children's behavior is selected by consequences, and we use relational frame theory to understand how their symbolic, or cognitive, world grows. We use applied behavioral processes from acceptance and commitment therapy to understand why young people often become distressed and stuck in unhelpful behavioral patterns, and understand how we can intervene to help them let go of unhelpful behavior and develop new, more functional overt and symbolic behavioral repertoires.

The full model is presented in Figure 1 and discussed below. It should be noted that DNA-V is not intended to be a trademarked package, but rather a model that captures empirically supported processes. For a detailed account of DNA-V and its intervention strategies, please see L. Hayes and Ciarrochi (2015). To review the evidence behind each of the processes implemented in DNA-V, please see Ciarrochi, Atkins, Hayes, Sahdra and Parker (2016). We will first provide an overview of each aspect of the model.

Values and vitality: We begin at the center of our DNA-V model. Here, connections with values and vitality define the consequences that, ideally, would select a particular behavior in a particular environment. Value becomes sophisticated verbal processes that grow with us. In infancy, meaningful contingent behavior is profound yet simple—we seek contact and connection with humans who provide nurturing and secure attachment (Ainsworth, 1989). Our evolutionary heritage attests to our need for others (Bowlby, 1979). Over time, however, as we learn to navigate symbolic language, we can broaden our repertoire of meaningful behaviors and our range of valued behavior expands.

By adolescence, values are a kind of verbal behavior that can transform experience and help young people move beyond immediate reinforcement. For example, a teen who values working with animals and wants to become a veterinarian can use language to transform his or her perceptions of studying behavior from "drudgery" to "the first step to becoming a veterinarian." Practically, we can shape young people to utilize verbal rules that link behavior to selection criteria; for example, by asking questions like, "What is this behavior for?" "What is important to you?" and "Does this behavior bring you vitality and well-being?" Of course, not all teens will develop the same language skill, and those who are unskilled at using language to build meaning in life may instead take action just for the immediate reinforcement an activity brings. We do not assume that all teens can connect their behavior *to value*, but we hope that, through optimal growth, we can shape behavior in this direction.

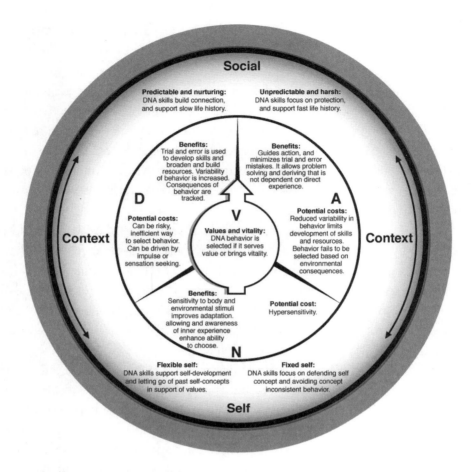

Figure 1: The DNA-V model of valued action. Discoverer (D), Noticer (N), and Advisor (A) are metaphors for three classes of behavior that can function to support an individual's values and vitality (V). Psychological flexibility occurs when people are able to persist in D, N, or A behaviors, or change them, in the service of values.

As can be seen in Figure 1, V ("values") is surrounded by D, N, and A. These represent three clusters of behavior that we call discoverer, noticer, and advisor. Notably, they are classes of behavior rather than "things." We use DNA as a metaphor to describe our model because it encompasses behaviors that all young people have, and that all can express in different ways—just as cellular DNA is expressed in different ways. The DNA acronym helps us imply that these skills are basic capabilities, a bit like our basic biology. Naturally, there is individual variation, but all young people are capable of developing their DNA skills. Context, adaptation, and selection by consequences matter here, too. Just like biological DNA, the DNA processes in our model need the right environmental contexts to find optimal expression.

Shaping a young person's D, N, and A behaviors is done in the service of increasing valued action and vitality. Research suggests that humans generally experience value across six domains: connecting with others, giving, being active, connecting with the present moment, challenging oneself and learning, and caring for oneself (L. Hayes & Ciarrochi, 2015). An evolutionary account would assume that each of these behaviors is associated with well-being (a proximate explanation) because they ultimately increase fitness. In addition, because we are a group species that need and depend on each other (E. O. Wilson, 2012), our valued behaviors tend to almost always be linked to other people. For example, connecting and giving are examples of cooperative behavior, and humans who work together to care for and protect their young and each other are more likely to survive than ones who do not (D. S. Wilson, Vugt, & O'Gorman, 2008; E. O. Wilson, 2012). Challenging oneself and being active are behaviors that promote fitness because they help one master the environment and compete with other humans for status and reproductive success. Caring for oneself promotes health and reproductive potential. Finally, connecting with the present moment is a behavior that undermines the harmful effects of language (e.g., ruminating about the past) and increases sensitivity to present-moment information that is being signaled via the body and the physical environment. All these aspects of value can flourish through our use of discoverer, noticer, and advisor behaviors.

Advisor, or our inner voice, is a metaphorical term we use to describe the verbal behavior that functions to efficiently guide action and avoid trial-and-error mistakes. It is an application of relational frame theory's account of human language and cognition, and the operant principle of "arbitrary applicable derived relational responding" (Tornke, 2010). The advisor is based on a person's history, which includes experiences, past problem solving, reasoning, and past teachings. The advisor is a large part, but not all, of our symbolic language use. Advisor includes evaluations of the self, others, and the world, and rules about how to live in the world. Our use of the term "advisor" relates most closely to how others use terms like automatic negative thoughts, self-concepts, hope, dysfunctional beliefs, and schema (Ciarrochi & Bailey, 2008). It also relates to interventions used in cognitive behavioral therapy, such as cognitive reappraisal and problem solving. One goal with DNA-V training is to help young people develop useful beliefs, and use those beliefs to guide effective action. DNA-V training also helps young people to notice when some beliefs are unhelpful and to mindfully "unhook" from them and not allow them to impact valued action. Essentially, DNA-V helps young people to evolve a more helpful advisor.

Noticer is the term for our ability to focus on present-moment events that are occurring both inside and outside our bodies. We are all born as noticers; seeing, hearing, touching, tasting, and smelling our world. However, our noticer behavior is rapidly shaped by experience. Like all mammals, we have evolved to experience danger through our bodies, primarily via our ventral vagal complex (the tenth cranial nerve). External messages of safety or danger bring changes to our heart rate, bronchi, vocalization, middle ear, and facial expressivity (Porges, 2011). If we grow in a harsh or socially unsupportive environment, our noticer skills make us hypersensitive to cues that relate to threat and safety (Szalavitz & Perry, 2010). In contrast, if we grow in a nurturing environment, our noticer skills make us relatively more sensitive to cues that indicate novelty and elicit curiosity and exploration (Ainsworth, Blehar, Waters, & Wall, 2015). If we are soothed by loving carers when upset, we learn how to "listen" to the signals in our bodies, to trust our bodies, and to trust that we can connect with others and seek their help if needed (Ainsworth et al., 2015; Ainsworth, 1989). A "safe noticer" supports the broadening of behavioral repertoires and skills (Fredrickson, Cohn, Coffey, Pek, & Finkel, 2008), and maps closely to the concept of mindfulness and nonreactive awareness, which is the ability to focus on the present moment, on purpose, and with curiosity (Kabat-Zinn, 1994). Skilled noticers are able to maintain focused attention, accepting and allowing negative experience to occur; are nonreactive to inner experience such as impulses and doubts; and can practice nonattachment, or nonclinging, to pleasant experiences and feelings (Sahdra, Ciarrochi, & Parker, 2016).

Discoverer is a term for behavior that utilizes trial-and-error learning in order to expand skills and resources. Children's discovery is commonly characterized by play and exploration, while adolescent discovery is commonly characterized by love of novelty, sensation seeking, and adaptive risk taking. Adolescent discoverers take risks in order to develop independence and learn adult roles. Discoverer skills are used to help young people find ways to live with vitality and to direct their life toward valued ends. To do this, they must continually examine the workability of current and old behaviors, by tracking the consequences of their actions. Our use of the term discoverer refers to a broad range of research streams, including (1) adaptive risk taking across species (Ellis et al., 2012; Hawley, 2011), (2) Broaden and Build of psychological theories (Fredrickson, 2001), (3) contingency shaping that begins in childhood (Forehand & McMahon, 1981), and (4) behavioral activation interventions (Cuijpers, van Straten, & Warmerdam, 2007).

Context forms the outer circle of our model (Figure 1) and refers to all the evolutionary, life history, and immediate events that have exerted, and are exerting, an organizing influence on a young person's D, N, and A behaviors. It is important to note that, from a functional contextualist perspective, there is no a priori way to specify what is "context" and what is behavior. Rather, the behavior refers to what is of interest to the interventionist at a particular point in time, and context refers to all the factors that may affect that behavior. Two contextual factors are important to single out for young people's development—social context and self-context.

Social context of DNA-V. Humans don't so much live *with* each other; they live *for* each other. Attachment to others is as important as nutrition (Bowlby, 1979; Hrdy, 2009). Our colleague Ellis discusses the importance of social context, and especially of the ways that a harsh versus nurturing social environment can shape how young people seek to get their needs met. In DNA-V, we consider how attachments and social forces surround and reciprocally influence the growing individual's D, N, and A behaviors. Thus, young people who are displaying antisocial behavior are defining their "anti" in terms of their "social" world. We would argue, based on studies showing the importance of human connection, that even the most disagreeable narcissists need the praise of other people and will use their D, N, and A behaviors to get it. Social connection increases our well-being, while social isolation is a risk factor for heart disease, obesity, impaired thinking, mental health problems, and death (Hawkley & Cacioppo, 2010; Heinrich & Gullone, 2006; Holt-Lunstad, Smith, & Layton, 2010). Thus, DNA-V puts a strong emphasis on understanding the extent young people are getting their social needs met, the extent they are surrounded by support from adults and peers, and the extent that this support builds D, N, and A skills.

Self context of DNA-V. In addition to the social context, a young person's sense of self can shape his or her D, N, and A behaviors. We discuss this further in the following section, but briefly here we refer not just to self-evaluations (advisor behavior), but also to how one relates to those self-evaluations (see Barnes-Holmes, Hayes, & Dymond, 2001, for a behavioral account of self). Two key ways of relating to oneself are viewing oneself with self-compassion and viewing oneself as capable of change, or having a "growth mindset" (Marshall et al., 2015; Yeager & Dweck, 2012). Within our framework, self-compassion involves the ability to notice negative self-evaluations, to recognize that those self-evaluations are normal, and to treat oneself with kindness in the service of valued activity. Young people with self-compassion experience higher well-being, and manage better than on occasions when their self-esteem is low (Marshall et al., 2015). Young people with a growth mindset believe that self-evaluations like "untalented" do not define them and believe that they can improve themselves (Yeager & Dweck, 2012). Together, growth mindset and self-compassion give young people what we call a "flexible self-view," or the ability to respond flexibly to setbacks and self-limiting beliefs and to utilize their D, N, and A behaviors to expand and grow. In contrast, young people who have an inflexible/rigid self-view tend to utilize their D, N, and A behaviors within the relatively narrow boundaries of how they define themselves. For example, consider a young man who believes rigidly "I am an athlete." He may utilize DNA-V well to improve his athleticism, but what happens when he gets injured and cannot play sports anymore? In that case, he may struggle to use D, N, and A effectively and be dominated by thoughts such as "If I am not an athlete, I am nobody."

The above summary demonstrates that DNA-V is contextual and considers adaptation, variation, selection, and retention across streams of research. Ideally, D, N, and A skills evolve and change based on the consequences they have for a young person's values and vitality. We now turn our focus to one critical aspect of DNA-V—symbolic adaptation and selection by consequences.

A CBS Approach to Evolving Symbolic Life

Symbolic activity is central to all aspects of DNA-V. For example, the advisor "uses" evaluations and rules to avoid trial and error, the noticer uses words to describe inner and outer experiences, and the discoverer uses words to imagine better worlds and new ways of being. Value may start out as nonverbal preferences, but becomes more symbolic as we learn to speak of what we want in life. Self-view is inherently symbolic as it involves a verbal "I" that gets related to evaluations and descriptions, and can be verbally placed into different perspectives (e.g., "I am different now than I was then"). Social-view involves someone experiencing another, at least in part, verbally (e.g., "In the future, he will reject me") and being able to place that other into verbal perspective (e.g., "I am sad, but she is angry").

We now turn to a discussion on how symbolic behavior, or language, evolves and can be selected and itself serve as selection consequences for adolescents.

Typically, evolutionary theories focus on the ways that observable behavior is shaped by environmental costs and benefits. For example, theories may seek to explain adolescent sexual behavior in terms of the dangerousness of the environment, the degree of match between ancestral and current environment, and the ratio of males to females in the environment (Ellis et al., 2012). These theories predict how the evolutionary environment of adaptation works through the genetic inheritance system to influence adolescent capacities and preferences in specific contexts. However, they do not present the full evolutionary story. CBS and some evolutionary scientists have argued that, in addition to the genetic inheritance system, there are epigenetic, behavioral, and symbolic inheritance systems (transmission through language and other forms of symbolic communication such as culture) (Jablonka & Lamb, 2006).

CBS can be seen as an extension of evolutionary principles to understanding and shaping private behavior such as symbolic thinking (see chapter 1). Here we argue that we will never reduce the prevalence rates of suffering in adolescents if we fail to incorporate the symbolic inheritance system into our models, along with the other inheritance systems. We need to understand not only what young people's external environment is like, but also how they use symbols to make sense of it, and how symbolic understandings are passed down from parent to child. This is the only way we will fully understand how humans transition from dependent children to adults who can assume adult roles and responsibilities.

From a CBS view, language is a type of behavior that (1) has evolved as a unique capacity in humans, and (2) is influenced by the principles of variation, selection, and retention (D. S. Wilson et al., 2014). Wilson and his colleagues refer to a network of symbolic relations that regulate behavior as a "symbotype"; here, we refer to this more simply as "beliefs." Like a genotype, beliefs evolve and "survive" based on what they cause[1] the organism to do. Direct correspondence between those beliefs and the physical world is not a requirement. For example, religious beliefs may help people cooperate and outcompete other groups, even if scientists believe they are untrue (D. S. Wilson, 2010).

Our task is to help teens evolve flexibly with language. Human babies begin life in the world of sensation, responding to physical stimuli with crying, reaching, smiling, laughing, and throwing up. As they approach 18 months, they learn to walk, play, sing (often badly), and prefer one type of cake over another. They do not yet feel self-conscious about how they play or sing. They do not yet feel guilty about eating cake. Then they develop language. They develop the ability to leave present-moment stimuli, and use words or symbols to "experience" the past ("I did something embarrassing") or future ("If I see a stranger approaching me in a white van, I need to run away"). Even when they are focused on the present moment, they can transform that moment with judgment ("I am unpopular"; "she is mean").

Human adolescents grow increasingly "inside" a vast symbolic world that may have scarcely any connection to physical reality. Throughout much of human history, this symbolic world incorporated supernatural agents, the sacred and profane, rituals, rites of passage, and counterintuitive religious concepts (bleeding statues, virgin births), all of which were probably in the service of building cooperation and provided clear definitions of roles and responsibility within a community (Alcorta & Sosis, 2005). The modern adolescent also lives inside a symbolic world, but it is unclear how much these symbols influence cooperation and prepare one for adult roles. For example, modern adolescents are often "living inside" media, movies, tweets, emails, websites, and social media—all filled with stories, gossip, and advertisements. They grow inside contexts where social worth can be determined by a small symbol on a sneaker.

---

1 By causal, we mean beliefs have a causal role in the subsequent action. We do not mean that belief is a sufficient cause.

This means that although we humans have gained the ability to extract ourselves from a physical jungle, through language we are now recreating the danger of the jungle in our heads again and again. For example, a teenage girl is excluded from playing with other girls and experiences shame. Later, she stands in front of her mirror, pinching her "fat," wishing she could cut it off, even though she is in fact very thin. She may do this well into adulthood, long after the schoolyard incident is over. Or a boy sits alone in the room doing his math homework when he recalls the teacher asking him, in front of the whole class, "Are you stupid or something?" He relives the shame, as if the teacher and the class were right there in the room with him. He closes the math book. What is striking in both these examples is that for humans, overt behavior is driven largely by fear created in symbolic events, rather than in danger within the immediate environment.

The reason symbols become so important to our life is because our context reinforces the idea that we should interact with symbols as if they are the "real thing." If a mother says, "Be careful of snakes," she wants her child to react to the word "snake" as if it were a real snake. It is clearly useful for the child to respond to the word "snake" as if there were a real snake because such a response not only gets the mother's approval, but helps the child survive. Indeed, symbols are such useful tools for survival that we want to use them all the time, in every aspect of life. This is where we start to see problems. Symbols are used even when they are not useful.

Generally, the overuse of symbols can be problematic in three ways. First, behavior can be selected based on the symbolic consequences, rather than physical world consequences. For example, instead of pursuing actual social connection, status, and autonomy, we can buy the symbols of these things. We buy an expensive car because it is advertised as giving us the freedom to "Go your own way," even though the debt it brings forces us to work longer hours at a job we hate. We avoid the doctor so we don't have to *think* about ill health, and in doing so increase our risk of ill health arising. We think ourselves into feeling superior to others, even when we are not (Ehrlinger, Johnson, Banner, Dunning, & Kruger, 2008) and even when feeling superior interferes with our ability to collaborate and connect (Nevicka, Velden, De Hoogh, & Vianen, 2011). It is as if we entirely lose contact with the physical world.

The second potential problem with language overuse is that it can elicit emotional reactions, even when such reactions have no link to what is happening in the present moment. For example, the mere thought of a past, shameful experience can lead a young person to act ashamed and withdraw despite a safe social context.

The third potential problem with language overuse is that it makes it possible for us to misapply rules learned in one niche (e.g., the family) to other niches (e.g., school) and contributes to environmental mismatch and insensitivity of present-moment costs and benefits (S. C. Hayes, Brownstein, Zettle, Rosenfarb, & Korn, 1986). For example, in an abusive family, a child may learn that nobody is to be trusted. When at school, this same child may fail to pick up on cues from a caring adult who can be trusted and wants to help the child (Ciarrochi, Deane, Wilson, & Rickwood, 2002). Thus, the child may fail to emit help-seeking behavior (lack of variation) and miss the opportunity to receive help/reinforcement as a result of that behavior (selection).

CBS Provides a Distinctive, Pragmatic Approach to Youth Development

CBS shares assumptions with evolutionary and behavioral science, which, in turn, differ quite dramatically from many assumptions typically found in other areas of psychology, particularly developmental psychology and the search for "normative" adolescence. What allows us to bridge evolution, learning, symbolic, and cultural streams is our CBS assumptions. Youth behavior is assumed to be an adaptation to a particular context (including evolutionary, life history, and current context), and the goal of CBS is to identify the *changeable* aspects of context that are subject to intervention. This is taking a functional contextualist view, which is the philosophical framework that underpins CBS (S. C. Hayes et al., 2012; S. C. Hayes, Hayes, & Reese, 1988).

A contextual behavioral scientist assumes that behavior is an ongoing act in context. Behavior in this view is anything that the organism does, including overt behavior (what we see) and also covert behavior (the unseen, thinking, feeling, and brain activity). Context involves all the evolutionary life history (comprising genetic, epigenetic, behavioral, and symbolic history), plus immediate environmental events that have exerted, and are exerting, an organizing influence on behavior (D. S. Wilson et al., 2014). This view is often contrasted with an elemental realist worldview, which seeks to understand how different aspects of the young person work together (e.g., beliefs influence feelings), identifies what parts are normal versus abnormal (e.g., absence or presence of rumination), and seeks to find and fix dysfunctional parts (e.g., reducing rumination) (Ciarrochi & Bailey, 2008). An elemental realist might, for example, compare "normal" and clinically anxious youth in order to identify how they differ (e.g., negative schema), and seek to make the anxious youth more similar to the normal youth (reduce negative schema). In contrast, a contextual behavioral approach never assumes that any behavior or inner experience is inherently dysfunctional. CBS scientists look for ways that behavior might be an adaptation to a distinctive context. They assume there is no "normal" adolescent.

These two views influence adolescents research and interventions. From an elemental realist view, adolescents are commonly accused of being self-absorbed or excessively worried about what others think, as if these are inherently dysfunctional states that need to be corrected (i.e., adult-like thinking is the expected standard of maturity). In contrast, functional contextualists assume that the adolescent tendency to think about themselves, and engage in social comparison or gossip, has an adaptive purpose. Their task is to discover the purpose of behavioral patterns for each adolescent and for us as a species. If, for example, self-absorption is present across cultures and environments, we look for how such absorption may help young people to acquire such things as social status, prestige, and dominance (Ellis et al., 2012).

The contextual view also applies to biology. Functional contextualists generally do not view biological patterns as sufficient causes of overt behavior. Rather, they view biological patterns as behaviors that need to be explained by evolutionary and current environmental influences. The reason for this viewpoint is practical. Functional contextualists want to identify aspects of the environment that can be changed to produce positive change in both physiology and overt behavior. An elemental realist explanation such as "Dopamine deficiency causes depression" is incomplete for the functional contextualist because it does not indicate what aspect of the changeable environment, such as social isolation, leads to both dopamine deficiency and depression. Functional contextualists examine how biology interacts with environmental costs and benefits to produce adaptations. For example, in nurturing environments, the onset of puberty may be associated with an increase in risk taking that helps a young person forge long-lasting relationships outside of the family. In harsh environments, risk taking may focus on immediate gains and a discounting of longer-term gains that are highly uncertain (Dishion, Ha, & Veronneau, 2012). Thus the goal of CBS-based interventions is not to eradicate adolescents' love of risk taking, but rather to create environments that allow their risk-taking behavior to be prosocial rather than antisocial (Ellis et al., 2012; Hawley, 2011).

Conclusions

Youth intervention research is impressive, but fragmented. As long as it remains so, we will never be able to develop optimal youth interventions. For example, what is the use of challenging negative thoughts if we do not consider the social and family context in which they occur? Negative thoughts can be highly adaptive. What is the use of teaching young people that mental health is a "brain problem" if we fail to consider how this message may demotivate young people? In many contexts, anxiety and depression are not the fault of a defective brain, but an adaptation within a deficient environment. What is the use of treating biological symptoms of depression with pills if we fail to consider the social factors that are causing the biological symptoms, as well as the long-term psychological and biological consequences of pharmacology? What is the use of teaching young people to have long-term goals and hope when we do not help them to escape a harsh environment and reduce uncertainty about their future? Even traits like mindfulness, which is often treated as a universal good, can have negative consequences in certain contexts (Farias & Wiholm, 2015; Shapiro, 1992). In short, there is no biological profile, internal experience, or external behavior that can be judged as good independent of individual, social, and evolutionary context. There is no psychological pill that is universally healthy to youth.

This chapter has provided one example of how to defragment youth intervention research, using well-defined evolutionary terms such as costs, benefits, context, variation, and selection. We have had little space to talk about retention and inheritance, but would like to note, in closing, that verbal inheritance means that we could see dramatic improvements in the human condition in just a few generations. Imagine if we could teach young people to become mindful of the ways that symbols can dominate our interpretations of experience and become unhelpful. They might then learn to use symbols like tools, and "put them down" when no longer useful. They might become less caught up in self-criticism, materialism, and prejudice. Could they pass these lessons on to their children? Or imagine if all young people learned to judge their behavior in terms of how it served their values, and especially how it helped them build connection and love? Or imagine young people who understood that they are not fixed, and the future is not fixed, and they can improve themselves and this world. What might they teach their children?

References

Ainsworth, M. S. (1989). Attachments beyond infancy. *American Psychologist, 44,* 709.

Ainsworth, M. S., Blehar, M. C., Waters, E., & Wall, S. N. (2015). *Patterns of attachment: A psychological study of the strange situation.* Hillsdale, NJ: Lawrence Erlbaum.

Alcorta, C., & Sosis, R. (2005). Ritual, emotion, and sacred symbols. *Human Nature, 16,* 323–359.

Barnes-Holmes, D., Hayes, S. C., & Dymond, S. (2001). Self and self-directed rules. In S. C. Hayes, Dermot Barnes-Holmes, et al. (Eds.), *Relational frame theory: A post-Skinnerian account of human language and cognition.* New York: Plenum.

Biglan, A., & Hayes, S. C. (2015). *The Wiley handbook of contextual behavioral science*. West Sussex, UK: Wiley-Blackwell.

Bowlby, J. (1979). *The making and breaking of affectional bonds*. New York: Routledge.

Bryant, G. (2006). On hasty generalizations about evolutionary psychology. *American Journal of Psychology, 119*, 481–487.

Chawla, N., & Ostafin, B. (2007). Experiential avoidance as a functional dimensional approach to psychopathology: An empirical review. *Journal of Clinical Psychology, 63*, 871–890.

Ciarrochi, J., Atkins, P., Hayes, L., Sahdra, B., and Parker, P. (2016). Contextual positive psychology: Policy recommendations for implementing positive psychology into schools. *Frontiers of Psychology, 7*, 1561

Ciarrochi, J., & Bailey, A. (2008). *A CBT-practitioner's guide to ACT: How to bridge the gap between cognitive behavioral therapy and acceptance and commitment therapy*. Oakland, CA: New Harbinger Publications.

Ciarrochi, J., Deane, F. P., Wilson, C. J., & Rickwood, D. (2002). Adolescents who need help the most are the least likely to seek it: The relationship between low emotional competence and low intention to seek help. *British Journal of Guidance & Counselling, 30*, 173–188.

Ciarrochi, J., & Forgas, J. P. (1999). On being tense yet tolerant: The paradoxical effects of trait anxiety and aversive mood on intergroup judgments. *Group dynamics: Theory, research, and practice, 3*, 227–238.

Ciarrochi, J., & Parker, P. (2016). Nonattachment and mindfulness: Related but distinct constructs. *Psychological Assessment, 28*, 819–829.

Cuijpers, P., van Straten, N., & Warmerdam, L. (2007). Behavioral activation treatments for depression: A meta-analysis. *Clinical Psychology Review, 27*, 318–326.

Dishion, T. J., Ha, T., & Veronneau, M. H. (2012). An ecological analysis of the effects of deviant peer clustering on sexual promiscuity, problem behavior, and childbearing from early adolescence to adulthood: An enhancement of the life history framework. *Developmental Psychology, 48*, 703–717. doi:10.1037/a0027304

Ehrlinger, J., Johnson, K., Banner, M., Dunning, D., & Kruger, J. (2008). Why the unskilled are unaware: Further explorations of (absent) self-insight among the incompetent. *Organizational Behavior and Human Decision Processes, 105*, 98–121.

Ellis, B. J., Del Giudice, M., Dishion, T. J., Figueredo, A. J., Gray, P., Griskevicius, V., … Volk, A. A. (2012). The evolutionary basis of risky adolescent behavior: Implications for science, policy, and practice. *Developmental Psychology, 48*, 598.

Farias, M., & Wiholm, C. (2015). *The Buddha pill: Can mediation change you?* London: Watkins Media Limited.

Ford, B., & Mauss, I. (2014). The paradoxical effects of pursuing positive emotion. In J. Gruber & T. Moskowitz (Eds.), *Positive emotion: Integrating the light sides and dark sides*. Oxford Scholarship Online. doi:10.1093/acprof:oso/9780199926725.001.0001

Forehand, R. L., & McMahon, R. J. (1981). *Helping the noncompliant child*. New York: Guilford.

Fredrickson, B. L. (2001). The role of positive emotions in positive psychology: The broaden-and-build theory of positive emotions. *American Psychologist, 56*, 218–226.

Fredrickson, B. L., Cohn, M. A., Coffey, K. A., Pek, J., & Finkel, S. M. (2008). Open hearts build lives: Positive emotions, induced through loving-kindness meditation, build consequential personal resource. *Journal of Personality and Social Psychology, 95*, 1045–1062.

Goldney, R., Eckert, K., Hawthorne, G., & Taylor, A. (2010). Changes in the prevalence of major depression in an Australian community sample between 1998 and 2008. *Australian and New Zealand Journal of Psychiatry, 44*, 901–910.

Haslam, N., & Kvaale, E. (2015). Biogenetic explanations of mental disorder: The mixed-blessings model. *Psychological Science, 24*, 399–404.

Hawkley, L. C., & Cacioppo, J. T. (2010). Loneliness matters: A theoretical and empirical review of consequences and mechanisms. *Annals of Behavioral Medicine, 40*, 218–227.

Hawley, P. H. (2011). The evolution of adolescence and the adolescence of evolution: The coming of age of humans and the theory about the forces that made them. *Journal of Research on Adolescence, 21*, 307–316. doi:10.1111/j.1532–7795.2010.00732.x

Hayes, L., & Ciarrochi, J. (2015). *The thriving adolescent: Using acceptance and commitment therapy and positive psychology to help teens manage emotions, achieve goals, and build connection.* Oakland, CA: Context Press.

Hayes, S. C., Barnes-Holmes, D., & Roche, B. (Eds.). (2001). *Relational frame theory: A post-Skinnerian account of human language and cognition.* New York: Plenum.

Hayes, S. C., Brownstein, A. J., Zettle, R. D., Rosenfarb, I., & Korn, Z. (1986). Rule-governed behavior and sensitivity to changing consequences of responding. *Journal of the Experimental Analysis of Behavior, 45*, 237–256.

Hayes, S. C., Hayes, L. J., & Reese, H. W. (1988). Finding the philosophical core: A review of Stephen C. Pepper's *World Hypotheses. Journal of the Experimental Analysis of Behavior, 50*, 97–111.

Hayes, S. C., Luoma, J. B., Bond, F. W., Masuda, A., & Lillis, J. (2006). Acceptance and commitment therapy: Model, processes and outcomes. *Behaviour Research and Therapy, 44*, 1–25.

Hayes, S. C., Strosahl, K. D., & Wilson, K. G. (2012). *Acceptance and commitment therapy, second edition: The process and practice of mindful change.* New York: Guilford.

Heinrich, L. M., & Gullone, E. (2006). The clinical significance of loneliness: A literature review. *Clinical Psychology Review, 26*, 695–718.

Holt-Lunstad, J., Smith, T., & Layton, J. (2010). Social relationships and mortality risk: A meta-analytic review. *PloS Medicine, 7.* doi.org/10.1371/journal.pmed.1000316

Hrdy, S. (2009). *Mothers and others: The evolutionary origins of mutual understanding.* Cambridge, MA: Harvard University Press.

Jablonka, E., & Lamb, M. (2006). *Evolution in four dimensions: Genetic, epigenetic, behavioral, and symbolic variation in the history of life.* Cambridge, MA: MIT Press.

Kabat-Zinn, J. (1994). *Wherever you go there you are: Mindfulness meditation in everyday life.* New York: Hyperion.

Kann, L., Kinchen, S., Shanklin, S. L., Flint, K. H., Kawkins, J., Harris, W. A., … Whittle, L. (2014). Youth risk behavior surveillance—United States, 2013. *MMWR Surveillance Summaries, 63*, 1–168.

Marshall, S., Parker, P., Ciarrochi, J., Sahdra, B., Jackson, C., & Heaven, P. (2015). Self-compassion protects against the negative effects of low self-esteem: A longitudinal study in a large adolescent sample. *Personality and Individual Differences, 74*, 116–121.

Martin, G., Swannell, S. V., Hazell, P. L., Harrison, J. E., & Taylor, A. W. (2010). Self-injury in Australia: A community survey. *Medical Journal of Australia, 193*, 506.

Milin, R., Kutcher, S., Lewis, S. P., Walker, S., Wei, Y., Ferrill, N., & Armstrong, M. (2016). Impact of a mental health curriculum for high school students on knowledge and stigma:

A randomized controlled trial. *Journal of the American Academy of Child and Adolescent Psychiatry, 55*, 383–391. doi:10.1016/j.jaac.2016.02.018

Nevicka, B., Velden, F., De Hoogh, A., & Vianen, A. (2011). Reality at odds with perceptions: Narcissistic leaders and group performance. *Psychological Science, 22*, 1259–1264.

Perkonigg, A., Kessler, R. C., Storz, S., & Wittchen, H. U. (2000). Traumatic events and post-traumatic stress disorder in the community: Prevalence, risk factors and comorbidity. *Acta Psychiatrica Scandinavica, 101*, 46–59.

Perry, B. D. (2009). Examining child maltreatment through a neurodevelopmental lens: Clinical applications of the neurosequential model of therapeutics. *Journal of Loss and Trauma, 14*, 240–255. doi:10.1080/15325020903004350

Porges, S. W. (2011). *The polyvagal theory: Neurophysiological foundations of emotions, attachment, communication, and self-regulation.* New York: Norton.

Sahdra, B., Ciarrochi, J., & Parker, P. (2016). Nonattachment and mindfulness: Related but distinct constructs. *Psychological Assessment, 28*, 819–829.

Seligman, M. E. (2004). *Authentic happiness: Using the new positive psychology to realize your potential for lasting fulfillment.* New York: Atria Books.

Shapiro, D. (1992). Adverse effects of meditation: A preliminary investigation of long-term meditators. *International Journal of Psychosomatics, 39*, 1–4.

Skinner, B. F. (1969). *Contingencies of reinforcement: A theoretical analysis.* Englewood Cliffs, NJ: Prentice Hall.

Steinberg, L. (2001). We know some things: Parent-adolescent relationships in retrospect and prospect. *Journal of Research on Adolescence, 11*, 1–19.

Szalavitz, M., & Perry, B. D. (2010). *Born for love.* New York: William Morrow.

Tornke, N. (2010). *Learning RFT: An introduction to relational frame theory and its clinical application.* Oakland, CA: New Harbinger Publications.

Twenge, J., & Campbell, W. (2008). Increases in positive self-views among high school students: Birth-cohort changes in anticipated performance, self-satisfaction, self-liking, and self-competence. *Psychological Science, 19*, 1082–1086.

Twenge, J., Campbell, W., & Freeman, E. (2012). Generational differences in young adults' life goals, concern for others, and civic orientation, 1966–2009. *Journal of Personality and Social Psychology, 102*, 1045–1062.

Wilson, D. S. (2010). *Darwin's cathedral: Evolution, religion, and the nature of society.* Chicago: University of Chicago Press.

Wilson, D. S., Hayes, S., Biglan, A., & Embry, D. (2014). Evolving the future: Toward a science of intentional change. *Behavioral and Brain Sciences, 37*, 395–460.

Wilson, D. S., Vugt, M., & O'Gorman, R. (2008). Multilevel selection theory and major evolutionary transitions. *Current Directions in Psychological Science, 17*, 6–9.

Wilson, E. O. (2012). *The social conquest of earth.* New York: Liveright.

Wittchen, H.-U., Nelson, C. B., & Lachner, G. (1998). Prevalence of mental disorders and psychosocial impairments in adolescents and young adults. *Psychological Medicine, 28*, 109–126.

Yeager, D., & Dweck, C. (2012). Mindsets that promote resilience: When students believe that personal characteristics can be developed. *Educational Psychologist, 47*, 302–314.

Dialogue on Development and Adolescence

Participants: Joseph Ciarrochi, Bruce J. Ellis,
Louise L. Hayes, and David Sloan Wilson

David Sloan Wilson: Let me get the ball rolling with a few observations about both chapters. They're strikingly similar in their emphasis on behavior in context. Bruce says, "What's in it for the kids?" Much of what we consider pathological is in fact adaptive in the evolutionary sense of the word. That's a subtle but essential point to make about evolutionary thinking in general. Then Joe and Louise made exactly the same point for behavior in context, which is the essence of contextual behavioral science and its evolutionary roots.

    At the same time, there are some striking differences between the chapters—not just between this pair, but the other pairs of chapters in the book. To explain, let me digress just a little and say that I've written that the human behavioral system should be thought of as like the immune system, with an adaptive component and an innate component. The innate component is a psychological and physiological architecture that evolved by genetic evolution and would be much the same in humans as in other mammals. We could also call this closed phenotypic plasticity. The adaptive component is more open-ended and represented by Skinnerian processes: things like operant conditioning, and also language as studied in CBS with relational frame theory and by authors such as Eva Jablonka and Marion Lamb in their book *Evolution in Four Dimensions*.

    Against this background, I see Bruce's chapter as largely about the innate component of the human behavioral system, much like in other mammals, and Joe's and Louise's chapter as concentrated on the adaptive component of the behavioral system, having much to do with selection by consequences and its extensions into symbolic thought and language. There is not very much overlap between the two chapters in this particular respect.

Bruce J. Ellis: I share your observation that there were similarities and also quite distinct differences in our chapters. E. O. Wilson said that what evolves is the directedness of learning: the relative ease with which certain associations are made and others are bypassed, even in the face of relatively strong reinforcement. I don't think of it so much as having an innate and a learned component. What evolves is a certain directedness or preparedness of learning. If you start thinking about how contexts affect adolescents, you're not starting with a blank slate by any means. You have certain types of information and certain contexts that are going to be extremely important to adolescents because they are trying to transition to a reproductive phase of a life span where they have to successfully mate, where they have to successfully attract partners. So, there are going to be lots of contingencies in the environment that will be really important to them—things like social inclusion, social centrality, relative status—these things are hugely influential in adolescence. So thinking about how the social and symbolic context and social media and all these things operate—it's going to be in a dramatically nonarbitrary way.

Louise L. Hayes: Thank you, Bruce. I agree that there are a lot of similarities and some big differences in our chapters. I would like to talk about how CBS can change what we think about context as one of those things. From a CBS perspective, context is not just the physical environment, but it also brings in this new aspect—our verbal environment becomes a context. Adolescence is, of course, that explosion of language and cognition and behavior, with the ability to begin to look outside your family and to understand your world and where you fit in. Relational frame theory helps us look at how verbal behavior is selected by consequences inside the individual. So context becomes not just the physical environment but also the verbal environment, and we need to look at how that evolves and develops. We can see how young people are shaping their own world by their memories, their history, how they use time, and how they use social interactions. This allows us to layer on the decades of research showing how children's behavior can be adapted and changed by selection of consequences—and allows us to get inside that symbolic verbal realm, where people's worth can be determined by what they think.

David: In the whole volume there is something very constructivist about CBS and relational frame theory—and that's not necessarily a bad thing. There's a baby in the bathwater of constructivism. And there's a baby in the bathwater in the kind of the innatism that we think of in at least

126

some schools of evolutionary psychology. The challenge is to keep the babies and throw out the bathwater.

Joseph Ciarrochi: It's a difference in emphasis. We accept everything Bruce says, but one focus of CBS is on the changeable aspects of the environment and how we can get a handle on those things to improve the condition of young people. To give you an example of how the two might link up, we certainly recognize how sensitive adolescents are to social evaluation and relative status. One of the functions of language is to help us avoid trial-and-error learning. Let me show where that comes in practically with young people. In modern culture, when a kid has a negative thought like *I am unloved* or *I am worthless*, adults often respond by saying, "No, you're a wonderful person. You're great. You're lovely." What we find, again and again, is that when people challenge negative thoughts, a lot of times a young person will resist and fight back. That's kind of weird. Shouldn't everyone want to "unlock the giant within" and have positive thoughts about themselves? It turns out that if you start to look at it functionally, language is there to keep us safe. And so when young people think something like *I'm no good at math*, then that verbal statement is there to protect them from embarrassing themselves, making mistakes, and losing social status in math. The reason that they resist your reassurance is because you're basically asking them to do something that makes them feel quite unsafe. That comes from a better understanding of how language has evolved to help us avoid trial-and-error learning. So when you seek to challenge so-called negative beliefs of young people, you're also challenging their safety and security. Those beliefs, however negative, often provide them with security.

David: Good point!

Bruce: Thinking about this issue of resistance, one of the species-typical changes that occurs in adolescence is that they become more resistant to parental advice and information. As someone once said, when your child reaches adolescence, you get fired as your kid's manager and have to do everything you can to try to get rehired as their consultant. This becomes very relevant to thinking about interventions with adolescents—you do not want adolescents to feel that they're being manipulated or stigmatized. Many interventions single out adolescents. If you're in a school context and you have someone identified as having behavioral problems or failing in their classes, they may get put in a special class, get targeted for an antibullying intervention, and so on. As soon

127

as you pick out an adolescent and target him or her for an adult-driven intervention, you've almost lost before you've begun, because adolescents don't want to be managed in that way. That presents challenges. Let me mention a little about the Meaningful Roles Intervention and how it tries to get around that problem. It is an antibullying intervention that I developed in collaboration with Dennis Embry, Anthony Volk, and José-Michael Gonzalez. We recently published our first paper on this, showing initial results. It's designed as a schoolwide antibullying intervention. Basically, it's a jobs program: every kid at the school gets a job to do that has some important role in running the school. But before you go in, you actually collect data in the school about who is engaging in more antisocial or bullying behaviors. Then everyone gets a job, but they get targeted for particular jobs. The bullies get jobs that give them an alternative, prosocial way to obtain status.

The typical intervention of zero tolerance, which just inflicts punishments to stop bullying, doesn't work at all. Why would kids give up a very effective strategy with nothing in return? It's remarkable to me, when you actually look at prevalent interventions, how common it is to use punishment to stigmatize kids, to target kids; and to try to use coercive methods to, for instance, stop bullying, which often have effects that are just the opposite of what's intended.

Joseph: That sounds like an amazing program, Bruce. Louise and I have been doing something similar where we empower young people to be the leaders in their own community. But one thing that I want to note was that when we work with young people—the whole model in contextual-based science is driven by their values, their needs, their vitality. To begin, we need to know their needs in that context. Would you agree, Louise?

Louise: I would absolutely agree. I know I said it earlier, but I think it's important to repeat that when we talk about changing their context, we can also talk about changing their verbal context, so it could be the context inside them. We'd have to measure it, but I imagine that Bruce's intervention is changing the verbal context inside the bullies. It's giving them more variation in how they see themselves and how they understand themselves. That's what Joe and I have been trying to do by working on all of those levels when we change context.

Joseph: I would like to see if Louise and I could combine Bruce's approach with what we do. Adolescents are viewed as if you put them in school for four or five years until they're ready to do something in the real world. Why not think of them as potentially being the force of change now?

Bruce: Right. The intervention tries to shift them away from a mode of just being passive vessels at school into a mode where every one of them has a meaningful job that's important to running the school. Some of them are the gardeners, others are taking care of animals. There's a sign painter, a PA announcer. When everyone is doing something to help the school run, it improves the school environment for everyone.

Joseph: That's fabulous.

David: Here is another major point I'd like to discuss. Much of this conversation is based on the premise: "Change the context and individuals (or groups) will adapt to that context," as if the process of adaptation is straightforward. But actually it's not straightforward, or at least sometimes it's not. Even to change in the direction that everyone wants to change will still require some kind of work. That's one place where ACT comes in, to accomplish that kind of work through things that are familiar to the ACT community—changing psychological flexibility, perspective taking, working around obstacles, and so on. It's a process that needs to be managed because our natural learning abilities, in true evolutionary style, just climb a local peak and are not capable of traversing a multipeak adaptive landscape.

Joseph: Right! One thing that we do with DNA-V is—this won't be controversial at all to you, David—is the same processes at a group level. We want groups, as well as individuals, to become more adaptive and more vital. So I think we can scale this up to different levels.

Louise: One important thing to add when you think about adolescents is that there's an assumption around valuing. That's why the vitality aspect is really important. As Bruce and Joe both said, we need to think about adolescents in context and to look at where they're getting their vitality. A young person may be getting his or her vitality in a different way, and not have the verbal faculties developmentally in a way that you might use it with adults. We need to think about what their needs are, which might not be your typical middle-class needs—that's a tricky aspect.

Bruce: David, what you just brought up is one of the big questions. To what extent are individuals going to be sensitive to context, or to what extent are they going to be resistant to that context? As a developmental psychologist, I think a lot about developmental programming. Early life experiences get under the skin. Based on your early exposure to microbes, your immune system gets programmed; based on your early exposure to threats and dangers, your stress physiology gets programmed, your HPA access gets programmed; based on your early exposure to energetics, your metabolism gets programmed.

David: And that's overlaid upon genetic differences coming out of the box.

Bruce: Yes, absolutely. All these things get under the skin; they all get epigenetically encoded. And so development has a lot of inertia. People often think, "Why aren't we just infinitely malleable? Why don't we just change perfectly with every situation?" Well, the fact is we don't. People have personalities and there's stability over time. The earlier you begin programming for a particular strategy, the better you tend to become at that strategy. At the same time, there is much sensitivity to context. Individuals have to read their contexts; they have to react to the environment that they're in. So ultimately, understanding individual differences is going to be looking at the interaction between developmental experiences and current contexts. More and more, what I'm seeing and what I'm engaging in are collaborations between, for instance, developmental psychologists and social psychologists, to study the interactions between variation in developmental experiences and variation in some sort of manipulated environmental experience or psychological state. These studies show that what early life experiences are doing is sensitizing you to react in different ways to current contexts. You have to measure both and think about both to really answer the question as to what extent your intervention, or this environmental change, is going to be effective, and for whom.

David: I do think that's something that the CBS folks can take on more than they have. There's a bit of a "blank slate" assumption. There's a quote in Louise's and Joe's chapter that says something to the effect that "naturally there's individual variation, but all young people are capable of developing their DNA skills." There's a sense in which that's true and can be justified. Yet, in accord with what Bruce was saying, there might also be some added value in contextualizing the training.

Joseph: The DNA skills are contextualized, David. When we say "DNA skills," we've defined it for each person, according to what they need in their environment. They're not skills that can be independent of how it's working in the context of that individual person, given his or her history. Louise and I spent a lot of time, for example, talking about trauma in *The Thriving Adolescent*, in our chapter on self-compassion, and how that changes people's Noticer skills. Maybe you want to talk a bit about this, Louise?

Louise: I think what we mean in that quote—and I understand your point, if it seems decontextualized—what we mean is that we need to step aside from the idea of looking at formal behavior and what it looks like from the outside, and start looking at how it's functioning for the individual. So you can't decontextualize it. It might look from a social perspective to be aversive, antisocial behavior, but function very well for the individual. When we talk about developing the DNA skills, we're looking at developing them in a way that helps them be more flexible for the environment that they're in.

David: Right. The context comes in when you actually go through the process, such as when examining the obstacles that get in the way of reaching valued goals.

Well, this has been awesome. Nothing but good can come from this kind of conversation. In closing, I also wanted to mention another common denominator for your two chapters, which is great science. Both are very evidence-based and represent the best of the behavioral sciences.

CHAPTER 8

Situated Empathy: Applying Contextualism to the Science of Recognizing Other People's Emotions

Kibby McMahon

M. Zachary Rosenthal
Duke University

A facial expression with wide, upturned lips and visible teeth may be universally understood as a smile, but its meaning can vary wildly. Americans tend to believe that when a person smiles, he or she is experiencing pleasant emotions paired with a desire to foster connection or trust. In contrast, Japanese people might interpret a smile as an indication of embarrassment, shame, or pain that is appropriately masked for social settings. Thus, the meaning of a smile is determined by the context in which it arises and who perceives it.

Determining how people perceive and attribute meaning to emotional expressions has been an important goal within the science of *empathy*, the ability to recognize, share, and understand another person's feelings (de Waal, 1996; Zaki, 2014). Empathy is critical to the survival of the human species because it drives altruistic behaviors (e.g., helping others or ensuring another's safety), promotes social cohesion, and allows individuals to form intimate bonds with others (de Waal, 2008). Given its importance within adaptive human functioning, empathy has been an important topic within the science and practice of psychology. This chapter will focus particularly on the empathy process of perceiving another person's emotion.

Previous investigations within the basic science of empathy have often treated the social context as a distinct feature that influences one's empathic abilities, but that approach implicitly undermines how fundamental context is to emotional perception and interpretation itself. Reintegrating context within our conceptualization of empathy requires a different approach. Steven Pepper's *World Hypotheses* (1942) provides alternative models of truth that can guide this different approach. In this chapter, we consider how emotional expressions are accurately recognized from the perspective of one such model, *contextualism*. We will first review the prevailing approach within empathy research and its limitations. Then we will introduce a model of contextualism and how it can bolster our understanding of emotions and empathy. Finally, we will discuss the implications of contextualism within psychological science and practice.

The Misguided Quest for Empathy's Universality

Empathy is a crucial skill for us as social beings because it is important to identify the emotional states of other people in our environment. Social theories of emotion go so far as to say that emotion is inherently a medium for communicating information about internal and external states to other people (Hareli & Hess, 2012). Therefore, empirical investigations of empathy have traditionally focused on studying how one person (i.e., "the perceiver") can identify which emotion another person (i.e., "the target") is feeling, based on his or her verbal or nonverbal cues (i.e., expressions). The roots of this approach can be traced back to Darwin's book *The Expression of the Emotions in Man and Animals* (1872), which introduced the idea that there are distinct, adaptive facial expressions associated with distinct subjective emotional experiences across the human species. Resting on the assumption that humans make facial expressions of emotion that others can perceive, the driving question of this early research was whether given facial expressions communicate the same emotions cross-culturally.

The most famous psychological researcher to tackle this question to date is Paul Ekman. Ekman ventured into New Guinea to gather evidence that people of both literate and preliterate societies can identify the same emotion from a photograph of a facial expression (Ekman, Sorenson, & Friesen, 1969; Ekman & Friesen, 1971). In their seminal 1971 work, he and Wallace Friesen conducted the study within the Fore culture of New Guinea's South East Highlands, which had little exposure to Western culture and the way in which members of Western culture display emotion. Participants read stories and were presented with a set of photographs of faces with different emotional expressions. They were instructed to judge which emotional expression best fit the emotion described in the story. Remarkably, there were significant commonalities among the Fore participants and Western adults in their judgments. Ekman and his colleagues widely published these findings as evidence suggesting that individuals of different cultures can identify the same basic emotions in other people (Ekman et al., 1969; Ekman, Friesen, & Ellsworth, 1972; Ekman & Friesen, 1971; Ekman, 1993), supporting a *universality theory of emotion*, wherein a person's emotional state gives rise to a distinct, corresponding expression that can be recognized by others across populations and contexts.

This work led to the proliferation of scientific research, using diverse methodologies and samples, that operationalized empathy as the classification of an emotion from a particular configuration of facial expressions. Critical to the success of early research on this topic was the development and validation of the Pictures of Facial Affect Series, a set of photographed faces displaying one of six basic emotional expressions: sadness, happiness, fear, anger, disgust, and surprise (Ekman & Friesen, 1976). Independent raters coded each photograph for the emotion they thought the face expressed, leading to the Facial Action Coding System (FACS) manual for coding facial expressions (Ekman & Rosenberg, 2005). Each photograph has a specific emotion that is the "correct" answer, as coded by these independent raters. Studies using this approach generally display the photographs to participants and measure the degree to which participants arrive at the correct answers. Although there are alternative sources of photographs, this set and coding system are widely used as standardized stimuli in research examining empathy. Emotion recognition *accuracy* is defined by the degree to which participants select the emotion label assigned.

The ecological validity of this paradigm has been widely debated (Ekman, 1994; Russell, 1994), however, because this artificial laboratory task is very different from the way people usually make sense of others' emotions in everyday naturalistic settings, such as during a conversation with a loved one or passing someone on the street. Some argue that there are rarely realistic circumstances in which people have to judge emotions in an offline, static context such as from photographs of faces (Dziobek, 2012). Suppose, for example, "that facial expressions rarely or seldom occur in the intensity and clarity portrayed in the photographs shown to subjects or that observers rarely or seldom have the time to study the face as they did when shown a still photograph. Then, their degree of recognition of specific emotions from the kind of photographs used in the experimental context might not tell us what happens in most everyday face-to-face encounters" (Russell, 1994, p. 130).

In the early literature supporting the universality of basic emotions, there was evidence that the basic emotions are recognized within different cultures at notably different rates among different populations (Biehl et al., 1997; Matsumoto, 1989). Russell (1994) pointed out that cross-cultural studies generally used the same set of photographs of Western faces with posed expressions. As a result, studies with populations that had more access to Western culture yielded high rates of performance, and thus were more supportive of the theory than findings from studies in more isolated cultures (Russell, 1994). For example, Ekman himself found that over 50 percent of his participants in the Fore population judged the sad facial expression as angry (Ekman et al., 1969). If culture of origin moderates performance on measures of accurate emotional classification, it calls into question the value of this analytic approach because such studies may actually only measure whether participants know the stereotypical way Western people express emotions, not how these participants feel, express their feelings, or recognize the feelings of others from their own community.

A major limitation in this research is that it overlooks how much information context can provide about the meaning of an emotional expression. Russell noted that these studies ask participants to judge emotions from expressions without contextual information, whereas "to refer to a specific facial expression as a *signal* requires that the expression communicate the hypothesized message not only when the face is seen alone, but when seen embedded in a reasonable range of naturally occurring contexts" (1994, p. 123). There is a wealth of evidence that context influences perception of emotional expression (e.g., Campos, Mumme, Kermoian, & Campos, 1994). Without this crucial contextual information, research findings may not apply to real-life social settings.

Empathy researchers have tried to improve the ecological validity of their research with more advanced tasks that assess how empathy happens within everyday social contexts. If emotions are adaptive methods of communicating inner states to other members of our species, social interaction is a primary setting in which empathy occurs (Hareli & Hess, 2012; McArthur & Baron, 1983). The Empathic Accuracy Paradigm (Ickes, Stinson, Bissonnette, & Garcia, 1990) is a prime example of rigorous experimental methodology that aims to reflect real social interaction. This task requires that two participants talk to each other for a few minutes while they are being videotaped. After their conversation, both participants independently review the tape and record what thoughts and feelings they were having in particular moments as well as what they thought their partner was thinking and feeling.

Although this task is a step toward studying empathy within real social contexts, the main outcome measure is *empathy accuracy* as defined by the correspondence of participants' predictions and self-reported experiences from their partner (Ickes, 1993). This defines accuracy as arriving at a consensus between the perceiver and the target (Zaki & Ochsner, 2011). Studies using these tasks have found that strangers have an average empathic accuracy of around 22% (Ickes, 2003). This rate is so low that it raises the question: is accurately guessing what another person will say they are feeling only one aspect of empathy? These studies also found that empathy accuracy with friends was 50% higher than with strangers, likely because their judgments included historical information from the friendship as part of the context for decision making. If social context (e.g., culture, the nature of the relationship between perceiver and target) has a large impact on emotional perception, then it should be included in the fundamental conceptualization and operationalization of empathy.

World Views and Emotional Perception

Integrating context into an understanding of empathy requires considerable philosophical and conceptual clarity about how to approach the issue of context itself. In his book *World Hypotheses: A Study in Evidence* (1942), Stephen Pepper describes four common sets of assumptions and methods that he argued are used to study and understand the world—what he termed "world hypotheses" or worldviews. Worldviews consist of an underlying commonsense *root metaphor*, the event under investigation, and a *truth criterion*, or the principle guiding our method of analyzing this event. Worldviews have been used previously to help understand and simplify philosophy of science issues in psychological research programs (e.g., Hayes, Hayes, Reese, & Sarbin, 1993). We will attempt in this chapter to apply two of Pepper's worldviews—formism and contextualism—to the context of emotional perception.

Formism has as its root metaphor the idea that the world consists of repeated forms, and it is our analytic task to discover what they are and how they are reflected in particular events. In this view, many different events in the world can have a common, underlying nature as they may be all manifestations of the same form. Scientists from the formist perspective would believe that to understand events, one must specify and name the form, and show that its identified characteristics correspond to the world.

Formism is the philosophy of science that best describes the conventional research approach to examining emotions and empathy. The universal theory of emotion is grounded in the assumption that there exist universal prototype emotions that can be recognized by people in a universal, prototypical way. Research in support of this theory studies how individuals can recognize prototype emotions from facial expressions. In this case, events (i.e., facial expressions) are reflections of underlying forms (prototypical emotions) and participants' judgments of the facial expressions are considered "true" if they recognize this underlying form correctly.

For example, Ekman's work took the perspective that "angry expression" is a form that is represented in a photograph of a face, which is identified and named by people who view the photograph. When someone names that photograph "an angry expression," that statement is true because they have named the event (e.g., the photograph) by its characteristic form (e.g., an angry expression) that is tied to a hypothesized underlying emotion. Then, to study the event of "Person A's ability to recognize expressions of anger," the formist scientist would study the nature of that event by evaluating if Person A can identify and name the angry expression in the photograph, which is considered true, or "accurate." In this view, contextual factors like culture can disguise the underlying truth, but are not inherently part of the truth of emotions or recognizing these emotions. As a result, science from the formist perspective defines true empathy as the ability to recognize the "correct" form of an emotional expression, *despite* variations due to culture and other individual differences.

Formism can be contrasted with contextualism, which is based on the root metaphor of purposive, historical, situated act: the act-in-context. An analysis of an event is considered "true" if the purpose of analysis is achieved within the context. This is a *pragmatic* or *functionalist* understanding of truth, in which what is true is what "works" toward an analytic goal in a given context (Hayes, 1993). Contextualism argues that the very nature of an event is dependent on the context (Pepper, 1942; Kohlenberg, 2000). From this perspective, understanding the true nature of an event also requires understanding its context, including culture and other relevant influencing factors, because these are the functions that make up the whole event.

There is a wide range of factors within a given social context, which can include both internal and external factors. Internal factors can be factors that primarily exert their influence "inside the skin" of a person, including thoughts, sensations, and other psychological events. On the other hand, external factors can be social contexts and the many facets of society that impact individuals' psychological processes. For example, Bronfenbrenner's (1979) ecological model of human development provides an account of the different interwoven layers of society that form the external context. Within this model, factors within an individual's immediate environment, such as the individual's family or workplace, are called *microsystems*. *Macrosystems* are overarching factors of a society, such as values, lifestyles, or socioeconomic status. Contextualist scientists consider these factors as interdependent and fundamental to the very nature of the phenomena they study. Therefore, emotion science from a contextualist perspective aims to identify and understand these contextual factors in order to fully understand emotional phenomena. Experimental research methodology is particularly useful for this purpose (Hayes, 1993). Reviewing previous experimental research on emotion and empathy can shed light on which factors have already been identified.

Contextualism and Emotion

Viewing empathy and emotion science through the lens of contextualism can lead to many complexities since context can fundamentally shape several different processes involving feeling, expressing, and recognizing emotions. For example, the nature of emotions themselves depends on the context in which they arise. Campos and his colleagues (1994) outlined a functionalist approach to the understanding of emotion, arguing that emotions primarily serve to fulfill a function within a given context (p. 284). From this perspective, these authors integrated the social environment within their definition of emotion: "one cannot understand emotion by examining the person or environmental events as separate entities … *Emotion* can be succinctly defined from a functionalist perspective as the attempt by the person to establish, maintain, change, or terminate the relation between the person and the environment on matters of significance to a person" (p. 285). In other words, our feelings motivate us to take certain actions within our environment to achieve a goal. For example, if defending our reputation were important to us, we would feel angry when someone insulted us and be moved to fight back against the slanderer. This would especially be the case if we lived in a society that emphasized the importance of defending our reputations aggressively. Furthermore, we might also be more prone to anger if we had confidence in our fighting skills and our heart raced from a large cup of coffee that morning. Therefore, there are external (i.e., outside one's skin) contextual factors determining our goals and interpretation of that situation, and internal (i.e., inside one's skin) contextual factors determining physiological and behavioral responding. If any of these factors vary, the emotional experience and expression may vary as well. Thus, the functionalist views emotions as inherently interpersonal phenomena that motivate people within their social environments (Campos et al., 1994).

According to Campos and his colleagues' review (1994), functionalists focus on contextual factors that characterize an emotion's function and the ability to fulfill that function within a given environment. Research that supported the universal theory of emotion defined emotion as a specific subjective state that is associated with a distinct facial expression. This takes a formist view that assumes that the same basic emotions and their expressions exist across all individuals and contexts. Instead, the functional contextualist view assumes that emotions are defined by their function within a certain context. Furthermore, the environment determines important aspects of how an emotion serves its function successfully.

First, the social environment (e.g., social norms, relationships with other people) can determine what the goal of the person is and how likely the person is to succeed in this goal (Campos et al., 1994). One important goal for emotions is to communicate needs or information about the environment to other people. Given this communication function, emotions will be felt and expressed in ways that most effectively get the message across. For example, physiological responses determine action patterns with functional consequences (a heart beating associated with fear may prime an individual to escape a dangerous situation). Furthermore, the way the individual interprets or makes sense of the context can lead to different appraisals of these physiological responses (e.g., whether a heart beating is a sign of fear or anger). Appraisals, or interpretations of a situation, can influence the goals and meaning of the social situation in which an emotion arises (Campos et al., 1994, p. 297). Finally, culture is a crucial macrosystem that may determine one's situational goals and the acceptable behavioral repertoires that could help achieve those goals. Therefore, the functionalist approach to emotion considers all the contextual factors that define all components of an emotion, including the physiology, subjective experience, appraisal, behavioral response, and goal. There is no one singular causal or defining factor because all of these factors and components are fundamental to the nature of emotion itself. While a formist scientist dismisses cultural differences as variations from a universal conceptualization of an emotion, a contextualist scientist regards those differences as central to how the emotion is conceptualized in the first place.

Contextualism and Research on Empathy

A contextualistic perspective illuminates not just how context can fundamentally shape what an emotion is and how it is expressed to fulfill a function in a specific environment, but also how it shapes others to perceive those emotional expressions. In his work on *pragmatic accuracy*, Swann (1984) states that "the mark of an optimal or accurate person perception is its ability to promote the interaction goals of perceivers. Simply put, an accurate belief is an instrumental belief" (p. 461). From this viewpoint, a person's perception is "accurate" if it helps him or her achieve a function within a certain environment. There are three components of pragmatic accuracy: (1) the environment, or *where* the recognition occurs; (2) the person's function within that environment, or *why* the recognition occurs; and (3) *what* needs to be recognized to fulfill that function. All three components help us understand if empathy is "accurate" from a contextualist point of view. It is not based on a universal criterion or "form," such as "if someone can judge the correct emotion from this facial expression." Instead, it's defined as "if someone can judge another person's emotion in a way that fulfills a certain function within that particular social interaction." This fundamentally changes how someone's empathy abilities are measured, since all three components are important in this definition.

As a contextualist views it, empathy research must take into account both the perceiver's and the target's characteristics and how they may influence each other. For example, a perceiver's empathy abilities predict how well that perceiver can guess a target's emotions, but only if the target is highly expressive (Zaki, Bolger, & Ochsner, 2008). The more the perceiver and target interact, the more the perceiver can access emotional features such as eye gazes or asking questions (Ickes et al., 1990), leading to higher empathy accuracy. Thus, empathy abilities depend on the interactions between individual differences and the social context, which is often dynamic and multifaceted.

Along these lines, other research demonstrates how the internal context of a perceiver can affect his or her empathy abilities. For example, the emotional state of a perceiver also influences his or her ability to recognize other people's emotions (Schmid & Schmid Mast, 2010; Niedenthal, Halberstadt, & Margolin, 2000). Psychopathology also has an impact on empathy, since borderline personality disorder, schizophrenia, depression, and social phobia are all linked to abnormal empathy performance (Roepke, Vater, Preißler, Heekeren, & Dziobek, 2012; Mandal, Pandey, & Prasad, 1998; Joorman & Gotlib, 2006). In addition, empathic responses can also be altered by appraisals, or what is known about the social context. For example, observers who believed that social targets are competitive or untrustworthy inhibited spontaneous empathic responses (Lanzetta & Englis, 1989; Singer et al., 2006; Cikara, Bruneau, & Saxe, 2011), whereas believing the target is a cooperative partner (Lanzetta & Englis, 1989) or attending to his or her painful experiences (Lamm, Batson, & Decety, 2007) enhanced empathic responses. A perceiver's physiology also has an impact on empathy processes, but even that effect can be context-dependent. One review demonstrated that oxytocin may impact certain social cognitive processes such as the ability to process social cues (Bartz, Zaki, Bolger, & Ochsner, 2011). However, this means that the context and the availability of the social cues in the environment will ultimately determine the outcome of oxytocin's impact on empathy. For example, oxytocin might increase a person's attention to a target's emotional expressions, but his or her response to that expression would depend on if the target is a friend or a foe (Bartz et al., 2011). Therefore, an individual's empathic responses can be shaped by internal factors, such as physiology, beliefs, or other related psychological processes.

On a broader level, research has also shown that macrosystems of the social context also influence an individual's empathic abilities (Kraus, Côté, & Keltner,

2010; Mendes & Kraus, 2014; Matsumoto, 1989; Schmid Mast, Jonas, & Hall, 2006). The original work of Ekman and colleagues (Ekman & Friesen, 1971) focused on their participants' abilities to recognize emotional expressions despite cross-cultural differences. Matsumoto (1989) instead brought attention to these cultural differences as clues to which aspects of society lead to such profound impacts on empathy. In his review, he found that judgments of emotional expressions are influenced by their culture's *power distance,* or the degree to which there are differences in power among individuals, and *individualism,* the degree to which the society values individual goals and independence from others. Each one of these aspects determines the society's social norms and values, which in turn influences how individuals express and perceive emotion. More recent experimental research supports these findings. For example, one study found that people lower in socio-economic status had higher empathy accuracy than others (Kraus et al., 2010). Furthermore, people wearing clothes that signaled low status (e.g., sweat pants) were more sensitive to the emotional cues of people wearing clothes that signaled high class (e.g., a suit) (Mendes & Kraus, 2014). These findings support the contextualist view that information about the social context is necessary to understand empathy that arises within it. Research needs to identify and study these factors instead of simply controlling for them or disregarding them.

Implications for Future Research Directions

While research has begun to lay the groundwork of a contextualist approach to empathy, it needs to take additional steps. First, research should go beyond a single standardized criterion for empathy accuracy, such as the consensus criterion. The contextualist approach to defining empathy would include: (1) the environment, or *where* the empathic response occurs; (2) the person's function within that environment, or *why* the empathic response occurs; and (3) *what* empathic response is needed to fulfill that function. For example, previous research has demonstrated that there are some circumstances in which people are often motivated to *avoid* recognizing another person's emotions (Zaki, 2014). The formist researchers would consider that avoidance as "poor performance" or a failure to empathize correctly. The contextualist researcher would consider how that avoidance helped the person achieve a certain goal in the environment, and what other factors contributed to that process. Therefore, contextualism can offer a more flexible definition of empathy that depends on the circumstance in which it arises.

Second, future investigations can make conscious efforts to identify how the context influences empathy in various naturalistic contexts. Previous research has traditionally studied empathy by standardizing the environment and function.

For example, in Ekman's research paradigm, the environment is the research setting in which participants view photographs of faces, the participant's goal is to guess the "correct" emotion, and the function of a participant's empathy skills is to complete the tasks as instructed (e.g., "select the emotion label that best describes the facial expression"). Such laboratory tasks limit the effects of motivations, goals, or needs a person might have in his or her own life. A broader consideration of environment and function is needed. Research needs to investigate why and how we empathize with others to achieve specific interpersonal goals within daily life. For example, we may recognize another person's emotions differently if we want to comfort the person or compete against the person. Ecological momentary assessment or other methods that capture people's behavior in real-life settings are potential tools that can be used for this purpose. Findings from this research can have increased ecological validity if they can generalize to a wider variety of environments and populations.

Finally, investigations can also study specific cases of empathy through a thorough examination of all the relevant contextual factors, conducting an in-depth analysis of empathy within the specific context in which it arises. Importantly, such a contextual analysis would aim to identify the relevant contextual factors that contribute to an empathy process, fulfilling its function within a given environment. Such analyses might start with a thorough examination of single cases, which may lead to new hypotheses on the effects of contextual factors (Biglan, 2004). Psychotherapy is a potentially rich resource for such data. Such investigations of empathy would more closely reflect how empathy occurs within real-life settings.

Treatment Implications

Psychotherapy is a setting that regularly requires empathy and the keen perception and understanding of emotions. Contextualism has already informed the field of psychotherapy and has given rise to a wave of contemporary cognitive behavioral therapies (CBTs). In behavioral therapies informed by contextualism, clients' emotions are examined and understood within context (Hayes, Villatte, Levin, & Hildebrandt, 2011). There is no "right" or "wrong" emotion objectively across contexts, but emotions are understood to serve a function (e.g., communicate, motivate action tendencies) within a particular environment. Using this approach, emotions are understood within the broader context of what matters to clients (i.e., their values) in any given context. The therapist and client collaboratively identify the temporal links in the chain of an antecedent and consequences associated with problem behaviors. This approach, commonly called functional analysis or behavioral analysis, is used to identify contextual factors associated with emotions and problem behaviors (Kohlenberg, 2000). In order to change emotions and related behavioral sequelae, functions and goals are contextualized according to the person's values. In sum, contextualist psychotherapists understand emotions as embedded within the context of the client's life. Therefore, emotions and their functions are inseparable from the context, which includes a clear emphasis on client values relevant to the context.

For a therapist to recognize and understand how a client is feeling, the therapeutic relationship is the context that gives rise to empathy. Contextualism encourages the perspective that a client's emotions do not occur in a vacuum, but instead are situated within the therapy setting. As the therapist perceives and understands how the client is feeling, contextual factors influence this process of empathy. Specifically, contextual factors influence the therapist's perspective when he or she perceives the client's emotions (Kohlenberg, 2000). First, the function of the therapist's empathy should be made explicit. Often this function is to help the client achieve his or her goals for therapy. A behavioral analysis can identify all the contextual factors that contribute to how the therapist perceives a client's emotions in service of this therapeutic function. Findings from contextual empathy research can be applied here. For example, the therapist's cultural background, state of physiological arousal, and subjective emotional state may all influence his or her ability to recognize and respond effectively (i.e., respond in a way that works given the client's values and goals for the context of therapy) to the client's emotions. Understanding the effect of these factors can inform how the client's emotions are perceived in similar contexts. Therefore, contextualism allows for a flexible approach to understanding how emotions are felt, expressed, and perceived in context. Contemporary CBT therapists draw upon this wisdom in an effort to deeply understand their clients' emotional experiences and associated behavioral responses in a manner that is guided by the clients' values in context.

Conclusion

Contextualism is an alternative to the formist view of emotions in general, and empathy in particular. Formism in emotion science leads implicitly to the idea that there is a "true" way of feeling, expressing, and perceiving an emotion. The goal of that approach is to find and label this "true" way and treat differences across contexts as noise that disguise recognition of the underlying form. What is noise to the formist is the phenomenon of interest to the contextualist. The contextualist believes that emotions or empathic processes allow a person to achieve a goal within a particular context. This affords a more flexible and pragmatic view of what is "true," a view that considers any relevant contextual information vital to the very understanding of an emotion's nature and function. When applied to psychological science and practice, contextualism expands the definition of emotions and empathy to include the myriad of moderating and mediating variables in the interest of a more functional approach.

References

Bartz, J. A., Zaki, J., Bolger, N., & Ochsner, K. N. (2011). Social effects of oxytocin in humans: Context and person matter. *Trends in Cognitive Sciences, 15,* 301–309.

Biehl, M., Matsumoto, D., Ekman, P., Hearn, V., Heider, K., Kudoh, T., & Ton, V. (1997). Matsumoto and Ekman's Japanese and Caucasian facial expressions of emotion (JACFEE): Reliability data and cross-national differences. *Journal of Nonverbal Behavior, 21,* 3–21.

Biglan, A. (2004). Contextualism and the development of effective prevention practices. *Prevention Science, 5,* 15–21.

Bronfenbrenner, U. (1979). *The ecology of human development: Experiments by nature and design.* Cambridge, MA: Harvard University Press.

Campos, J., Mumme, D., Kermoian, R., & Campos, R. (1994). A functionalist perspective on the nature of emotion. In N. A. Fox (Ed.), *The development of emotion regulation: Biological and behavioral considerations.* Chicago: Society for Research in Child Development.

Cikara, M., Bruneau, E. G., & Saxe, R. R. (2011). Us and them: Intergroup failures of empathy. *Current Directions in Psychological Science, 20,* 149–153.

Darwin, C. (1872). *The expression of the emotions in man and animals.* London: Murray.

de Waal, F. B. M. (1996). *Good natured: The origins of right and wrong in humans and other animals.* Cambridge, MA: Harvard University Press.

de Waal, F. B. M. (2008). Putting the altruism back into altruism: The evolution of empathy. *Annual Review of Psychology, 59,* 279–300.

Dziobek, I. (2012). Comment: Towards a more ecologically valid assessment of empathy. *Emotion Review, 4,* 18–19.

Ekman, P. (1993). Facial expression and emotion. *American Psychologist, 48,* 384–392.

Ekman, P. (1994). Strong evidence for universals in facial expressions: A reply to Russell's mistaken critique. *Psychological Bulletin, 115,* 268–87.

Ekman, P., & Friesen, W. V. (1971). Constants across cultures in the face and emotion. *Journal of Personality and Social Psychology, 17,* 124–129.

Ekman, P., & Friesen, W. V. (1976). *Pictures of facial affect.* Palo Alto, CA: Consulting Psychologists Press.

Ekman, P., Friesen, W. V., & Ellsworth, P. (1972). *Emotion in the human face: Guidelines for research and an integration of findings.* New York: Pergamon Press.

Ekman, P., & Rosenberg, E. L. (2005). *What the face reveals: Basic and applied studies of spontaneous expression using the Facial Action Coding System (FACS).* New York: Oxford University Press.

Ekman, P., Sorenson, E. R., & Friesen, W. V. (1969). Pan-cultural elements in facial displays of emotion. *Science, 164*, 86–88.

Hareli, S., & Hess, U. (2012). The social signal value of emotions. *Cognition and Emotion, 26*, 385–389.

Hayes, S. C. (1993). Analytic goals and the varieties of scientific contextualism. In S. C. Hayes, L. J. Hayes, H. W. Reese, & T. R. Sarbin (Eds.), *Varieties of scientific contextualism*. Oakland, CA: Context Press.

Hayes, S. C., Hayes, L. J., Reese, H. W., & Sarbin, T. R. (Eds.). (1993). *Varieties of scientific contextualism*. Oakland, CA: Context Press.

Hayes, S. C., Villatte, M., Levin, M., & Hildebrandt, M. (2011). Open, aware, and active: Contextual approaches as an emerging trend in the behavioral and cognitive therapies. *Annual Review of Clinical Psychology, 7*, 141–168.

Ickes, W. (1993). Empathic accuracy. *Journal of Personality, 61*, 587–610.

Ickes, W. (2003). *Everyday mind reading*. New York: Perseus Press.

Ickes, W., Stinson, L., Bissonnette, V., & Garcia, S. (1990). Naturalistic social cognition: Empathic accuracy in mixed-sex dyads. *Journal of Personality and Social Psychology, 59*, 730–742.

Joorman, J., & Gotlib, I. H. (2006). Is this happiness I see? Biases in the identification of emotional facial expressions in depression and social phobia. *Journal of Abnormal Psychology, 115*, 705.

Kohlenberg, B. S. (2000). Emotion and the relationship in psychotherapy: A behavior analytic perspective. In M. J. Dougher (Ed.), *Clinical behavior analysis*. Reno, NV: Context Press.

Kraus, M. W., Côté, S., & Keltner, D. (2010). Social class, contextualism, and empathic accuracy. *Psychological Science, 21*, 1716–1723.

Lamm, C., Batson, C. D., & Decety, J. (2007). The neural substrate of human empathy: Effects of perspective-taking and cognitive appraisal. *Journal of Cognitive Neuroscience, 19*, 42–58.

Lanzetta, J. T., & Englis, B. G. (1989). Expectations of cooperation and competition and their effects on observers' vicarious emotional responses. *Journal of Personality and Social Psychology, 56*, 543–554.

Mandal, M. K., Pandey, R., & Prasad, A. B. (1998). Facial expressions of emotions and schizophrenia: A review. *Schizophrenia Bulletin, 24*, 399–412.

Matsumoto, D. (1989). Cultural influences on the perception of emotion. *Journal of Cross-Cultural Psychology, 20*, 92–105.

McArthur, L. Z., & Baron, R. M. (1983). Toward an ecological theory of social perception. *Psychological Review, 90*, 215–238.

Mendes, W. B., & Kraus, M. W. (2014). Sartorial symbols of social class elicit class-consistent behavioral and physiological responses: A dyadic approach. *Journal of Experimental Psychology: General, 143*, 2330–2340.

Niedenthal, P. M., Halberstadt, J. B., & Margolin, J. (2000). Emotional state and the detection of change in facial expression of emotion. *European Journal of Social Psychology, 30*, 211–222.

Pepper, S. C. (1942). *World hypotheses: A study in evidence*. Berkeley, CA: University of California Press.

Roepke, S., Vater, A., Preißler, S., Heekeren, H. R., & Dziobek, I. (2012). Social cognition in borderline personality disorder. *Frontiers in Neuroscience, 6,* 195.

Russell, J. A. (1994). Is there universal recognition of emotion from facial expression? A review of the cross-cultural studies. *Psychological Bulletin, 115,* 102–41.

Schmid Mast, M., Jonas, K., & Hall, J. A. (2006). Give a person power and he or she will show interpersonal sensitivity: The phenomenon and its why and when. *Journal of Personality and Social Psychology, 97,* 835-850. doi.org/10.1037/a0016234

Schmid, P. C., & Schmid Mast, M. (2010). Mood effects on emotion recognition. *Motivation and Emotion, 34,* 288–292.

Singer, T., Seymour, B., O'Doherty, J. P., Stephan, K. E., Dolan, R. J., & Frith, C. D. (2006). Empathic neural responses are modulated by the perceived fairness of others. *Nature, 439,* 466–469.

Swann, W. B. (1984). Quest for accuracy in person perception: A matter of pragmatics. *Psychological Review, 91,* 457–477.

Zaki, J. (2014). Empathy: A motivated account. *Psychological Bulletin, 140,* 1608–1647. doi.org/10.1037/a0037679

Zaki, J., Bolger, N., & Ochsner, K. (2008). It takes two: The interpersonal nature of empathic accuracy. *Psychological Science, 19,* 399–404.

Zaki, J., & Ochsner, K. (2011). Reintegrating the study of accuracy into social cognition research. *Psychological Inquiry, 22,* 159–182.

CHAPTER 9

The Social and Contextual Nature of Emotion: An Evolutionary Perspective

Lynn E. O'Connor
Wright Institute

Jack W. Berry
Samford University

It has been suggested that no one can clearly define the word "emotion." According to Joseph LeDoux, "there are as many theories of emotions as there are emotion theorists." Even so, we all know in a general way what an emotion is: a brief psycho-physiological event that includes the subjective capacity to experience what we think of as a feeling. We know what emotions are because we share them. We're utterly dependent upon our emotions. They provide us with information and direction as we react to changes in the physical and social environment; our feelings tell us what's going on.

Traditionally—at least in academic and clinical circles—emotions have been considered intrapsychic mental phenomena (Keltner & Haidt, 1999). In the sway of our individualistic culture, with its consistent focus on the decontextualized, isolated individual, we see ourselves, first and foremost, as individuals separated enough to believe that the emotions we experience are occurring to us alone. We live in a culture in which children often grow up sleeping alone in their own bedroom, not even shared with siblings. Alone. Our western individualistic lifestyle has constrained our theorizing about emotions, and sent us off in numerous wrong directions. For most of human history, and in much of the world today, babies are rarely out of physical contact with their mother (Konner, 2005). The idea of putting infants to sleep in a bed, let alone a room, by themselves, is unimaginable.

In the era of evolutionary adaptation (EEA; Cosmides & Tooby, 2000), perhaps best understood by examining the lifestyles and customs of the few remaining hunter and gatherer cultures, people were seldom alone, and the idea of privacy was entirely foreign. Considered as social phenomena, emotions are interpsychic, serving to connect and influence members of social groups—the family, the tribe, the nation. From this perspective, the social world in which we live determines how we think about—and even how we experience—emotions.

The varied experiences we think of as emotions evolved as part of the highly complex mental apparatus of the mammalian brain, exquisitely designed and organized for nursing babies who come into the world entirely dependent upon their mother. Reptiles are born fully developed; their only feature close to mammalian emotions may be something, perhaps a sensation, that signals danger or safety, recommending approach or avoidance. The nursing mother, the maternal mammal, needs the extensive information provided by emotions to tell her what is happening in the world as she moves from moment to moment, making it possible to successfully adapt to changing physical, psychological, and social conditions. This is especially important when she needs to understand the communication of affect from her baby. Without the capacity to feel and interpret emotions, a mother would fail to know if her baby was hungry, wet, tired, or cold. Mothers with an emotional connection to their babies are better able to care for them, and babies able to relate to the emotions felt and expressed by their mothers are better able to learn. The intimacy of the relationship between mothers and babies is fostered by that emotional connection, and it affects everything in a newborn's life, including the regulation of breathing, heart rate, and temperature (Feldman, 2007).

In human groups, it makes most sense to consider emotions as social events that evolved to protect, nurture, and aid the social structure and communication among people within groups. The interactive connection between mother and infant is a theme carried throughout life, as we never stop needing person-to-person connection; it forms the basis even for adult emotion regulation in mammals (Lewis, Amini, & Lannon, 2000). Beginning with the mother-infant connection, in our highly social species, emotions form a basis for human connection, without which we couldn't survive. While letting us in on what's going on in our world, emotions serve as guides for behavior, in part because of the speed by which they are processed (Ohman, Flykt, & Lundqvist, 2000). If we waited for more fully conscious cognitions to instruct us in any given situation, we wouldn't be able to adapt to a rapidly changing environment in which speed is of the essence. Emotions allow us to act when we have to—offering a clear reflection of danger and safety. One remarkable feature of emotions is that they are largely unconscious and able to take over before we know what hit us (Scherer, 2005). This makes them difficult to study, and sets them apart from conscious cognition, although the two are inevitably interrelated. We may be slow thinkers, but we're awfully fast reactors, thanks to our emotions. When the baby is crying, the mother is flooded with chemicals propelling her into action. Without conscious thought, she performs the actions she knows will bring comfort. We may consciously recognize and be able to label our emotions later, even long after taking action, although often they remain hidden from overt awareness. In the course of daily life, we experience numerous emotions but rarely stop to articulate their appearance.

People tend to categorize emotions as negative and positive, healthy and unhealthy, prosocial and antisocial, not because, objectively, some are more useful than others, but because, subjectively, some are more unpleasant, or even more disturbing or disruptive, than others. Buddhists speak of less pleasant emotions as "afflictive" while other religions may consider them "immoral." Some unpleasant emotions, when exaggerated, may render people dysfunctional in a behavioral sense. Biological disruptions by way of pathogens, neurotoxins, malnutrition, developmental disasters, or end-of-the-line random mutations, may lead to emotions that do not have clear adaptive functions. Someone who fails to read or express emotions in a way that's culturally appropriate may be socially excluded and isolated, a deadly situation for a member of our social species. Prosocial emotions, such as empathy and compassion, are far more socially acceptable, and certainly more favorable in how we experience them. Love feels better than hate. Pride feels better than shame. We'd rather feel happy and joyful than be hit by a wave of guilt, shame, or remorse. Regardless of categorization, however, all emotions are the product of evolutionary pressures and of the environmental conditions to which they are adaptations, and thus are important.

The interpersonal nature of emotions makes it imperative that they be communicated, person-to-person. Because emotions are so often unconscious at the moment in which they arise, their expression is initially seen in nonverbal behavior, although they may quickly become conscious and conveyed through language. When a newly in love couple gets "swept off their feet" with some feeling, it's the feeling first, then one or both of them label it "love." Before either of them can get to the point of naming the feeling, unbeknownst to either of them they've signaled one another with widened pupils, demonstrating intense interest. Emotions allow us to communicate our feelings to others in our social group, before we know it. An emotion may signal danger or safety. When someone expresses fear, members of the social group know there's danger. Everyone knows what it is to feel anxiety and fear, the emotions warning us that something bad may happen to us, or to others we care about. When someone falls in love, the potential partner knows it. The communication of affect may be a proximal purpose of emotions. Despite the disputes that exist between emotion researchers, in our view, the evidence suggests that many emotions are universal, going back in evolutionary history.

Many Emotions Are Universal

Darwin, noting the universality of core emotions in mammals, described the continuity of their expression as they evolved. In *The Expression of Emotions in Man and Animals* (1872) he demonstrated the emotional expressions in mammals preceding us, with exquisite illustrations of emotions in dogs, cats, and other mammals, including our species. Despite the centrality of emotions in human life, it was only in the early 1960s that psychologists began to study them in earnest. Inspired by Darwin, many researchers, such as Tomkins (1962), Plutchik (1962, 1963), Izard (1971), and Ekman (1972), began to investigate what are often called "basic emotions" in humans, which are emotions that we share with other primates and possibly other mammals. Basic emotions are commonly thought to have distinct physiological profiles and response tendencies aimed at helping us adapt to environmental demands, and also to have reliable, hardwired communication signals. Presenting photographs of people expressing six basic emotions (happiness, sadness, fear, anger, surprise, and disgust), Ekman and Friesen (1971) demonstrated that across cultures—including people with minimal contact with the modern world—these emotions are universally recognized. Since then, an increasing number of emotions, beyond the initial six, have been included among the basic emotions by at least some researchers, along with compound emotions, or the combination of emotions.

Emotions in Context: Culture, Sex, Social Position, and Class

While the capacity for emotions is universal, when and how people experience them depends on the culture, environment, and social context. Potential variability is a characteristic of all psychological traits; whether we're speaking of personality or proneness to moodiness, genetic variability is interacting with environmental and developmental factors. The huge range of possible variations is what makes us unique as individuals, and what creates the group differences between us. Thus, while our capacity to feel emotions is universal—that is, we are all able to feel every emotion, as well as to read and respond to emotions felt and expressed by others—we exhibit wide differences between groups, and between individuals within groups. Variability provides the basic units upon which evolution is working to create the adaptations we know as emotions.

After the initial studies demonstrating the universality of emotions and linking them to fundamental biological capacities, years of research have followed, highlighting the relevance of culture to emotions. Cultures differ markedly in the value they place upon emotions, the conditions under which they are expressed, and even in the frequency with which they are experienced. The expression of emotions may be culture-specific and context-dependent. Furthermore, even given differences between cultures, within every culture there are differences in the expression of emotions dependent upon social context, such as gender/sex, social status and class, and the nature of relationships. The acceptable expression of emotions in any given situation is regulated by what we refer to as "display rules." Display rules provide instruction as to when, with whom, and how we express or hide our emotions. For example, the rules for expressing emotions differ when someone is communicating with a supervisor than when speaking to a family member. While cross-cultural research has attempted to make clear distinctions in emotions between cultures, these studies may, in quite a few cases, be primarily capturing "public behavior." Context matters. This may be similar to studies of personality. Women may appear agreeable and introverted in the workplace with male (or female) bosses, while being quite the opposite when at home with their mothers, siblings, or partners. Tsai (2007) has noted that Asian Americans tend to find more calming emotions "ideal" compared to the European American preferences for more excitement-related emotions. This may or may not be equally true for public and private behavior.

Culture-specific differences in the expression of emotions were illustrated by Briggs (1970) in her discussion of the unusual degree to which people from Inuit society avoid the expression of anger, and in fact seem to experience anger less often. Surviving in the arctic is exceptionally difficult. Scarcity in the environment demanded that groups remain small. The expression of anger might incur a serious dispute, resulting in exclusion from the group. In many parts of the world, an angry dispute leads to a family or group member splitting away and moving elsewhere. This kind of separation of individuals from their group would be lethal in the harsh arctic environment. Therefore, it is clearly adaptive to have rules against the expression of anger in an environment where walking away permanently is out of the question. As we've moved into a global economy, differences between regions and nations have become less obvious and thus less predictive. Matsumoto (2002), after first finding significant differences in emotion studies comparing Japanese Americans and European Americans, later noted that with globalization, the differences have largely disappeared.

Within every network of relationships in any specific culture, there are found elaborate rules and regulations for emotional expression. In Western culture, anger expressed by women is heavily frowned upon, whereas anger expressed by men may be a source of social dominance. The angry male CEO may, in part, have risen by way of his temper. Here, anger is employed by men in asserting dominance in the ranking of group members in a work place. In the same work setting, however, women are not advanced by the overt expression of anger—quite the opposite. In one study, it was found that a face when appearing with a woman's hair was considered seriously angry while the same face, given the hair typical of a man's, was seen as relatively neutral (Barrett & Bliss-Moreau, 2009). In a recent American election, negative commentaries about a female candidate supported research findings in education: a woman in Western culture who appears strong and confident may be perceived as angry, as a "bitch," or even worse for her social ranking, she may be classified as "dishonest." This is still a considerable problem for women in academia. Overall, students rank female professors below male professors even if the latter's educational outcomes are poorer (Boring, Ottoboni, & Stark, 2016), or if the gender of virtual professors in online courses is randomly assigned irrespective of actual gender (MacNell, Driscoll, & Hunt, 2015), suggesting that student evaluations are more measures of gender bias than of teaching ability. The bias against intellectual and emotional confidence expressed by women, speaking in public, illustrates the effect of cultural rules related to the expression of emotions, while also demonstrating the connection between culturally specific rules for displays of emotions and politics.

Individualism and Collectivism: The Evolution of Culture-Specific Emotionality

Cultures differ in their rules of behavior, with associated culture-specific emotionality. Applying evolutionary concepts to the emergence of individualistic and collectivistic societies provides an example of how and why this might happen.

A strength of approaching social phenomena from an evolutionary perspective is that it invites us to ask broad questions such as why and how did one culture become collectivistic while another became individualistic? Why might one form of social organization, with specific behaviors and emotional tendencies, emerge in the Western nations and another in regions of Asia?

The answers to these questions begin with concrete environmental factors that create selection pressures, leading to particular group differences. An evolutionary analysis of the conditions explaining the emergence of collectivism has identified at least one pressing problem for which collectivism was a partial solution, namely, the prevalence of pathogens. Collectivism is marked by behaviors and attitudes such as strong social cohesion, conformity, suspicion toward strangers, and social sensitivity, with attending emotionality, all of which serve as defensive strategies against pathogens (Fincher, Thornhill, Murray, & Schaller, 2008). In regions with pathogens such as typhus, malaria, and leprosy, people living in a closely knit, cooperative, collectivistic culture are more likely to survive and thus a collectivistic form of social organization is a successful adaptation to these environmental problems.

Cultural neuroscience has begun to look to biological factors interacting with behavioral, social, and emotional variations that are correlated with differences between cultures. Specific genetic variants are being identified that potentially explain differences in emotionality between individualistic and collectivistic cultures, which are likely driven by gene-culture coevolution. A collectivistic society and culture provides the socially supportive environment needed for emotionally sensitive people. A variety of the serotonin transporter gene, 5-HTTLPR, has been identified that is associated with emotional sensitivity, but the outcomes of this genetic fact vary by culture (Chiao & Blizinsky, 2010). In China, about 70% of the population carries the gene variant, with no ill effect. In the West, however, the approximately 40% of the population who carry the same 5-HTTLPR polymorphism are vulnerable to mood and anxiety disorders. This make sense, because in the context of a socially supportive collectivistic environment, Chinese people carrying the variant, with its associated sensitive emotional makeup, were likely to survive and thrive. In the West, however, an individualistic culture emerged due to the lesser need to respond to the challenge of pathogens. In the context of that culture, emotionally sensitive people can face greater emotional and psychological challenges. Such findings demonstrate the complexity of cross-cultural differences in emotions, due to the process of gene-culture coevolution.

Another contemporary example of gene-culture coevolution possibly affecting emotionality or emotional tendencies may be found by following the current changes we're seeing in "female choice," relevant to mate selection in primates. In many primate species, including our own, females generally choose with whom to engage in reproduction. Human females are looking for partners who will best provide for their infants and children as they go through the very long, multiyear process of development from infancy to adulthood. Until the last forty (or so) years, females—when they were able to choose their reproductive partners—were likely to select men who were strong, muscular, athletic, and highly "masculine" in personality and emotional tendencies. The "macho" male was likely to be a more reliable provider for the offspring. However, cultural changes are afoot; with escalating speed, technology is changing the way we make things, and males who are skilled at new technologies are becoming the more effective providers. In reaction, female choice is changing; the contemporary woman is now more likely to select a technology-savvy mate with whom to raise her children (Konner, 2015). The Bill Gates/Steve Jobs/Barack Obama model of manhood is emerging, with changes in men's emotionality. The new female choice of mate is more "feminine" in terms of emotional sensitivity; the new man may be more socially sensitive, higher in empathy and compassion, and more willing to demonstrate the softer emotions. Today, the new man is likely to be a more effective provider, and for better or worse, he is even more likely to cry. Like the emergence of collectivism in China with associated genetic factors, the male children of the new man are, like their fathers, likely to exhibit greater social and emotional sensitivity. Today, changes in culture, as seen in the rise of technology, are likely to be influencing the genetics of the new man, that is, the new female choice of partner, initiating the process of coevolution leading to more sensitive, empathic, and overtly altruistic men. It seems likely that complex neurotransmitter variations are coevolving, helping to support the changes in social emotions. The changes in display rules for the male and for the female are increasingly obvious, and likewise, the very nature of contemporary men may be shifting.

Altruism, Empathy, Compassion, and Guilt

Despite cultural, gender, or social-class differences in emotional displays, it seems likely that the experience of emotion is far more uniform than its expression. The experience of happiness or the experience of fear may be quite similar for all of us, although we may differ in exactly where and how we're allowed to express them. As highly social mammals living in relatively large groups, we are fundamentally motivated by altruism, albeit unconsciously much of the time. Despite the competition going on between us, within our groups, this basic nature—what Buddhists refer to as "Buddha nature"—works to hold our families and larger social groups together. Empathy, the capacity and tendency to feel what others are feeling, is derived from a neural architecture including mirror neurons (Gallese & Goldman, 1998). If you see a conspecific in pain, you feel pain. If you are close to someone who is happy, you will find yourself happy. Laughter is contagious, as are anxiety, depression, and anger. We may be competitive, but most often we cooperate. The human nervous system has evolved for cooperation based on multiple levels of selection (Wilson, 2015). Cooperation within our groups allows us to be successful at the most fundamental tasks in life, including work and reproduction. We cooperate in raising our children and in caring for our families, the young and the elderly alike. We organize our groups by way of cooperation. We accept leaders, we're loyal followers, we tolerate hierarchies and sometimes a less-than-fulfilling division of labor. Furthermore, we cooperate for the success of our group as a whole; in between-group competition, groups with more cooperators outdo groups with fewer cooperators.

Emotions, as adaptations to particular problems, are selected for at multiple levels, and what appears beneficial for the individual within the group may be detrimental to the group as a whole, in competition between groups. We may analyze almost any emotion from the perspective of multilevel selection theory (Wilson & Wilson, 2007). Anger may serve many positive functions at the level of the individual, such as enforcing the loyalty of friends or the faithfulness of one's partner, or in establishing a cohabiting relationship with a group member. Parents, responding to a child who has walked heedlessly into the street without looking when he or she knows better, may react quickly with anger in order to save their child's life. When someone threatens a child's life, parents appropriately grow fiercely angry. Anger may also serve a positive function for the group. By expressing anger and hatred, we may be conveying a message to members of another group, to the "out-group," about our willingness to fight for our own "in-group." Anger and hate may thus be prosocial when directed toward the "other" (the out-group, the enemy in between-group competition), meaning that it may help to support your group.

Group members may react with anger when another group member fails to follow culturally accepted rules. Anger may appear within a group of friends or colleagues as a way to punish a group member when cultural norms have been violated. This is considered altruistic punishment, in that it is motivated by altruistic intentions, for the good of the group. It can be powerful in protecting the group from intruders if group members all get angry together. Anger is a frequent factor in both within-group and between-group competition; the specifics of evolutionary context matter. The Dalai Lama, speaking of political tyranny and oppression—perhaps with Tibet in mind firstly, but also in reaction to the threat to democracy perceived in recent American politics—said that anger was not only appropriate, but it must continue unabated until the oppressors are overthrown.

Survivor guilt, a pervasive though usually unconscious emotion, describes the feeling of uneasiness and guilt when someone believes that by pursuing and succeeding at ordinary life goals, including good relationships, work, and happiness, they will harm another who is less successful. While it might be selected against at the level of individual fitness, its persistence suggests that it is selected for at the level of the group. It can help groups to have many people with high levels of survivor guilt. People high in survivor guilt try to cooperate with others, which clearly promotes group effectiveness. They may avoid leadership and be more comfortable being followers, supporting the social hierarchies needed in organizing the socio-economic world of large groups of people, despite our strong preference for equity. In smaller groups, such as groups in the European Economic Area, survivor guilt may have been helpful in maintaining equality; if few were comfortable surpassing another, survivor guilt could have played an important role in fostering a democratic and equality-based social structure. Cosmides and Tooby (2005) have described a human capacity to "detect cheaters" in order to avoid being "taken in" by others. Survivor guilt is, essentially, the capacity for cheater detection turned inward, upon oneself. When suffering from survivor guilt, people are suspicious of their own happiness, afraid that it comes at the cost of someone else's misery.

In our mass culture, however, people who suffer from high levels of survivor guilt may inhibit ordinary drive and ambition to fulfill normal human aspirations. Guilt-ridden people may sacrifice themselves, engaging in pathological altruism. For example, a woman may remain in a destructive relationship because she believes that it will harm her partner should she leave him. Someone else who has a drinking problem and wants to stop drinking may believe it will make his family members, most of whom are heavy or even alcoholic drinkers, feel inadequate by comparison. The examples are everywhere; survivor guilt seems to underlie many of the problems that bring people to therapy (O'Connor, Berry, Lewis, & Stiver, 2012). Destructive survivor guilt appears to be connected to underlying, unconscious, pathogenic cognitions—that is, beliefs that one is omnipotently responsible for the well-being of another, when such responsibility is impossible. This emotion—which most of us can relate to when we stop and think about it—persists, despite how damaging it may be to individuals. It seems likely that there are heavy group selection pressures for this emotion, given that it is an overwhelming problem for so many people.

Psychotherapy: New Emotional Experiences May Make Life Better

The two principles threading through this discussion about emotions are that emotions are more interpersonal than is often thought, and that context matters. These principles apply to when and how people experience emotions, how they are expressed, and how they work as one of the mechanisms holding people together. Context may refer to geography, ethnicity, class and status, relationships, and even the level of selection at which they are being analyzed. Extending this further into particular human activities, psychotherapy and the therapeutic relationship itself are contexts. When successful, psychotherapy is a social context specifically aimed at personal change, and the emotional connection between the therapist and client is the medium in which change happens, whether or not emotions per se are a topic or something directly or indirectly expressed.

What happens in psychotherapy, no matter what "school," style, or type, involves the therapist providing a different interpersonal, social context—one with a positive emotional tone—while teaching the client about the relevance of contexts, or situations, to their current condition. Clients begin treatment with a set of problems to solve. In many cases, they don't know exactly what those problems are and furthermore, they believe that they themselves are "to blame." Even those who appear to be externalizing blame for all of their problems are falling in line with an under-the-surface belief indicating how they are at fault. Some current therapies include careful review of clients' conscious beliefs, followed by efforts to retell their life stories, removing the sense of blame clients feel. The tendency for clients to internalize blame for events that have occurred in their lives may be a cultural bias of Westerners, according to many Tibetan Buddhist teachers, including His Holiness the Dalai Lama. These highly respected Asian teachers of Buddhism were surprised, even startled, when they began teaching Western students and found them so full of self-blame and what we have described as "self-hate guilt" in many studies. Therefore, at least in psychotherapy with Westerners, both explicit and implicit efforts to modify this propensity to internalization, to reduce the sense of shame and guilt clients suffer from, may be the most important work of therapy. And while this may take place through "treatment by attitude" (Weiss, 1993; see also Sampson, 2005) with no explicit mention of emotions, it is treatment almost entirely focused on changing long-held disturbing cognitions and emotions.

Emotions may, in fact, be the ever-present heartbeat of therapy, and here the context is this very private, intimate, interpersonal dance between two people whose conscious aim is to help clients overcome difficulties in managing their lives with some degree of comfort. While clients may not be entirely conscious of the problems for which they are seeking solutions, they are hoping to gain success in their pursuit of quite ordinary, normal life goals, such as happy relationships with intimate partners, friends, and family members; success in parenting if they have children; work they enjoy; and the ability to find ways in which to financially support themselves and those they love. Clients also aim to be helpfully contributing to their whole social group. When therapists understand the fundamentally altruistic motivation driving their clients, albeit under the surface of conscious awareness, they are able to convey an acceptance and even admiration that by itself creates the conditions for healthy emotion regulation. This overview describes any successful psychotherapeutic experience, even one in which part of the treatment involves the prescription of psychotropic medications.

When people are burdened by difficulties and swimming in dark emotions, they feel bad. The important point is that negative interpersonal interactions, or the absence of social connections, are the most important sources of unhappiness. Conditions that make it difficult to have or maintain relationships create a context for negative emotions. This might be the outcome of a disturbed psycho-social background, such as when children are raised by parents with untreated mental illnesses, or exposed to harmful environmental factors such as malnutrition, neurotoxins, or pathogens. Traumas faced by groups as a whole—such as wars over scarce resources, or natural disasters such as earthquakes, weather events, and other random kinds of misfortune—tend to intensify positive socio-emotional interactions because people seem prone to help one another more than usual when the whole group is suffering. Whatever the cause of systematically disrupted interpersonal relationships, the emotional connection between client and therapist presents a way to have a new social experience, one in which clients find themselves feeling closely connected to another.

Some contemporary therapies focus consciously and explicitly on emotions. These treatments involve overt discussions of emotions, including those experienced and felt between the therapist and client in the treatment itself. Other therapies take place in which there is never any direct discussion of feelings, either within the therapy or even in describing prior experiences. Cure may still happen, by way of the unconscious work of interpersonal emotions. One might wonder, if emotions are interpersonal and treatments depend upon the therapist/client relationships, how, then, could some of the new computerized treatments be so effective. Here it seems likely that the words and values expressed by the computer come to represent a person in clients' lives, and they find themselves having conversations "in mind" with an imagined person. A common observation is that therapy is having an impact when clients, in their imagination, start to have conversations with their therapist, and likewise, when therapists are also starting to talk silently to their clients. Although research is needed on this point, when exchanges are with a computer program, it seems likely that a person emerges anyway. So great is our need for interemotional experiences, we try to find them in our dreams and in our imagination, when we're alone in our very alone culture.

Conclusion

The story we have tried to tell in this discussion of emotions is one describing their interpersonal nature, and their integration with context. From our own "Buddha nature," each one of us, client and therapist alike, is creating the link between us with our emotions, allowing us to live and flourish all over our planet.

References

Barrett, L. F., & Bliss-Moreau, E. (2009). She's emotional. He's having a bad day: Attributional explanations for emotion stereotypes. *Emotion, 9,* 649.

Boring, A., Ottoboni, K. & Stark, P. B. (2016). Student evaluations of teaching (mostly) do not measure teaching effectiveness. *ScienceOpen Research*. doi:10.14293/S2199–1006.1 .SOR-EDU.AETBZC.v1

Briggs, J. L. (1970). *Never in anger: Portrait of an Eskimo family (Vol. 12)*. Cambridge, MA: Harvard University Press.

Chiao, J. Y., & Blizinsky, K. D. (2010). Culture-gene coevolution of individualism-collectivism and the serotonin transporter gene. *Proceedings of the Royal Society of London B: Biological Sciences, 277*, 529–537.

Cosmides, L., & Tooby, J. (2000). Evolutionary psychology and the emotions. In M. Lewis & J. M. Haviland-Jones (Eds), *Handbook of the emotions* (2nd ed.). New York: Guilford.

Cosmides, L., & Tooby, J. (2005). Neurocognitive adaptations designed for social exchange. In D. M. Buss (Ed.), *Handbook of evolutionary psychology*. Hoboken, NJ: John Wiley & Sons.

Darwin, C. (1872). *The expression of the emotions in man and animals*. London: Murray.

Ekman, P. (1972). Universal and cultural differences in facial expression of emotion. In J. R. Cole (Ed.), *Nebraska symposium on motivation, 1971* (pp. 207–283). Lincoln, NB: Nebraska University Press.

Ekman, P., & Friesen, W. V. (1971). Constants across cultures in the face and emotion. *Journal of Personality and Social Psychology, 17*, 124–129.

Feldman, R. (2007). Parent-infant synchrony and the construction of shared timing: Physiological precursors, developmental outcomes, and risk conditions. *Journal of Child Psychology and Psychiatry, 48*, 329–354. doi:10.1111/j.1469–7610.2006.01701.x

Fincher, C. L., Thornhill, R., Murray, D. R., & Schaller, M. (2008). Pathogen prevalence predicts human cross-cultural variability in individualism/collectivism. *Proceedings of the Royal Society of London B: Biological Sciences, 275*, 1279–1285.

Gallese, V. & Goldman, A. (1998). Mirror neurons and the simulation theory of mind-reading. *Trends in Cognitive Sciences, 2*, 493–501. doi:10.1016/S1364–6613(98)01262–5

Izard, C. E. (1971). *The face of emotion*. New York: Appleton-Century-Crofts.

Keltner, D., & Haidt, J. (1999). Social functions of emotions at four levels of analysis. *Cognition and Emotion, 13*, 505–521.

Konner, M. (2005). *Hunter-gatherer infancy and childhood: The !Kung and others*. In B. S. Hewlett & M. E. Lamb (Eds.), *Hunter-gatherer childhoods: Evolutionary, developmental and cultural perspectives*. New Brunswick, NJ: Aldine Transaction.

Konner, M. (2015). *Women after all: Sex, education and the end of male supremacy*. New York: Norton.

Lewis, T., Amini, F., & Lannon, R. (2000). *A general theory of love*. New York: Random House.

MacNell, L., Driscoll, A., & Hunt, A. N. (2015). What's in a name: Exposing gender bias in student ratings of teaching. *Innovative Higher Education, 40*, 291–303. doi:10.1007/s10755 –014–9313–4

Matsumoto, D. R. (2002). *The new Japan: Debunking seven cultural stereotypes*. London: Nicholas Brealey.

O'Connor, L. E., Berry, J. W., Lewis, T., & Stiver, D. (2012). Empathy-based pathogenic guilt, pathological altruism, and psychopathology. In B. Oakley, A. Knafo, G. Madhavan, & D. S. Wilson (Eds.), *Pathological altruism*. Oxford: Oxford University Press.

Ohman, A., Flykt, A., & Lundqvist, D. (2000). Unconscious emotion: Evolutionary perspectives, psychophysiological data and neuropsychological mechanisms. In R. D. Lane, L. Nadel, & G. Ahern (Eds.), *Cognitive neuroscience of emotion.* New York: Oxford University Press.

Plutchik, R. (1962). *The emotions: Facts, theories, and a new model.* New York: Random House.

Sampson, H. (2005). Treatment by attitudes. In G. Silberschatz (Ed.), *Transformative relationships: The control-mastery theory of psychotherapy.* New York: Routledge.

Scherer, K. R. (2005). Unconscious processing in emotion: The bulk of the iceberg. In L. F. Barrett, P. M. Niedenthal, & P. Winkielman (Eds.), *Emotion and consciousness.* New York: Guilford.

Tsai, J. L. (2007). Ideal affect: Cultural causes and behavioral consequences. *Perspectives on Psychological Science, 2,* 242–259.

Tomkins, S. S. (1962). *Affect, imagery, and consciousness (Vol. 1).* New York: Springer.

Weiss, J. (1993). *How psychotherapy works.* New York: Guilford.

Wilson, D. S. (2015). *Does altruism exist? Culture, genes, and the welfare of others.* New Haven, CT: Yale University Press.

Wilson, D. S., & Wilson, E. O. (2007). Rethinking the theoretical foundation of sociobiology. *Quarterly Review of Biology, 82,* 327–348.

Dialogue on Emotions and Empathy

Participants: Jack W. Berry, Steven C. Hayes, Kibby McMahon, Lynn E. O'Connor, and M. Zachary Rosenthal

Steven C. Hayes: Maybe we could talk first about what we mean by emotion—just to get on the same page—and why is it so hard in the field to answer that question?

M. Zachary Rosenthal: To my thinking, part of why this is such a challenge to define is that I think the way we talk about emotions really intersects with some of the fundamental aspects of human experience. It's an access point of so many things fundamental to being human. Emotion touches learning, touches memory, touches thinking, touches arousal and attention and sensation and perception. All of our basic processes at some level intersect with what we call "emotion": survival, reproduction, mating, trust, intimacy. And then there are other fundamentals of human experience, like imagination, inspiration—all are caused by, consequences of, or co-occur with what we call emotions.

Jack W. Berry: It is also difficult to define because it's so complex. There are a lot of philosophical and theoretical assumptions that impact people who are proposing definitions. And if you have widely different philosophical perspectives, it's going to create a breakdown in consensus. Just for an example, a methodological behaviorist, I assume, is not going to include feelings in any definition of emotion. In contrast, for William James feelings are essential to his definition: emotions are the feelings of bodily changes in response to stimuli and conscious experience is the key.

Lynn and I talk a lot about the role of consciousness in emotions, and that's partly based on our philosophical and theoretical perspectives. This reflects a commitment to an evolutionary perspective where what is ultimately important in evolution is that some kind of behavior has to make a difference to fitness. There is research showing that there are very rapid physiological changes and behavior responses that occur before feelings are felt, before you evaluate the situation. But when you respond quickly like that, this can get you out of trouble, or your child out of trouble, so even though consciousness isn't involved, it's quite functional. Because of that, we're willing to include in any definition of emotion unconscious processes as well. Others may simply not acknowledge that.

Kibby McMahon: I like the intersection between our chapter and yours, the idea of the survivalist function. The goal is fitness. And we think of it in a contextual way: whatever responses fulfill a certain goal or help us survive. What's tricky about emotions is that it's such a collection of different processes: behavioral, subjective experience, communication, physiological. There are so many different aspects. Emotion scientists tend to focus on one and say it's the physiological response, and that's what the emotion is. But we view it as a collection of these things— states of being that motivate humans or other organisms in their environment in a way that fulfills a goal, nested into the bigger goal of survival. That's a very broad definition of emotion, but that's how we tend to think about it.

Steve: Maybe we could go from there to dive into some of these assumptions that were talked about. So, you raised the issues of assumptions and Jack even mentioned William James. I think people should know one reason that we speak in the title of an "integrated framework" is that you can track the more contextualistic wing of behavioral science back to James, and James definitely you can link very tightly to Darwin and him trying to deal with the implications of evolution. So "integration" seemed to be a really good way of talking about how these two wings can relate to each other. But there are differences in your chapters having to do with assumptions, and they penetrate down beyond strategy to how we think about emotion.

Lynn E. O'Connor: We think of emotions as a package—they include biology, feelings, cognitions. Everything is sort of merged together. And you can't really separate the components; it really is a package. And we do think of emotions as relevant in all mammalian behavior. So that would be the evolutionary perspective.

Jack: We both emphasize that emotions actually do tend to serve interpersonal functions. These mechanisms evolved in part for group living. That is why we share our emotions, we express them, and not just through the face. There are other ways of getting information across.

This is why empathy is important. Emotions evolved for group living, so at some level there's going to be this interpersonal function.

Lynn: We think of emotion, we think of empathy, as really being able to understand what someone else is feeling; it doesn't necessarily mean that the emotion is shared. We make a distinction between being accurate about being able to read somebody's emotion and sharing emotions.

Kibby: I think what you're also raising is that part of emotions and empathy is conscious: Are you actually aware of what's going on or not? And could you share an emotion and be consciously aware of it or not? Could you feel an emotion, and be consciously aware of it or not?

Lynn: Most of the time we're *not* aware of our emotions. You have to stop and actually deliberately focus on it: *What is it that I was feeling?* Emotions are mostly unconscious.

Kibby: That is the interesting point of taking the dual-systems approach. So much of what we perceive and understand and even think about can be unconscious. But what's interesting about that is that people tend to call the unconscious part the "automatic system." When you become aware of it you can start to act with more intention, with more executive function and planning. When you feel an emotion and are not aware of it, you act—you may tend to act on it automatically, based on learning history or habits or associated patterns.

Lynn: Conscious or unconscious, emotions drive behavior.

Zach: I agree. They really are the engine that drives behavior, and perhaps more to the specific point here, they drive our social connectedness. My emotional expression is understood or not understood well by others who are listening. And my feeling of being understood is a function of how well someone is empathically attuned—are they able to classify the emotion I'm experiencing in a way that it's consistent enough with what I believe I'm experiencing. That helps me make sense of my own motivational tendencies. The clearer I am with my motivational needs—my behavioral response needs in context—the more efficient I can be; and the more efficient I can be, the more able I can be in attaining prosocial goals in context. When I'm emotionally attuned, empathic, aware, and able to make sense of somebody—it has an impact. It can decrease

165

autonomic arousal, or open up intentional systems, to be able to be more effective, to be more likely to learn, to be more open to new experiences. So simply just understanding others' emotional expression can have this cascade of effects—whether that's psychotherapy, whether that's me at home with my wife or my kids, or whether that's really in any social context.

Steve: One of the things that I saw in your two chapters is this issue of universality versus cultural specificity. There were some differences there.

Lynn: We think that emotions are all universal, across the board. But how they're expressed is different. There are cultures where it's not okay for women to show any anger. The same computer-drawn face is interpreted as being absolute rage if it's a woman, where for a man it wouldn't be. What's allowed to be expressed is very culture-dependent, it's bound by context.

Kibby: Maybe the universal point of the commonality between the way we see emotions is "survival as function." So, what could be common, universal, about emotions is that they help us as a species to survive, either individually or on a group level.

However, it's interesting that emotions and selves' experiences of emotions themselves can be very context-dependent. And when you have so much variability in how people express their emotions and even feel their emotions, what is universal? I'm half-Chinese and half-American and there are some emotions that my Chinese family experience completely differently. Sadness is a stomach ache. Anger is a headache. So I tend to believe we have less as universal than what was previously considered as universal.

Lynn: Do you really think giving it a different name makes it different? I mean, my experience cross-culturally with clients is that they may use a different verbal word but the feeling is the same. We say "survivor guilt," for example, and that was not talked about in most of the history of psychology. But I think it's an experience universally, because we really prefer equity—dogs prefer equity, we're not the only animal. So when things seem unequal it bothers us. Cheater detection turned inward is really what survivor guilt is about. If I say "forlorn" versus "sad," are those different things? Is it the same thing? What part is the same? What do we consider universal? What defines "universal"?

Jack: Part of our emphasis on universality just has to do with our evolutionary focus here. Emotions are evolved mechanisms; they're intended to help us solve problems and react successfully in recurring situations. So we have evolved capacities and these recurring situations are cross-cultural. Child raising is going to be important. Cooperation is going to be important. Everyone, regardless of culture, is going to encounter threats to their lives or their safety. There is going to be interpersonal conflict that might call forth aggression and anger. So we have these capacities evolved to respond to the situations. And I don't know if the situations have changed that much.

Zach: The physiological experience—there may be some shared universalities there. But I would argue that that may not be quite as nuanced as the affective-state terms that we use in our language. Capacities to approach, capacities to avoid or move away from, capacities to experience—the input capacities are universal. The output language that's used and expression, I think that's more nuanced than purely universal.

Jack: We often develop new terms that describe the situation in which we feel. How would you explain the definition of "schadenfreude" or something like that?

Kibby: It makes sense to see it as more the dimensional than the categorical. If you think about how the context would shape emotion, schadenfreude is just pleasure, it's just happiness. But it has a contextual part of "being happy because of someone else's pain," which can confer benefit to you and your family or whomever you want to survive. So I think it's interesting to think about what part of that is the universal part.

Jack: You can find a use for new words to describe happiness in various contexts.

Kibby: Is it a different kind of happiness? Does schadenfreude feel different than me getting a birthday present?

Steve: What are the standards for the categories and distinctions, when we say "the emotions"? Do you have ones that you think are particularly important?

Kibby: I think we talked about thinking about what works in the given context—what helps someone survive in the given context. Take threat, for example. It is a contextual category that leads to many different kinds of emotions—different action tendencies in order to succeed in overcoming that threat. So if we think that way, we keep coming back to *What about function?* What if function was the way we could categorize emotions?

Steve: Jack and Lynn: when you say all emotions are universal—it sounds like there is a kind of a categorical set.

Jack: You have to start somewhere. You don't want to say that every person in every different context has a very different emotion. Ekman started with the idea, "Let's focus on cross-cultural emotions. All of us have recognizable facial expressions." That enabled him to get started. You have to start somewhere.

Zach: Let me just jump in and try to offer maybe a way to think about it. The more efficient we can be with understanding what each other is expressing, the better, for lots of reasons, scientifically and in terms of more applied ways of thinking. What is the context for how we talk about emotion and its function? I think that's really important. In therapy I'm going to use words that I think will likely resonate, and give purpose and meaning and value and clarity, and help that individual change longstanding patterns of ways of thinking, relating, feeling. Some of those may be categorical emotional terms. But that's very different from the science I might do the next hour. There I might be using a dimensional framework that allows me to try to predict and control certain variables in a very particular context. The point I'm making is that the parameters of how we define emotion are a function of the context in which we do the defining.

Steve: Interesting. You just took the contextual nature of emotion and then you applied it to the contextual nature of analyzing emotion in the scientific culture versus the clinical culture. But let's move over to that issue of culture. How do your different perspectives land, when you consider the issue of culture? Is it just a matter of expression? Is it more than that?

Lynn: Which emotions are more acceptable or less acceptable does seem to depend on what's functional in the given setting. For example, there is restricted expression and feelings of anger among remote Inuit because they can't disperse. If there's a real fight, people can't just disperse or they will die. So there are a lot more rules against anger. And I think in fact people there probably experience less anger, because it's so frowned upon.

Zach: The list of what we call emotions might be very consistent across culture, across time. We might come up with unique terms to define the situations. But ultimately, survival threat cues are survival threat cues are survival threat cues. And cues that signal opportunity for mating or reproduction are those kinds of cues.

Lynn: We asked ourselves why thinking about it cross-culturally mattered and it is because increasingly we are dealing with a global world. There's been so much focus on differences, that it is really important that we do understand also the universality of our makeup and our propensity to adapt to local conditions.

Jack: It's true. Extreme relativism makes other people seem alien. And we know that empathy is fostered by similarities.

Zach: I agree—and maybe looking for a synthesis could do something to help us relate better to each other in our increasingly globalized world. Not because we assume universality is there at every level, because we know that maybe some things are universal and some are not.

Kibby: This is the first time in history we have so much contact with so many different cultures. We need a large vocabulary of understanding other people's emotional expressions. Either that, or we find a common language to express emotion. We all have text messaging and around the world, a smiley emoji is the same thing! That makes all these questions even more relevant now when—how do we understand emotional expression for someone else across the globe?

Steve: What would you want an applied professional who is reading these chapters and perhaps watching this discussion to take away?

Zach: What's important is not to assume that you are right about emotional expression or to teach your patients to believe that they are right or that they have the Truth. Instead, it is to really be thinking about emotional empathy and the understanding of other human beings in a more pragmatic, contextualistic way: what works.

Lynn: Sometimes when patients come and they're depressed they'll say, "That's the way I am." And I teach patients to look more deeply: "Well, when did your mood drop?" When they begin to be able to link their feelings to particular contexts, that really makes a difference. Teaching the idea of context and emotions to patients may be one the most important things I do.

Kibby: It's very important to teach patients that their emotions make sense in context. That takes away a lot of the shame, blame, and stigma of "I'm feeling sad." We need to say, "Let's look at all what's going on in your life and what's going on now, right here, in this session. It makes sense what you're feeling."

Zach: It is important to help clients make sense of their learning history, their values, their goals, where they need to go, and what's getting in the way. If a narrow response repertoire, as it relates to emotion, is one of those factors—and often it is—it is worth exploring that emotional area at a pace that makes sense for that client or patient.

CHAPTER 10

Relational Models, Leadership, and Organization Design: An Evolutionary View of Organizational Development

J. W. Stoelhorst

University of Amsterdam

Mark van Vugt

VU University, Amsterdam

Introduction

In this chapter, we explore how evolutionary theory can help managers guide organizational development—taking business firms as our main example. From an evolutionary perspective, that is, based on the explicit recognition that humans are a biological species with evolved behavioral dispositions, firms are best understood as historically and culturally specific solutions to the problem of achieving and sustaining large-scale cooperation among genetic strangers (Johnson, Price, & van Vugt, 2013; Stoelhorst & Richerson, 2013). This means that an understanding of the evolved dispositions that underlie the unique human ability to solve this problem should be helpful in successfully managing firms. This claim is in line with the recent suggestion that theories of, and attempts at, intentional change can benefit from being grounded in evolutionary theory (Wilson, Hayes, Biglan, & Embry, 2014).

In the management literature, the term "organizational development" refers broadly to the general phenomenon of organizational change, and more specifically to an approach to planned organizational change originating in the work of Kurt Lewin (Burnes & Cooke, 2012). Our view of organizational development bridges these two meanings of the term in the sense that we are concerned with organizational change in general, while focusing on the role of managers in planning this change. The explicit link from our arguments to the more specific approach to organizational development that emanated from Lewin's work lies in his famous dictum: "There is nothing so practical as a good theory" (Lewin, 1951). Our main purpose is to show the practical relevance of evolutionary theory for an understanding of organizational change and development in modern firms.

Below, we use evolutionary theory to conceptualize firms as cultural contexts that allow their members to reap the benefits of large-scale cooperation. Standing in the way of these benefits are social dilemmas: situations involving tensions between (short-term) individual and (long-term) collective interests. From this perspective, management is about shaping the cultural context of a firm in ways that sustain cooperation in the face of social dilemmas. Managers can help sustain cooperation by appealing to psychological mechanisms that originally evolved to sustain cooperation in the small-scale and egalitarian societies in which humans originally evolved. Central among these mechanisms is how a firm's members cognitively frame their social relationships with each other in terms of four elemental "relational models": Communal Sharing, Authority Ranking, Equality Matching, and Market Pricing (Fiske, 1991, 1992). Managers can guide organizational development by influencing the specific combination of relational models that the members of a firm adopt to structure their social interactions. Managers exert this influence through their leadership style (managers lead in part by example) and their power to change formal rules (managers lead in part by organization design). We give examples of each.

Modern Organizations from an Evolutionary Perspective

In the context of evolutionary theory, modern organizations such as firms are puzzling phenomena because they typically involve very large scale cooperation among individuals who are not directly related genetically. Cooperation in general, not to mention the human ability to sustain large-scale cooperation among non-kin, seems to fly in the face of the basic logic of evolutionary theory, which is that natural selection can only reward behaviors that increase an organism's individual fitness. Given this logic, the evolution of cooperation requires a solution to a social dilemma, that is, a solution to the fact that there is a tension between the narrow interests of individual organisms and the broader interests of the groups in which they may live. Because the cooperative behaviors necessary to sustain groups, by definition, benefit others at a cost to oneself, the basic logic of evolution leads to the prediction that cooperative behaviors will lose out to purely self-interested behaviors.[1]

While the logic underlying this prediction is inescapable, a number of highly successful cooperative species, such as ants, termites, and humans, have nevertheless evolved. Explaining this fact has been a central puzzle in evolutionary theory since Darwin (Cronin, 1991), but convincing explanations have only emerged relatively recently (for summaries, see Nowak, 2006; Barclay & van Vugt, 2015). *Homo sapiens* presents evolutionary theory with an especially interesting puzzle, because we are the only species that is able to sustain large-scale cooperation among non-kin using cultural transmission, that is, the ability to learn socially and build upon others' ideas (Tomasello, 2009). From an evolutionary perspective, modern firms are simply a historically and culturally specific manifestation of this ability.

The most convincing explanation of the human ability to sustain large-scale cooperation among non-kin invokes a combination of multilevel selection (Sober & Wilson, 1998; Wilson & Wilson, 2007) and gene-culture coevolution (Boyd & Richerson, 1985; Richerson & Boyd, 2005).[2] According to this explanation, the specific nature of human cooperation is the result of a long history of

---

1 To see this logic clearly, consider a group of cooperative organisms in which a genetic mutation gives rise to a self-interested individual. This individual will have a reproductive advantage because it will benefit from the cooperative behaviors of its group members without incurring the costs of matching their contributions. As a result, the genes underlying self-interested behavior will increase in relative frequency, and, over time, self-interested individuals will replace cooperative individuals within a group. (Note that this example puts aside the question how cooperative behaviors would have evolved in the first place. Nevertheless, it illustrates well the basic free-riding problem in both the evolution of cooperation and in social dilemmas more generally.)

2 Explanations for the evolution of cooperation include kin-selection, different types of reciprocity (direct, indirect, and network), and multilevel selection theory (Nowak, 2006). All these mechanisms likely played a role in the evolution of human cooperation, but only the combination of multilevel selection with a gene-culture coevolutionary logic seems to fully explain the specific nature of human cooperation (cf. Henrich, 2004; Stoelhorst & Richerson, 2013). Multilevel selection recognizes groups as well as individuals as units of selection and explains the evolution of cooperative behaviors as the result of a vector of selection pressures, with within-group selection among individuals favoring self-interested behaviors (such as free riding on the cooperative behaviors of others) and between-group selection favoring cooperative behaviors (such as rewarding others' cooperative behaviors and punishing others' uncooperative behaviors, even at a cost to oneself). In cultural group selection, it is not just the genetic behavioral dispositions of individual group members that are selected for, but also the group's cultural norms that influence members' behaviors. Over evolutionary history, groups with more cooperative norms would, ceteris paribus, have been more successful in between-group competition, and the evolution of these more cooperative norms would, in turn, have provided within-group selection pressures favoring more cooperative genetic dispositions.

between-group competition that favored cooperative cultural norms, which, in turn, provided within-group selection pressures favoring individuals with more cooperative genetic dispositions. This ultimate explanation of human cooperation is in line with a host of empirical findings about the proximate mechanisms that drive human behaviors in the face of social dilemmas (for a review, see Van Lange, Balliet, Parks, & van Vugt, 2014). In fact, it has been argued that most of our social and moral psychology has evolved to sustain within-group cooperation in the face of between-group competition (Greene, 2013; Haidt, 2012).[3]

Thus, the evolutionary approach to human organizational behavior revolves around our evolved ability to put the long-term group interest above our short-term self-interest, or, in other words, our ability to overcome social dilemmas (Van Vugt & Ronay, 2014). While we are endowed with behavioral dispositions to manage social dilemmas, this does not mean that these dilemmas no longer exist: human organizations are always at risk of being undermined by within-group competition (Campbell, 1994). Moreover, there is a problem of "evolutionary mismatch": our cooperative dispositions originally evolved in the context of small, egalitarian, familial hunter-gatherer groups (Boehm, 2012). As a result, our evolved psychology is not necessarily well adapted to the specific challenges of establishing cooperation in large-scale modern organizations involving frequent interactions with genetic strangers in the context of bureaucratic, hierarchical structures (Van Vugt & Ronay, 2014).

The Managerial Challenge from an Evolutionary Perspective

Given the above, managers face two key challenges. The first is to provide an environment that helps overcome social dilemmas, while taking into account the evolutionary mismatch between the large-scale hierarchical environment of modern firms and the small-scale egalitarian environments in which our ability to solve social dilemmas originally evolved (Von Rueden & van Vugt, 2016). The second is to overcome social dilemmas in ways that make the firm successful in competition with other firms. It is not enough to find just any cooperative solution; given the fact that firms compete with other firms, it is their relative

---

3 For instance, one way in which our evolved cooperative dispositions express themselves is through a range of emotions that reinforce cooperative behaviors, such as guilt and shame (related to observing group norms oneself), anger and contempt (related to punishing others that do not observe group norms), and feelings of reward when punishing noncooperators (cf. Boehm, 2012; Greene, 2013).

efficiency that matters (Johnson et al., 2013). What does an evolutionary perspective suggest that managers should do to face up to these challenges?

Following Wilson's and his colleagues' (2014) vision of an evolutionary science of intentional change, the first suggestion is that managers should think of their firms as cultural contexts. As Wilson and his colleagues emphasize, the evolutionary success of *Homo sapiens* derives in large part from our unique capacity for symbolic thought. This capacity has resulted in a second, cultural, inheritance system that sits on top of our genetically inherited behavioral dispositions. Culture essentially serves as a store of functional rules of behavior that are passed on through imitation and learning. The crux of our capacity for symbolic thought is that it allows for endless combinatorial possibilities, making culture a very versatile tool to help groups adapt to different environmental conditions. This tool is at work at many different levels of analysis, from small groups to entire national cultures, and is also central to the success of firms.

To this first suggestion we add a second, which is that managers can think of organizational cultures as implementations of specific combinations of four elemental "relational models" (Fiske, 1991, 1992). Fiske (2004, p. 3) describes relational models as cognitive frames that people use, automatically and often unconsciously, "to plan and to generate their own action, to understand, remember, and anticipate others' action, to coordinate the joint production of collective action and institutions, and to evaluate their own and others' actions." The four relational models identified by Fiske are Communal Sharing (a relationship of unity, community, and collective identity), Authority Ranking (a relationship of hierarchical differences, accompanied by the exercise of command and complementary display of deference and respect), Equality Matching (a relationship among equals manifested in balanced reciprocity), and Market Pricing (a relationships where people compute cost/benefit ratios and pursue their self-interest) (Fiske, 1991).

Relational models theory holds that these four models suffice to account for the high diversity in social relationships observed across cultures (Fiske, 1991). There are two reasons for this. First, individuals relate in different ways when interacting in different domains of their relationship, which generates variety across relationships.[4] Second, the specific cultural context (e.g., ethnic, national, or organizational) determines the specific implementation rules of the relational models—that is, when, how, and with whom to implement a particular relational

---

4 For example, a manager and subordinate may interact through AR when it comes to work assignments, through MP when it comes to salary negotiations, and through CS when visiting a client.

model (Fiske, 1991, 2004, 2012).[5] Seen within the broader evolutionary view on human behavior, the four relational models may be understood as the elemental building blocks that enable and constrain the symbolic relations that define cultures. The relational models enable symbolic relations because of their combinatorial potential and flexibility with respect to the specific implementation rules they allow, but they also constrain symbolic relations because these relations ultimately are the result of only four elemental models that derive from deeply ingrained aspects of our evolved social and moral psychology.

If we think of firms as cultural contexts, then the managerial challenge translates into shaping a firm's symbolic relations, that is, the formal and informal rules that shape the collective behaviors of its members, in ways that (1) sustain within-group cooperation in the face of social dilemmas; and (2) make the firm successful in the face of between-group competition with other firms. Moreover, if we think of organizational cultures as specific implementations of the four relational models, then the managerial challenge translates into triggering implementations of these models that lead to symbolic relations that help achieve these two objectives. The logical follow-up question, then, is how can managers influence the relational models that the members of a firm adopt to frame their social relationships? Below, we highlight two particularly powerful mechanisms through which managers can exercise such influence: leadership style and organization design.

The Role of Leadership Style in Organizational Development

The first implication of our arguments so far is that the success of firms depends to a large degree on the cooperative actions of their members, and that managers play a significant role in shaping these actions (Spisak, O'Brien, Nicholson, & van Vugt, 2015). For instance, managers can galvanize cooperation through leadership-by-example, serving as prestigious role models that team members want to imitate (Smith et al., 2015). They can also lead in a more directive way, for instance by rewarding cooperative efforts and by punishing free-riders (O'Gorman, Henrich, & van Vugt, 2008). Leadership in firms can be more democratic and participatory, involving many stakeholders in the decision-making process, or it can be relatively authoritarian, even despotic. The second implication of our

---

5 For example, the implementation rules of AR in Japanese firms, with their traditional focus on seniority, will be very different from the implementation rules of AR at Intel, where the late Andrew Grove promoted a practice of constructive confrontation in which expertise could (and, in fact, should) trump hierarchical position.

arguments is that firms' competitive contexts determine to some degree which "style" of leadership works best. For instance, human groups prefer leaders with more aggressive, dominant personalities when competition with other groups is salient, whereas they select more trustworthy and less dominant individuals as leaders to manage conflict within groups (Spisak, Homan, Grabo, & van Vugt, 2011).

The leadership style of managers serves as a strong signal of the prevailing relational model within the firm. Communal Sharing (CS), for instance, coincides with a leadership style promoting trust and a collectively shared identity. Managers adopting a leadership style based on CS treat team members as if they are part of the same family (much like in our evolutionary past). In such a relationship, resources are distributed on the basis of need, and people are expected to show a high degree of commitment, even altruism, toward each other and their firm. Organizational citizenship behaviors (OCBs), that is, voluntary contributions to the organization beyond members' contractual tasks, are routinely expected. The leadership style that characterizes CS is caring and considerate and there is a high degree of concern for the team as a whole, which is more important than the welfare of individual members. The concept of transformational leadership (Bass, 1990) fits management by CS as managers attempt to transform individual members' interests to an overarching team or firm interest by showing leadership-by-example and through displaying a servant, self-sacrificial leadership style.

Managers who adopt an Authority Ranking (AR) model display a leadership style that operates via "noblesse oblige." Here, managers must provide security to team members, and in return they are bestowed with prestige and receive a larger share of the resources, (e.g., in salary or in other privileges). Furthermore, team members are treated individually by managers based on the hierarchical position they occupy in the team, which may be based on seniority, competence, or talent. The service-for-prestige model explains leadership according to AR. AR is akin to the way Big Men leaders operate in traditional societies (Price & van Vugt, 2014; Von Rueden & van Vugt, 2016). Big Men achieve status by generously providing public services to the group. In this relational frame, managers have power over team members, but it comes with a duty to protect the interests of the team. AR is characterized by an authoritarian, paternalistic leadership style, but one that is based on prestige rather than on coercion (cf. Henrich & Gil-White, 2001).

Managers who operate via Equality Matching form individual exchange relationships with team members and they operate via tit-for-tat mechanisms to induce cooperation. Team members who contribute more to the team can expect to receive rewards, whereas individuals who are undercontributing can expect punishment. In EM, managers will match the cooperative efforts of team members by investing more in their relationships with high-contributing members. Leadership according to EM shares similarities with Leader-Member Exchange (LMX) theory (Graen & Uhl-Bien, 1995), as the key relationship is the interpersonal relationship between manager and team member. The quality of this relationship determines the extent to which managers trust each member to cooperate. Leading via EM also bears similarities to a transactional leadership style characterized by a system of contingent rewards and punishment for individual worker performances (Bass, 1990).

A more elaborate form of transactional leadership is applied by managers who use Market Pricing (MP) as the dominant relational model. Here, leadership operates through a negotiated contract between the firm and its workers in which the tasks and roles of each team member are clearly laid out, as are the incentives they receive in return. Money is the currency and managers can incentivize cooperative efforts by providing salary increases or bonuses for extraordinary performances. In MP, team members operate strategically, doing just what is needed to receive monetary rewards. OCBs are generally not expected within this model, nor is a high intrinsic motivation to stay with the firm and meet the job demands.

As illustrated above, different leadership styles will trigger different relational models, and in doing so galvanize cooperation in different ways. As a result, organizational cultures will vary depending upon the dominant relational model. Of course, managers may apply a hybrid of relational models in their interactions with team members. For instance, the distribution of office space may follow an AR model with the more senior members getting a larger space, whereas salaries and bonuses might follow an MP model. A CS model may be applied to the distribution of extra work, with team members with young families having to do less overtime, whereas EM may prevail in dealing with free-riders. Nevertheless, based on their personalities, a particular model is likely to dominate in the leadership style of individual managers. For instance, CS relations may be encouraged by managers with a highly charismatic personalized style.

Importantly, relational models theory can be used to develop hypotheses about cooperation failures within firms (cf. Bridoux & Stoelhorst, 2016; Giessner & van Quaquebeke, 2010). We see at least two possible reasons for such failures. First, cooperation failures may result from a mismatch between the dominant relational model adopted by managers and the model adopted by team members. For instance, team members may want to operate via CS, pooling their efforts and distributing resources according to need. Yet managers may be instructed by those higher up in the hierarchy to operate via an EM or MP model, whereby team members are rewarded individually for their cooperative efforts. Second, cooperation failures may result from managers being inconsistent in the kind of relational model they apply. An example is when managers who generally operate via an MP model make a special request to team members to engage in OCB's (such as doing unpaid overwork) because the company is in a difficult financial position. Mismatches like these can account for perceived violations of ethical and fairness considerations, which may cause a breakdown in cooperation (Giessner & van Quaquebeke, 2010).

The Role of Organization Design in Organizational Development

Having considered the role of managers in guiding organizational development at the micro-level of their interactions with team members, we now turn to the role of managers at the macro-level of organizational design. Modern firms epitomize the unique human ability to sustain large-scale cooperation among non-kin. Firms are among the largest and most successful forms of human organization, allowing their members to overcome the social dilemmas that stand in the way of reaping the benefits of cooperation—and doing so despite the evolutionary mismatch between their large-scale and hierarchical nature and the small-scale egalitarian societies in which human social psychology originally evolved. How are firms able to achieve this?

To help answer this question, consider the struggle of large firms with the advantages and disadvantages of the "multi-divisional form" (M-form), which "[i]n terms of its impact, not just on economic activity, but also on human life as a whole, … must rank as one of the major innovations of the last century" (Roberts, 2004, p. 2). The emergence of the M-form in the first half of the twentieth century and its subsequent diffusion were described by Chandler (1962), who noted that it arose because firms realized that the more traditional functional organization, whereby firms organized themselves on the basis of functional specialization (i.e., around, for instance, manufacturing, sales, and R&D departments), stood in the way of their growth strategies. Instead, firms began organizing themselves around divisions, which were headed by general managers responsible for the growth and profitability of their divisions. This organization form became an extremely successful vehicle for the further increase in the scale of human cooperation.

The M-form makes use of the principle of "near-decomposability" (Simon, 2002): each of its divisions is a relatively self-contained cooperative unit (itself often organized along functional lines). In addition to its divisions, an M-form has a corporate headquarters (HQ), which is charged with managing coordination across the different divisions (Chandler, 1991). This is typically done on the basis of a combination of MP and AR. HQ has authority over the general managers of the divisions (AR) and operates an internal capital market that achieves coordination by allocating cash flows among the divisions based on their past performance and future profit potential (MP). The combination of near-decomposability with MP and AR was a recipe for success in the munificent economic environment of the 1950s and 1960s, and the M-form came to dominate the landscape of economic organization—accounting for a very large share of economic activity.

Notwithstanding its initial success, from an evolutionary perspective it is easy to see that even if the M-form may be well suited to exploiting the benefits of large-scale cooperation, it also introduces an additional layer of social dilemmas. Whereas in the traditional functional organization the problem was "merely" to have functional departments cooperate with each other, the M-form introduced the additional problem of cooperation among divisions. This problem was exacerbated by the changing economic landscape of the 1970s and 1980s, with mounting global competition and increased influence of financial markets. These developments put strong selective pressures on firms and increasingly forced them to look for competitive advantage in exploiting unique knowledge synergies across divisions rather than in economies of scale alone. But the need to look for synergies across divisions went against the core design principle of the M-form: that divisions can be managed as stand-alone units (Strikwerda & Stoelhorst, 2009).

Strikwerda and Stoelhorst (2009) describe how some large organizations have evolved toward a new organizational form that may be better able to meet the exigencies of economic selection pressures in the twenty-first century. In contrast to near-decomposability, MP, and AR, this new "multidimensional form" is based on interdependence, CS, and EM. In the traditional M-form, the core design principle of near-decomposability means that divisions should be self-contained units in which the general manager has the responsibility for performance in a specific market, as well as hierarchical control over all the resources needed to serve that market. In contrast, in the new multidimensional form, managers are deliberately made dependent on each other by giving one manager responsibility for a certain market and another manager control over the resources to serve that market. In fact, this principle of designing interdependence into the organization can be taken much further than designing along merely two dimensions—hence the term multidimensional.[6]

Bridoux and Stoelhorst (2016) argue that when production involves strong interdependencies, CS and EM frames are better able to sustain cooperation than are MP and AR frames. In line with this, the multidimensional form appeals

---

6 For instance, IBM is organized along four overlapping dimensions: products/solutions, regions, accounts, and distribution channels (Strikwerda & Stoelhorst, 2009).

much more explicitly to CS and EM frames. This does not mean that MP and AR no longer play any role at all. Managers in the multidimensional form typically are rewarded for their performance on the dimension for which they are responsible (MP), and in case of conflicts of interest that managers of the different dimensions cannot solve amongst themselves, corporate management can still step in (AR). But to make the mutual dependence among managers in the multidimensional form work, the principle of reciprocal interactions among equals (EM) and the principle of contributing to a collective goal (CS) are much more central organizing principles than in the M-form.

In the multidimensional form, EM is triggered because of managers' dependence on each other. Whereas general managers of a division in an M-form simply stand in a hierarchical relation to their superiors at corporate headquarters and their subordinates within the division, interdependent managers in the multidimensional form need to develop EM relations with each other to be successful. CS is triggered in at least two interrelated ways. The first is through a trusted information system controlled by corporate headquarters that makes the organization's performance on the various dimensions transparent. The second is through a much stronger emphasis on the collective performance of the organization, typically through a focus on how well the organization as a whole is able to serve its clients. In fact, this client orientation is also used to adjudicate any remaining conflicts of interests among managers of the different dimensions, giving priority to the course of action that the trusted information system suggests will maximize success with customers.

Conclusion

In this chapter, we have only made a first start in considering organizational change and development from an evolutionary perspective—focusing our attention on the role of managers in guiding intentional change in modern firms. We have argued that, from an evolutionary perspective, firms are historically and culturally specific solutions to the problem of reaping the benefits of large-scale cooperation. Large-scale cooperation in modern firms is made possible because of psychological mechanisms that originally evolved to sustain cooperation in the relatively small and egalitarian societies in which humans lived for most of their evolutionary history. The main managerial challenge in modern firms is to provide a cultural context that is able to sustain within-group cooperation, despite this evolutionary mismatch, and to do so in ways that make a firm successful in competition with other firms.

We have discussed how managers can guide the evolution of organizational culture through organization design and leadership style. In doing so, our main contribution has been to integrate relational models theory into the emerging evolutionary theory of human behavior. This is especially relevant for the project of developing a science of intentional change grounded in evolutionary theory (Wilson et al., 2014), because relational models theory offers a concrete way to understand the building blocks of cultural diversity and organizational development. Different combinations of the four models combined with specific implementation rules allow humans to construct cultural contexts in endless variations. In that sense, the four relational models can, perhaps, be understood as the cultural equivalent to the four letters of the genetic alphabet. If so, then intentional cultural change may be thought of as the science of guiding the evolution of specific implementations of the four relational models.

Let us finally link back to Lewin's original views on organizational development. Some seventy years ago, Lewin (1951) did not just emphasize the value of theory for planned change, he also wanted such change to tackle social conflict and promote democratic values (Burnes & Cooke, 2012). Applying Lewin's approach is as relevant today as it was in his day, when the M-form was in its ascendance. For many, modern firms are synonymous with MP and AR, and while the emphasis on these two relational models has no doubt contributed to economic welfare, some would argue that this may have come at the expense of physical and psychological well-being. However, as the emergence of the multidimensional form described above illustrates, given that the creation of economic welfare in knowledge-driven economies increasingly depends on production processes involving strong interdependencies, EM and CS are likely to be more conducive to creating economic welfare than MP and AR (Bridoux & Stoelhorst, 2016). Moreover, as firms organize themselves along these lines, their cultures may not just make them more successful in the competition with other firms, but, given the fit between EM and CS and the small-scale and egalitarian origins of our evolved psychology, these organizational cultures may also hold the promise of increased well-being.

References

Barclay, P., & van Vugt, M. (2015). The evolutionary psychology of human personality: Adaptations, mistakes, and byproducts. In D. Schroeder & W. Graziano (Eds.), *Oxford handbook of prosocial behavior*. Oxford: Oxford University Press.

Bass, B. M. (1990). *Bass and Stogdill's handbook of leadership: Theory, research, and managerial applications* (3rd ed.). New York: Free Press.

Boehm, C. (2012). *Moral origins: The evolution of virtue, altruism and shame*. New York: Basic Books.

Boyd, R., & Richerson, P. J. (1985). *Culture and the evolutionary process*. Chicago: University of Chicago Press.

Bridoux, F. M., & Stoelhorst, J. W. (2016). Stakeholder relationships and social welfare: A behavioral theory of contributions to joint value creation. *Academy of Management Review, 41*, 229–251.

Burnes, B., & Cooke, B. (2012). The past, present and future of organization development: Taking the long view. *Human Relations, 65*, 1395–1429.

Campbell, D. T. (1994). How individual and face-to-face group selection undermine firm selection in organizational evolution. In J. A. Baum & J. V. Singh (Eds.), *Evolutionary dynamics of organizations*. Oxford: Oxford University Press.

Chandler, A. D. (1962). *Strategy and structure: Chapters in the history of American enterprise*. Cambridge, MA: MIT Press.

Chandler, A. D. (1991). The functions of the HQ unit in the multibusiness firm. *Strategic Management Journal, 12*, 31–50.

Cronin, H. (1991). *The ant and the peacock*. Cambridge, UK: Cambridge University Press.

Fiske, A. P. (1991). *Structures of social life: The four elementary forms of human relations*. New York: Free Press.

Fiske, A. P. (1992). The four elemental forms of sociality: Framework for a unified theory of social relations. *Psychological Review, 99*, 689–723.

Fiske, A. P. (2004). Relational models theory 2.0. In N. Haslam (Ed.), *Relational models theory: A contemporary overview*. Mahwah, NJ: Lawrence Erlbaum.

Fiske, A. P. (2012). Metarelational models: Configurations of social relationships. *European Journal of Social Psychology, 42*, 2–18.

Giessner, S. R., & van Quaquebeke, N. (2010). Using a relational models perspective to understand normatively appropriate conduct in ethical leadership. *Journal of Business Ethics, 95*, 43–55.

Graen, G., & Uhl-Bien, M. (1995). Relationship-based approach to leadership: Development of leader-member exchange (LMX) theory of leadership over 25 years: Applying a multi-level multi-domain perspective. *Leadership Quarterly, 6*, 219–247.

Greene, J. D. (2013). *Moral tribes: Emotion, reason, and the gap between us and them*. New York: Penguin.

Haidt, J. (2012). *The righteous mind: Why good people are divided by politics and religion*. New York: Pantheon.

Henrich, J. (2004). Cultural group selection, coevolutionary processes and large-scale cooperation. *Journal of Economic Behavior and Organization, 5*, 3–35.

Henrich, J., & Gil-White, F. J. (2001). The evolution of prestige: Freely conferred status as a mechanism for enhancing the benefits of cultural transmission. *Evolution and Human Behavior, 22*, 165–196.

Johnson, D. D. P., Price, M. E., & van Vugt, M. (2013). Darwin's invisible hand: Market competition, evolution and the firm. *Journal of Economic Behavior and Organization, 90* (Supplement), S128–S140.

Lewin, K. (1951). Problems of research in social psychology. In D. Cartwright (Ed.), *Field theory in social science: Selected theoretical papers*. New York: Harper & Row.

Nowak, M. A. (2006). Five rules for the evolution of cooperation. *Science, 314*, 1560–1563.

O'Gorman, R. O., Henrich, J., & van Vugt, M. (2008). Constraining free-riding in public goods games: Designated solitary punishers can sustain human cooperation. *Proceedings of the Royal Society B, 276*, 323–329.

Price, M., & van Vugt, M. (2014). The evolution of leader-follower reciprocity: The theory of service-for-prestige. *Frontiers in Human Neuroscience, 8*, 363.

Richerson, P. J., & Boyd, R. (2005). *Not by genes alone: How culture transformed human evolution*. Chicago: University of Chicago Press.

Roberts, J. (2004). *The modern firm: Organizational design for performance and growth*. Oxford: Oxford University Press.

Simon, H. A. (2002). Near decomposability and the speed of evolution. *Industrial and Corporate Change, 11*, 587–599.

Smith, J. E., Gavrilets, S., Borgerhoff Mulder, M., Hooper, P. L., El Moulden, C., Nettle, D., … Smith, E. A. (2015). Leadership in mammalian societies: Emergence, distribution, power, and pay-off. *Trends in Ecology and Evolution, 31*, 54–66.

Sober, E., & Wilson, D. S. (1998). *Unto others: The evolution and psychology of unselfish behavior*. Cambridge, MA: Harvard University Press.

Spisak, B. R., Homan, A. C., Grabo, A., & van Vugt, M. (2011). Facing the situation: Testing a biosocial contingency model of leadership in intergroup relations using masculine and feminine faces. *Leadership Quarterly, 23*, 273–280.

Spisak, B., O'Brien, M., Nicholson, N., & van Vugt, M. (2015). Niche-construction and the evolution of leadership. *Academy of Management Review, 40*, 291–306.

Stoelhorst, J. W., & Richerson, P. J. (2013). A naturalistic theory of economic organization. *Journal of Economic Behavior and Organization, 90 (Supplement)*, S45–S56.

Strikwerda, J., & Stoelhorst, J. W. (2009). The emergence and evolution of the multidimensional organization. *California Management Review, 51*, 11–31.

Tomasello, M. (2009). *The cultural origins of human cognition*. Cambridge, MA: Harvard University Press.

Van Lange, P. A. M., Balliet, D., Parks, C. D., & van Vugt, M. (2014). *Social dilemmas: The psychology of human cooperation*. Oxford: Oxford University Press.

Van Vugt, M., & Ronay, R. D. (2014). The evolutionary psychology of leadership: Theory, review, and roadmap. *Organizational Psychology Review, 4*, 74–95.

Von Rueden, C., & van Vugt, M. (2016). Leadership in small-scale societies: Some implications for theory, research, and practice. *Leadership Quarterly, 26*, 978–990. doi:10.1016/j.leaqua.2015.10.004

Wilson, D. S., Hayes, S. C., Biglan, A., & Embry, D. D. (2014). Evolving the future: Toward a science of intentional change. *Behavioral and Brain Sciences, 37*, 395–460.

Wilson, D. S., & Wilson, E. O. (2007). Rethinking the theoretical foundation of sociobiology. *Quarterly Review of Biology, 82*, 327–348.

Yukl, G. (2014). *Leadership in organizations* (8th ed.). Upper Saddle River, NJ: Prentice Hall.

CHAPTER 11

Organizational Flexibility: Creating a Mindful and Purpose-Driven Organization

Frank W. Bond

Goldsmiths, University of London

Psychological flexibility[1] is a primary determinant of mental health and behavioral effectiveness, as hypothesized by one of the contextually and empirically based theories of psychopathology: acceptance and commitment therapy (ACT; Hayes, Strosahl, & Wilson, 1999). It refers to people's ability to focus on their current situation, and based upon the opportunities afforded by that situation, to take appropriate action toward pursuing their values-based goals, even in the presence of challenging or unwanted psychological events (e.g., thoughts, feelings, physiological sensations, images, and memories) (Hayes, Luoma, Bond, Masuda, & Lillis, 2006).

A key implication of this concept—and hence its name—is that, in any given situation, people need to be flexible as to the degree to which they base their actions on their internal events or the contingencies of reinforcement (or punishment) that are present in that situation. ACT maintains, and research suggests, that people are more psychologically healthy and perform more effectively when they base their actions on their own values and goals (Bond et al., 2011). Thus, if a person values being an effective leader, she will take action on a difficult task even if doing so is anxiety provoking; in another situation, however, she might refrain from taking action (e.g., not having a cross word with a colleague) even if she strongly feels like doing so, in order to pursue her personally meaningful goal

---

1 For historical reasons, psychological flexibility has also been referred to as psychological acceptance, and psychological inflexibility has been referred to as experiential avoidance. Bond and colleagues (2011) discuss these reasons.

of being an effective leader. In short, people demonstrating psychological flexibility base their behavior, in any given situation, more on their values and goals and less on the vagaries of their internal events or current situational contingencies (Bond et al., 2011).

An implication of acting flexibly is that people will experience, at times, unwanted psychological events (e.g., anxiety) while pursuing their values-based goals. Thus, a great deal of ACT theory and practice emphasizes the use of *mindfulness* strategies for experiencing these events, so that they have less of a negative impact on individuals' psychological health and their ability to pursue their values-based goals. When people are mindful of their psychological events, they deliberately observe them on a moment-to-moment basis, in a nonelaborative, open, curious, and nonjudgmental manner (Brown & Ryan, 2003; Kabat-Zinn, 1990; Linehan, 1993; Marlatt & Kristeller, 1999).

Thus, psychological flexibility emphasizes both committed action toward meaningful goals and mindfulness. It is this combination of mutually enhancing processes that is likely to account for the many mental health and performance benefits associated with this individual characteristic (e.g., see Bond, Lloyd, & Guenole, 2013; Lloyd, Bond, & Flaxman, 2017). Research, however, does not just show an association between these variables. Randomized controlled intervention trials in the workplace (and the clinic) have shown that an increase in psychological flexibility is the mechanism, or mediator, by which ACT has improved: employee mental health (Bond & Bunce, 2000; Flaxman & Bond, 2010; Lloyd, Bond, & Flaxman, 2013), innovation potential (Bond & Bunce, 2000), emotional burnout levels, and unhelpful attitudes toward client groups (Hayes et al., 2004).

This chapter posits that the combination of a commitment to values-based actions and mindfulness (i.e., psychological flexibility) that is so beneficial to individuals can be designed into organizations (and teams), in order to produce similarly beneficial outcomes in those organizations. Theorists have long emphasized the importance of flexibility for organizational effectiveness, noting that, in order to pursue their key goals, organizations need structures, processes, technologies, and strategies (together called "organizational characteristics") that can help them to adapt across dimensions such as time, range, intention, and focus (Golden & Powell, 2000). These theorists have rarely, however, tended to discuss how the psychological events of employees can inhibit, or even undermine, the optimal identification and implementation of these characteristics, thus inhibiting organizational flexibility and, as a result, effectiveness.

One notable exception is Jaques (1955), a Kleinian psychoanalyst who helped to establish the Tavistock Institute of Human Relations in London; he postulated that leaders and workers unconsciously collaborate to design organizational characteristics, in order not only to achieve the company's primary aim (e.g., manufacturing a product), but also to defend employees against unwanted thoughts and feelings (e.g., through heavily standardized—and, thus, probably dehumanizing—processes). If such defense mechanisms are not "made conscious" to leaders, they produce a rigid organization that is less effective in achieving its goals. To this day, work psychologists still use the Tavistock's T-group intervention as part of their efforts to make leaders aware of group processes that may lead to unhelpful organizational characteristics and cultures. (See De Board, 1978, for an account of psychoanalysis and organizations.)

As has been done with psychoanalytic principles, I believe that we can scale up the concept of psychological flexibility to the organizational level. In so doing, we can provide a contemporary, empirically based account of how to create and maintain flexible organizations that help employees approach difficult psychological events in ways that promote their mental health, as well as the values-based goals of organizations. To do so, I will detail how we can design organizational characteristics that produce an establishment that is both "mindful" and committed to pursuing its values. I will argue that we can construct this model of organizational flexibility using extant principles, models, and strategies of organizational behavior (OB). OB is a field of study that investigates the impact that individual (e.g., personality, mental health), group (e.g., leadership, teams), and organizational characteristics (e.g., structure, processes) have on organizational effectiveness (including the health of individuals) (Robbins & Judge, 2007). I will seek to develop this model by specifying structures, processes, technologies, and strategies that are consistent with the aim of contextual behavioral science (CBS; Hayes, Barnes-Holmes, & Wilson, 2012), upon which psychological flexibility (and hence ACT) is based.

Contextual Behavioral Science and Organizational Behavior

The aim of CBS is to develop science and practices that predict *and influence* human behavior, and specifically, behavior that serves to achieve goals that are related to a set of articulated values (Hayes et al., 2012). For the organization that promotes CBS, the Association of Contextual Behavioral Science, these values are to ease human suffering and promote human well-being. (Consistent with its functional contextual philosophical roots, CBS conceives of *behavior* on an organismic level: everything that a human does, including acting, blinking, thinking, feeling, remembering, etc.; Hayes et al., 2012).

Having the goal of predicting and influencing human behavior necessitates that we analyze the behavior of the entire person in the context in which it is occurring. According to CBS, this context is defined by the interaction of the current situation and the consequences of one's behavior in that situation, given one's history (genetic, biological, and historical). This analysis of an "act-in-context" is a pragmatic one, in that it permits us to understand actions in a functional sense; it allows us to know *why* they are occurring, or *what* they are trying to achieve. It guides us in designing interventions that can let us change behavior—by changing the context in which it is occurring—so as better to promote human well-being and effectiveness in the workplace.

CBS's aim of predicting and influencing behavior, which necessitates a functional analysis of the act-in-context, also requires us to identify, in that analysis, a variable *that we can actually manipulate* (or influence). Thus, consider the following functional analysis: "In order to increase productivity in a given department, we need to increase staff engagement and motivation." This analysis may represent an accurate account of events; indeed, research shows that staff engagement and motivation are excellent *predictors* of productivity (e.g., Barrick, Stewart, & Piotrowski, 2002). From a pragmatic perspective, however, it is not very useful to us, since it provides no indication as to how we can actually *influence* productivity, because we cannot directly increase the engagement and motivation of staff. (Sadly, perhaps, people do not have such switches on their backs.) Instead, we need to establish a functional contextual analysis of this situation, such as: "In order to increase productivity in a given department, we need to give staff more control over how they do their work and ensure that we make them aware of the significance of what they are doing." Research indicates that increased levels of control and task significance significantly *predict* productivity (e.g., Hackman & Lawler, 1971), and research shows that we can *influence* these variables, as well (e.g., Bond, Flaxman, & Bunce, 2008); thus, this latter analysis is consistent with our functional contextual goal of predicting and influencing human behavior so as to promote better mental health and behavioral effectiveness.

By pursuing this goal of prediction-and-influence, we can ensure that we develop a model of organizational flexibility that consists of OB constructs, strategies, and techniques that we can manipulate in order to influence individual and organizational effectiveness. Indeed, such a pragmatic model can help all OB researchers and practitioners stay laser-focused on affecting change, unencumbered by superfluous constructs that no OB practitioner could directly influence (e.g., motivation, meaningfulness of work[2]).

---

2   This is not to say that psychological events, such as motivation, are unimportant—quite the contrary, they bring energy and vitality to our lives—it is simply that we cannot directly impact them, and our theories and models may be more useful if we treat those internal events as outcomes, or dependent variables.

Functional Twins: Psychological Flexibility and Organizational Flexibility Models

The hexagon (colloquially referred to as the hexaflex; see Figure 1) is a graphic representation of the six core psychological processes (discussed below) that constitute psychological flexibility—processes that we can influence through various ACT techniques (Hayes et al., 2006). The processes on the left of the hexaflex (acceptance and defusion) constitute the mindfulness processes, while those on the right (values and committed action) promote commitment to values-based action processes. The two at the center of the hexaflex (present moment and self-as-context) facilitate both types of processes. In the following section, I will discuss OB characteristics that I believe are, at the organizational level, functionally equivalent to the six hexaflex processes. We can influence these OB characteristics (shown in Figure 2) and, in so doing, I maintain that we can design organizations that are "mindful" (or open and aware) and committed to furthering their values; in other words, organizations that are flexible.

As discussed in the following section, the six organizational characteristics that constitute the *orgflex* serve a related function to their (spatially) corresponding psychological process in the hexaflex (compare Figures 1 and 2); thus, "purpose and goals" is located in the orgflex in the same position as "values" is in the hexaflex, and I maintain that they serve similar functions in terms of setting a meaningful course in the life of the organization or individual, respectively. Furthermore, it is hypothesized that each organizational characteristic on the orgflex can promote, to varying degrees, in individual workers, the corresponding psychological process on the hexaflex; thus, an organization that is open to discomfort (e.g., ambiguity and conflict) can model and reinforce those characteristics of acceptance in relation to workers' own psychological events.

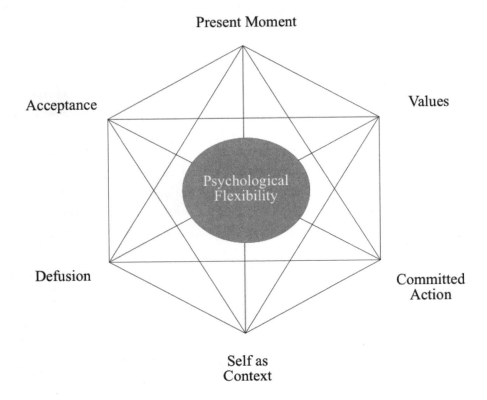

Figure 1. ACT's psychological flexibility model

The Six Characteristics of Organizational Flexibility

Purpose and Goals

For individuals, *values*, in ACT's hexaflex, refer to a direction of travel that people choose to take; they give meaning to their lives, and people need constantly to work toward them, as they can never be forever (if ever) achieved or sustained (Hayes et al., 2012); for example, a person has to work constantly on being a loving partner: it cannot be achieved in perpetuity. Similarly, the OB literature notes the importance of the *purpose* of an organization. Like a value, it guides an organization's goals (or vision) and day-to-day actions (or mission) (Marquis, Glynn, & Davis, 2007). An organizational purpose has three characteristics: (1) it meets a need in the world that will function to make the world a better place

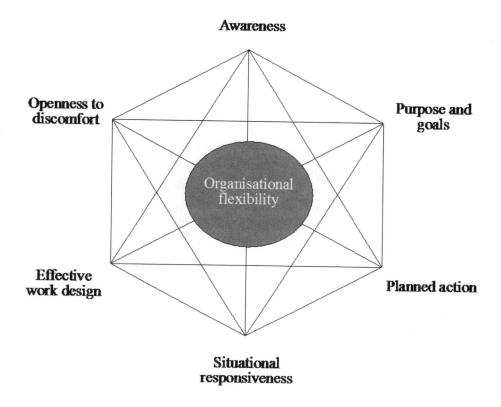

Figure 2. Organizational flexibility model

(e.g., anything from making machines for cancer treatment to providing entertainment to people); (2) it meets a need in society (e.g., providing transport for Londoners); and (3) as with ACT's definition of a value, it is aspirational but not sustainable (e.g., "preserve and improve human life"—Merck pharmaceuticals) (Marquis et al., 2007).

Planned Action

For the hexaflex, the term *committed action* involves the specification of actions, or goals, that individuals commit to take, in order to move toward their values (Hayes et al., 2012). Similarly, for the orgflex, organizations need to take *planned action* in order to further their purpose and goals. I use the term *planned*, instead of *committed*, for the orgflex, in order to highlight the difficulty of planning and implementing the interdependent steps (perhaps across many departments) needed to achieve an organizational goal. Indeed, the risk of goal failure, due to lack of sufficient planning, is reflected in the OB literature, in which one can find many different approaches to project management; they all have in common, however, planned strategies and processes to ensure that the project (i.e., goal) is actually delivered. The one that I advocate for promoting planned action in the orgflex is called "project definition" (Martin, 2009). This process specifies the "project" (or goal) toward which, and the related processes by which, teams and organizations work in order to further the organization's purpose. Thus, just as committed action is based on an individual's values, the process of project definition is very explicitly linked to the purpose of an organization. This emphasized link is a key reason why I believe this approach to planned action is similar to the hexaflex's committed action.

For individuals, taking committed action will likely involve creating an action plan that specifies the goal, how it will be achieved, psychological and external barriers that may get in the way of achieving the goal, and perhaps even a time frame in which sub-goals and the goal itself will be met (Bond, Hayes, & Barnes-Holmes, 2006). For organizations, project definition is a well-defined process (see Martin, 2009) that should result in the following: the problem definition (e.g., where are we now and where do we need to go); the specification of the project outcome (or goal); the project plan that outlines the approach to analysis, design, and completion of the plan, as well as the timeline for all of these processes; and, finally, it should clearly specify the project team, project leader, and reporting structure. This approach to project definition attempts to establish a clear framework that will help to ensure the successful and on-time fulfillment of an organization's goal.

In addition to an emphasis on identifying values/purpose-based goals, a second similarity between committed action and project definition is that, for both, problems are seen as an inevitable part of working toward goals, and they should be expected and addressed (Hayes, Villate, Leven, & Hildebrandt, 2011; Martin, 2009). In ACT, (psychological) "problems" such as anxiety and other unwanted internal events are considered "normal," and not something that needs to be changed or gotten rid of, in order to achieve one's goals. Similarly, in a project-definition approach, problems are not viewed as signs of trouble, undesirable, blameworthy, or even threatening to goal achievement; rather, they are accepted as an inevitable part of the process and should be revealed and addressed as quickly as possible. Thus, both committed action and the planned action approach of project definition view problems as a normal part of working toward goals; they need to be accepted and addressed, but they do not need to get in the way of achieving those goals.

Situational Responsiveness

For the hexaflex, *self-as-context* (SAC) is a complex process that has a wide range of psychological implications for matters ranging from mental health and autistic spectrum disorder to cognitive ability (Hayes et al., 2012). For the purposes of the orgflex, one particular function of SAC is of greatest relevance to organizations: SAC is a psychological space from which people can observe their self-conceptualizations (e.g., "I am a shy person," "I am an effective leader") without having such conceptualizations overly determine their actions (Hayes et al., 2012). Instead, from a perspective of SAC, people are better able to take actions, in a given context, that are more consistent with their values (e.g., intimacy) than with their thoughts as to who they are (e.g., "I am an unlovable person") and who they are not ("I am not confident"). In the orgflex, this function of SAC is seen in the concept of situational responsiveness, whereby organizations take strategic and operational decisions, in a given context, based more on their organizational purpose and goals and less on their own conceptualizations of what they are (as reflected in their culture or brand).

Situationally responsive organizations are those that keenly pursue and react to feedback from their task environment (Leavitt, 1965): their customers, competitors, suppliers, government regulators, and unions. They take operational and strategic decisions based more on that feedback (which includes market research, customer feedback, union engagement), and less on their brand (e.g., safe and reliable) and culture ("This is the way we do things around here"). As with people who overly base their actions on their self-conceptualizations, organizations that overly base their actions on their brand or culture/history are likely to be less flexible and adaptable, and hence less able to pursue effectively their purpose/values and goals. As Leavitt (1965) noted, for organizations to remain effective—to pursue their purposes and goals—they have to adapt successfully to the changes in their task environment. Those that are overly entangled in their brand or culture find it hard to make those necessary changes.

Organizational theorists have long noted how difficult it can be for organizations to change their cultures and how they present themselves to the public (i.e., their brand). The reasons for this self-defeating rigidity can be numerous, but Bion (1948) highlighted a particularly invidious one: employees at all levels have the tendency to nonconsciously collude to design and maintain rigid work structures and processes that function to help them to minimize unwanted psychological events (e.g., by requiring needlessly complex approval processes that discourage innovation and, crucially, the unease that can accompany it). Change is difficult enough, but it is even more so if the work characteristics that require change are functioning to prevent employees from experiencing discomfort of some kind.

The organizational development literature is replete with strategies to overcome this psychological "resistance to change" (e.g., French & Bell, 1999). One such way is to "design out" (or at least minimize) the rigid structures and processes that can inhibit situational responsiveness. This is accomplished, in part, through effective work design.

Effective Work Design

Cognitive *defusion* is an approach that people take toward their internal events that alters the undesirable functions of those events without changing their form, frequency, or situational sensitivity (Hayes et al., 2012). Put another way, defusion involves changing the way that people interact with their private experiences so that, while those experiences may still be present, they no longer have detrimental psychological/behavioral effects on them. Likewise, organizational researchers have long hypothesized that various forms of *work design*—that is, ways that people interact with their work tasks—can limit the impact that work demands have on people's physical and mental health. Karasek's (1979) demands-control model perhaps most explicitly makes this prediction. It maintains that highly demanding jobs will only have detrimental effects on people, such as coronary heart disease and psychological distress, if people have to approach their work without sufficient job control. If organizations provide people with some influence over how they carry out their work (even if it is demanding), they will not only experience fewer and less deleterious effects, but they will also perform their work more effectively and be more motivated in carrying it out. A comprehensive review of the work-control literature largely supports this hypothesis (Terry & Jimmieson, 1999; see also Bond et al., 2008).

Other well-established and empirically supported OB theories also posit beneficial effects of job control—for example, the job characteristics model (Hackman & Lawler, 1971), the sociotechnical systems approach (e.g., Emery & Trist, 1960), and the job demands-resources model (e.g., Bakker & Demerouti, 2007). Furthermore, these theories hypothesize that other work-design characteristics can have advantageous impacts on the health, performance, and attitudes of workers; these include support in carrying out one's work, the opportunity to do a variety of tasks, and the ability to do a complete job from start to finish (e.g., a team that builds a car from start to finish, or guiding a customer complaint from the time it is made to the time it is resolved). As for job control, there is long-standing and considerable research that shows the health and performance benefits of these and other work-design characteristics (Humphrey, Nahrgang, & Morgeson, 2007).

Thus, the concepts of both defusion and well-designed work maintain that unwanted thoughts and demanding work, respectively, do not necessarily have to lead to detrimental consequences: they only do so when they are approached from a fused stance or in a context of poor work design. As the hexaflex indicates, individuals can change the context in which they experience their internal experiences, but as the orgflex highlights, only organizations can create the context in which people carry out (or approach) their work.

Openness to Discomfort

It will be to no one's surprise that organizations can evoke challenging, unwanted emotions in people. The hexaflex shows how it is useful for individuals to *accept* and be open to those emotions in the pursuit of their values; the orgflex advocates this same open stance at the organizational level, and the OB literature champions many different structures, processes, strategies, and leadership approaches that require such *openness to discomfort*. Providing workers with job control is one such design principle that many managers find anxiety provoking; others include allowing employee participation in decision making; clearly, openly, and honestly communicating with employees in a timely manner; the project-definition process, noted above; and a transformational approach to leadership, which requires a personal, open, and "lead by example" leadership style (Bass, 1998).

If managers are unwilling to experience such discomfort, they may not redesign their organizations so that they reflect these OB characteristics. Such avoidance could compromise both the effectiveness of an organization and the health of its employees. Furthermore, leaders would miss an opportunity to serve as a role model to their staff as to how to experience discomfort and still pursue meaningful work, and indeed, personal goals.

Awareness

As the hexaflex advocates the need for individuals to be in the *present moment* and to be aware of, and open to, their internal events, so the orgflex maintains the same advice for teams and organizations. This advice is consistent with a whole field within OB that focuses on maintaining system *awareness*: human resource management. Most organizations of any size will have a human resources (HR) department that will develop policies and practices that function either to understand what is happening within the organization (e.g., through performance evaluation and staff surveys) or to train employees (essentially) to be aware of their actions (e.g., through diversity training and career development planning).

The role of maintaining awareness does not lie only in HR, however. For organizations to have flexible and high-performing individuals, teams, and departments, consistent monitoring needs to occur at each of those levels. For example, one useful technique that leaders can use with their teams or departments is referred to as "decision tracking" (Martin, 2009). This approach to maintaining awareness aims to obtain almost immediate feedback on results, in order to improve both learning and the decision-making process. Decision makers—even if they are not formal leaders—record a decision they have just made, along with the outcomes they anticipate, and they later read that document to reflect on and learn from the decision's consequences.

Leadership and Flexible Organizations

The orgflex highlights OB characteristics that can produce a flexible, and hence effective and healthy, organization, but senior and mid-level leaders need to design, maintain, and strategically change these characteristics. It is maintained here that, in order to do so effectively, these leaders need to be psychologically flexible. Avoidant, nonmindful leaders are less likely to be willing to experience the potential discomfort, ambiguity, or uncertainty that can result from providing their workers with job control, considering alternative solutions to problems, and taking a committed approach to completing a project. Furthermore, just as mindfulness facilitates the identification of values for individuals, it is thought that mindful—or open and aware—leaders are more clearly able to identify the purpose, vision, and mission for their organizations in the context in which they operate.

Research has yet to examine the relationship between leaders' psychological flexibility and the flexibility of their organizations, but it is hoped that this chapter may spur such investigations. If such a relationship were found, worksite ACT interventions would be useful to increase the psychological flexibility of leaders. As noted above, research has shown that ACT interventions in the workplace are effective in improving workers' productivity, mental health, and attitudes (Bond et al., 2013); and an increase in psychological flexibility is often seen as the mechanism by which those ACT interventions produce their beneficial effects. Thus, by increasing the psychological flexibility of leaders, they may be in a better position (with expert help, perhaps) to design, implement, and maintain the OB characteristics specified in the orgflex.

Conclusions

The orgflex highlights organizational characteristics that, it is hypothesized, will produce a more flexible, and hence, more productive and healthy, organization. Research has indicated that these characteristics are, individually, associated with better-performing organizations and, in the case of effective work design, better mental health (Humphrey et al., 2007). Research has not yet investigated the hypotheses that (1) the six characteristics in the orgflex combine to impact productivity and health, and (2) that they do so by increasing organizational flexibility. Even if future research supports these two hypotheses, this chapter does not maintain that the characteristics specified in the orgflex are the only ones that can enhance organizational flexibility. Indeed, Hayes (2010) developed an organizational flexibility model that was not explicitly informed by the OB literature, but rather by Ostrom's (1990) core design principles for group efficacy. As highlighted by the hexaflex, however, it is felt that flexibility will be best promoted when organizations are not only actively pursuing their purpose, vision, and mission, but when they are doing so "mindfully"; that is, when they are open to, and aware of, their internal and external processes and environments. It is hoped that the orgflex serves to emphasize, in particular, this latter point.

References

Bakker, A. B., & Demerouti, E. (2007). The job demands-resources model: State of the art. *Journal of Managerial Psychology, 22*, 309–328.

Barrick, M. R., Stewart, G. L., & Piotrowski, M. (2002). Personality and job performance: Test of the mediating effects of motivation among sales representatives. *Journal of Applied Psychology, 87*, 43.

Bass, B. M. (1998). *Transformational leadership*. Hillsdale, NJ: Lawrence Erlbaum.

Bion, W. R. (1948). Experiences in groups. *Human Relations, 1*, 314–320.

Bond, F. W., & Bunce, D. (2000). Mediators of change in emotion-focused and problem-focused worksite stress management interventions. *Journal of Occupational Health Psychology, 5*, 156–163.

Bond, F. W., Flaxman, P. E., & Bunce, D. (2008). The influence of psychological flexibility on work redesign: Mediated moderation of a work reorganization intervention. *Journal of Applied Psychology, 93*, 645–654.

Bond, F. W., Hayes, S. C., Baer, R. A., Carpenter, K. M., Guenole, N., Orcutt, H. K., ... Zettle, R. D. (2011). Preliminary psychometric properties of the Acceptance and Action Questionnaire—II: A revised measure of psychological inflexibility and experiential avoidance. *Behavior Therapy, 42*, 676–688.

Bond, F. W., Hayes, S. C., & Barnes-Homes, D. (2006). Psychological flexibility, ACT, and organizational behaviour. *Journal of Organizational Behavior Management, 26*, 25–54.

Bond, F. W., Lloyd, J., & Guenole, N. (2013). The work-related acceptance and action questionnaire (WAAQ): Initial psychometric findings and their implications for measuring psychological flexibility in specific contexts. *Journal of Occupational and Organizational Psychology, 86*, 331–347.

Brown, K. W., & Ryan, R. M. (2003). The benefits of being present: Mindfulness and its role in psychological well-being. *Journal of Personality and Social Psychology, 84*, 822–848.

De Board, R. (1978). *The psychoanalysis of organisations: A psychoanalytic approach to behaviour in groups and organisations.* London: Tavistock.

Emery, F. E., & Trist, E. L. (1960). Sociotechnical systems. In C. W. Churchman & M. Verhulst (Eds.), *Management sciences: Models and techniques* (Vol. 2). Oxford: Pergamon.

Flaxman, P. E., & Bond, F. W. (2010). A randomised worksite comparison of acceptance and commitment therapy and stress inoculation training. *Behaviour Research and Therapy, 48*, 816–820.

French, W. L., & Bell, C. H. (1999). *Organization development: Behavioral science interventions for organization improvement.* Englewood Cliffs, NJ: Prentice Hall.

Golden, W., & Powell, P. (2000). Towards a definition of flexibility: In search of the Holy Grail? *Omega, 28*, 373–384.

Hackman, J. R., & Lawler, E. E. (1971). Employee reactions to job characteristics. *Journal of Applied Psychology, 55*, 259–286.

Hayes, S. C. (2010, June). *Values, verbal relations and compassion: Can we do a better job of facing global challenges?* Paper presented at the ACBS Annual World Conference, Reno, NV.

Hayes, S. C., Barnes-Holmes, D., & Wilson, K. G. (2012). Contextual behavioral science: Creating a science more adequate to the challenge of the human condition. *Journal of Contextual Behavioral Science, 1*, 1–16.

Hayes, S. C., Luoma, J. B., Bond, F. W., Masuda, A., & Lillis, J. (2006). Acceptance and commitment therapy: Model, processes and outcomes. *Behavior Research and Therapy, 44*, 1–25.

Hayes, S. C., Strosahl, K. D., & Wilson, K. G. (1999). *Acceptance and commitment therapy: An experiential approach to behaviour change.* New York: Guilford.

Hayes, S. C., Strosahl, K. D., Wilson, K. G., Bassett, R. T., Pastorally, J., Toarmino, D., & McCurry, S. M. (2004). Measuring experiential avoidance: A preliminary test of a working model. *Psychological Record, 54*, 553–578.

Hayes, S. C., Villatte, M., Levin, M., & Hildebrandt, M. (2011). Open, aware, and active: Contextual approaches as an emerging trend in the behavioral and cognitive therapies. *Annual Review of Clinical Psychology, 7*, 141–168.

Humphrey, S. E., Nahrgang, J. D., & Morgeson, F. P. (2007). Integrating motivational, social, and contextual work design features: A meta-analytic summary and theoretical extension of the work design literature. *Journal of Applied Psychology, 92*, 1332.

Jaques, E. (1955). Social systems as a defence against persecutory and depressive anxiety. In M. Klein, P. Heimann, & R. Money-Kyrle (Eds.), *New directions in psychoanalysis.* London: Tavistock.

Kabat-Zinn, J. (1990). *Full catastrophe living: Using the wisdom of your mind and body to face stress, pain, and illness.* New York: Delacorte.

Karasek, R. A., Jr. (1979). Job demands, job decision latitude, and mental strain: Implications for job redesign. *Administrative Science Quarterly, 24*, 285–308.

Leavitt, H. J. (1965). Applied organizational change in industry: Structural, technological and humanistic approaches. In J. G. March (Ed.), *Handbook of organization.* Chicago: Rand McNally.

Linehan, M. M. (1993). *Cognitive-behavioural treatment of borderline personality disorder.* New York: Guilford.

Lloyd, J., Bond, F. W., & Flaxman, P. E. (2013). The value of psychological flexibility: Examining psychological mechanisms underpinning a cognitive behavioural therapy intervention for burnout. *Work and Stress, 27*, 181–199.

Lloyd, J., Bond, F. W., & Flaxman, P. E. (2017). Work-related self-efficacy as a moderator of the impact of a worksite stress management training intervention: Intrinsic work motivation as a higher order condition of effect. *Journal of Occupational Health Psychology, 22*, 115–127.

Marlatt, G. A., & Kristeller, J. L. (1999). Mindfulness and meditation. In W. R. Miller (Ed.), *Integrating spirituality into treatment.* Washington, DC: American Psychological Association.

Marquis, C., Glynn, M. A., & Davis, G. F. (2007). Community isomorphism and corporate social action. *Academy of Management Review, 32*, 925–945.

Martin, R. L. (2009). *The design of business: Why design thinking is the next competitive advantage.* Boston, MA: Harvard Business Press.

Ostrom, E. (1990). *Governing the commons: The evolution of institutions for collective action.* Cambridge, UK: Cambridge University Press.

Robbins, S. P., & Judge, T.A. (2007). *Organizational Behaviour* (12th ed.). New Jersey: Pearson Education, Inc.

Terry, D. J., & Jimmieson, N. L. (1999). *Work control and employee well-being: A decade review.* In C. L. Cooper & I. T. Robertson (Eds.), *International review of industrial and organizational psychology 1999.* West Sussex, UK: Wiley-Blackwell.

Dialogue on Organizational Development

Participants: Frank W. Bond, J. W. Stoelhorst,
Mark van Vugt, and David Sloan Wilson

David Sloan Wilson: Let me start by focusing on monitoring for groups working well, because that's one of the core design principles according to Elinor Ostrom. If you can't monitor agreed-upon behavior, then forget about it. But there are bad forms of monitoring, and there's overmonitoring. Maybe you'd like to elaborate on that a little bit.

Frank W. Bond: Take call-center workers. Maybe they have to call up people who are late on their payments; or call up people who are processing money; and then something else. Often the leader says, "Okay, I want you all to do it in this order. I'm going to check up on you at lunchtime to see how things are going." If we give people more control over how they do their work, then we're going to do better. I'm a horrible morning person, and there are certain tasks that I do not want to do in the morning. If can I do them in the afternoon, I am going to like that better.

David: Basically, providing elbow room for people to do the job. They'll be monitored; but do it their way.

Frank: Right. And if they need to step outside in the morning to call the school because the headmistress wants to talk about little Johnny, then she or he is going to be able to make up that time later on in the afternoon, when he or she tends to work faster. So you still have the monitoring, but you've also given people control to work in a way that's going to be more desirable for them. We find that it decreases absenteeism; performance basically remains the same; and turnover decreases as well. So the monitoring is necessary, but the degree of monitoring and the frequency of monitoring needs to be commensurate with the goal of the job. In many call centers, people will even know when you stand up from your chair to go to the bathroom, and things like that. So you have to balance that out.

Mark van Vugt: People bring to organizations a set of evolved preferences, which are part of our evolutionary legacy. A need for autonomy, for example, a need to be valued and respected, a need to be able to develop your competence and to be able to get prestige. People bring to their jobs needs and motives. The question is, is the modern organizational environment a way in which these needs and values and motives can thrive? Structures need to be aligned to people's motives and needs and they also need to be responsive to environments. Is there an organizational mismatch whereby the kind of structures that we have created after the Industrial Revolution are no longer best fitted to what people need and want? And how should we retailor and redesign them into the kind of cooperative communities that our ancestors thrived in for probably hundreds of thousands of years?

David: Let me nudge the conversation in this direction. The business and management organizational literature is a sprawling literature, so much work is being done; and also, of course, business is a highly competitive process, so between-group competition in the business world is intense. Businesses don't always work well, and there's something very new about both of these perspectives, the evolutionary perspective and the CBS perspective. So, help me understand the added value. Why is it that your perspectives are new, against the background of both the sprawling academic study of organizations and the organizations themselves?

J. W. Stoelhorst: Organizations tend to be designed along the lines of classical economic theory, which makes assumptions about us as humans that have largely been disproved. All four of us agree that values are important—but the question is, which values? If it's just about generating shareholder value, is that going to be commensurable with our evolutionary heritage?

Interestingly, economic theory also gets us a system that works along the lines of what Frank was suggesting. Not too much monitoring—nobody's telling executives how to do it; just as long as they are profitable, they're going to get a huge reward. But the problem is that one particular value is taking over. It's all about paying people on the basis of their performance toward a profitability goal, and any other type of value is going out the window.

I think you see this even more strongly in nonprofit firms. If you go to schools, if you go to the police, if you go to the healthcare sector, things are being managed on what we can measure—and what we can measure is all about productivity. So we start pushing more and more on productivity, and we push out all the intrinsic motivation that, especially in these nonprofit sectors, is probably what drove people to work there in the first place.

David: So, with these other types of groups, it's no longer financial profitability, but you're saying that something else similar takes hold.

JW: Yes, managing organizations in general has become a game of financial constraints, of measuring what you can measure, and rewarding people on that basis. I think that we've gone overboard in having this one particular perspective, in which the starting point is the idea that we're all *Homo economicus*.

Mark: That is the appeal of nonhierarchical organizations, which you see blossoming left and right—"bossless" organizations. People indicate that they enjoy working in them: productivity is high, satisfaction is high. And the question is, does this reflect something deeper about human nature and human motivation? Some organizational design folks are saying, well, these nonhierarchical organizations are nice, but they are also limited in scale. And as soon as you scale up organizations and give them more complex tasks, then you have to scale up the heirarchy. And that is when, they think, hierarchies start to emerge almost naturally.

David: I would think you have to have structure as you get larger, but whether this structure is hierarchical, and/or a bottom-up control in addition to top-down control, is a different matter. Take Toyota, for example, which succeeds by putting the executives on the shop floor, rather than on the top floor.

Frank: You read books on the East India Company, and on the major successes of the Industrial Revolution, and those weren't nice places to work. You know, the problem wasn't, "I feel stressed"; it was, "I died." And whether or not an employee had a problem didn't really matter; it was all about the money. It still is, for organizations, about the money—about shareholder value and all of that.

So you have some place like SalesForce.com, which is probably the most successful consumer resource management cloud-computing firm in the world, and they make a load of money. But what they also do is—they put a lot of money into the Gay Pride movement; they give people a half a day off per week to pursue a charitable goal. I think the smarter organizations are acknowledging: "Yes, we need to make more money, but how can we also fulfill the values that our employees have?"

David: Yeah, so I want to end with two big questions. Let me list both—they're related to each other. Often a company is organized for the benefit of the elites within that company. It's working great for the CEO, or the CEO and the shareholders, just extracting, extracting, extracting from the company. Good for them, not necessarily good for the company. And, of course, they're not going to want to give up control. *Homo economicus* is an ideology that serves those purposes, and these folks are not going to go lightly into the night. There is a kind of multilevel view of business evolution, which notes this disruptive within-group component. There is something almost revolutionary—I mean, literally so—about members of organizations that do not have much power wresting back that power. What may make that work is if organizations that do that in fact out-compete the companies that are handicapped by disruptive within-group processes.

The second question is how we do scientific research with groups like this. Often, companies and nonprofit organizations really don't have a scientific mindset, and the idea that they can be treated like living laboratories is not as simple as it might seem.

JW: The question you are asking is: "Who is being served by the shareholder-value ideology?" In fact, the CEOs that you mentioned are historically a relatively recent group to be co-opted into this financial market ideology: much of the shareholder-value thinking was directed against CEOs doing what they want. Historically, CEOs were not being controlled enough by shareholders—at least that is what the people pushing the shareholder-value ideology thought. In the end, CEOs, through their big bonuses, have been co-opted into the interests of the financial markets. If you look at this through an evolutionary lens, then you want institutions to keep competition peaceful and leaders honest. Capitalism has been fairly good at making competition more peaceful. But the current system is not very good at keeping leaders honest.

Frank: Government laws and rules are extremely important. You can't let a free market just do what it wants. That's not going to work. How you control that will vary from culture to culture. And then, within organizations, the leaders are extraordinarily important, and selecting those leaders is very important too. One thing that we haven't talked about, which organizational psychologists are very keen to talk about, is the whole process of selection. You know, how do we select the leaders, and indeed anyone, that we have in the company? What are the values against which we select those people? What are the competencies against which we select those people? We're making progress in getting people to change, but people don't like change. British sailors died of scurvy from about 1400 through 1600-something, and there was a clever chap at Cambridge who in 1400-something published this paper saying that if they just suck on a citrus for a minute, they're going to be fine. The Royal Navy implemented that two hundred years later. You know—"I'm an admiral, I know what's best." The move toward some evidence-based ways of doing things is perhaps making things change a little bit faster. But I think that's always going to be a difficulty.

David: Change is difficult, as you said, and we need a technology for just getting people to do what they want to do, either individually or collectively. That, I think, is a special contribution of ACT and relational frame theory. We get into realms that most human evolutionists don't talk about much, like Stephen Jay Gould on equilibrium about plans, constraints, and things that interfere with the process of just adapting to an environment.

Frank: And that's something that a number of us academics are working on at the moment. We want to identify what those problems are and how we can actually improve the performance and mental health of people.

David: I work with groups a lot and have really plunged into the business literature, so I regard myself as a colleague of yours in this subject area. And it's my experience that the groups that I approach don't have a scientific mindset, and they're so damn busy that the idea of getting them to step back and to reflect leads to "I can't do it! I'm too busy!" So how can we actually treat these groups as laboratories, and do good scientific research in this setting?

Mark: We're experimenting a little bit with that in the leadership lab, where we invite teams of leaders to subject themselves to a scientific approach. They're very curious what methodologies we can offer; but at the same time, (a) they don't have the time, and (b) they don't really like to be analyzed as groups. They like to be analyzed as individuals. The coaching of the CEOs and the senior management, executives, that's all well and good; but to actually study them as they are functioning as teams, it's quite scary to them, we find. So what we're trying to do is to basically combine a sort of individual approach, where we look at individual needs and motives of these leaders—could be in an executive board, for example, or in a supervisory board of a company or a non-profit organization—so, we give them individual feedback, but we also, what we do is, we observe a team meeting, then we analyze the team meeting in terms of the kinds of communications, all relatively anonymized. So we say something at the group level—for example, is there one individual who's talking the whole time and the others are passively listening but not actively contributing? Sometimes they don't realize how discussions are being monopolized by one individual, and so we give them feedback, which they cherish. The other thing is, we need, of course, that data for scientific purposes. What we enter in is a sort of quid pro quo. So far, that seems to be working well.

Frank: One thing we do, is we try to target individuals in organizations who we think can be key—but not too high. We literally wine them and dine them—I have a line in my budget for doing this. We ask them what they themselves need: how can we help you? And if they are a leader, inevitably they need to fulfill certain functions and deliverables. And we say we'll do it for free and we do that, usually it works, and that is our entry into the organization. So we find that that is a nice way of doing things, kind of turning the tables on them—if they wine and dine their customers, we try to do the same for them. That can be an effective way of doing things, especially in large organizations.

One thing that contextual science wants to do, is not just predict outcomes, but actually change outcomes as well. One of the great things about psychological flexibility as a psychological characteristic is we know that we can change it. It's been done in many, many studies. And we've seen that change mediates change in outcomes we like. We've long known about personality variables that predict very well, but we can't change them: neuroticism, conscientiousness, agreeableness, openness to experience, all of those. Psychological flexibility is different. We can go to HR and say, if you just take this seven-item questionnaire, we can help you figure out how best to help your people.

JW: Let me be frank on the challenges of working with organizations. It is very difficult to do controlled experiments at the level of organizations. As a result, there is not a lot of evidence-based substantiation of management practices. That's a problem in all of the social sciences—for ethical reasons, reasons of access, or just getting enough of a sample. So we tend to work across levels of analysis, trying to piece together a story from what we know from the lab and what we know from the field. One of the ways in which evolutionary reasoning can help is because it gives you this paradigmatic view of things, where you can begin to make meaningful connections between, say, an individual case of one firm and a whole lot of interesting findings in the lab. You can then start piecing together the puzzle of what this means for what we are as humans when we cooperate on a large scale, and how to best organize this.

David: Elinor Ostrom won the Nobel Prize by creating a database of common-pool resource groups from a very diffuse literature. Almost all of the accounts were descriptive, and nevertheless that did not prevent her from coding them, analyzing them, and then coming up with these core design principles, which could then be validated with other kinds of studies. There's a whole lot that can be done with data to mine, which can then be supplemented by more careful controlled studies.

CHAPTER 12

Evolutionary Mismatch: A Framework for Understanding Health and Disease in the Modern World— "Better Living Through Evolution"

Aaron P. Blaisdell

UCLA Department of Psychology

Despite their reputation as slimy, slithery creatures that litter sidewalks and driveways after a rainstorm, earthworms are amazing creatures. They chemically alter the soil in which they live and feed, resulting in a fitness benefit not only to themselves, but also for the plants and even microbes that live in the soil (Odling-Smee, 2009).

Ants, too, alter their world in amazing ways. On a trip to Colombia in 2015, I had the good fortune to witness firsthand a trail of leafcutter ants. These ants make cuttings from plant leaves which they carry back to their underground nests. Leaf matter then serves as a food source for fungus that the ants harvest for food. Other ants tend and protect aphids that feed on certain plants, milking the aphids for a nutritive secretion that feeds the ants.

Beavers build incredible domes to live in and dams to create ponds that protect them from predators. Environmental modifications created by beaver dams have profound effects on the ecosystem, ultimately enriching biodiversity. Thus, beavers are ecological engineers and are seen as a keystone species (Rosell, Bozsér, Collen, & Parker, 2005).

These are but three of many examples of what biologists call *niche construction*. Niche construction is the process by which an organism alters its environment, typically resulting in increased fitness. While many vertebrate and invertebrate species provide examples of niche construction, humans have taken the process to an extreme that goes far beyond the capacities of nonhuman animals (Stutz, 2014). As a result of taming fire, making tools of stone, wood, bone, and leaves, the invention of clothing and housing that protects against the elements, and ultimately the invention of language, humans have been able to leave their ancestral African birthplace and colonize much of the world's land-mass, including extreme environments within the Arctic Circle. We have been able to hunt large game, fish, and even large whales for food.

Modern humans present the ultimate case of niche construction, with the creation of agriculture, cities, large-scale, stratified societies, scientific and tech-nological feats such as lasers, particle accelerators, and gene editing, and ulti-mately by transporting a breathable earth atmosphere into the depths of the ocean and into space.

Despite producing amazing advancements and achievements, human niche construction is a double-edged sword. It certainly has enriched the human condi-tion in incredible and wondrous ways: art, music, stories, dance, science, math, and philosophy—all providing a deeply enriched human experience. At the same time, however, human niche construction is also the primary contributor to the majority of human disease and suffering worldwide, especially since the Industrial Revolution and the changes it has wrought on modern human society. Human ingenuity is remarkable at the individual level, but it has become incredibly pow-erful with the cultural accumulation of knowledge, skills, and technology. This has allowed human societies to make advances and changes in lifestyle and habitat faster than the pace of biological evolution can keep up with. As a result, we've created a living environment that is often far different from that to which our bodies and minds are adapted. Thus, many modern-day ills and diseases can be viewed as cases of *evolutionary mismatch*. Like the foolish painter who unwittingly paints himself into a corner, so too has humankind placed itself and the world we inhabit in a precarious condition.

Before turning to the various maladies brought on by evolutionary mismatch, let's first compare in very general terms the prototypical ancestral environment to the typical modern environment as experienced by people living in the modern age. The ancestral environment in which humans existed for over 99% of the time since the origin of *Homo sapiens* (including *H. sapiens neanderthalensis*) around 500,000 years ago consisted of living primarily in outside environments. Such environments are characterized as being nonlinear, with lots of fluctuations in light, temperature, elevation, activity, and sensory input; daily interactions with a small number of familiar individuals ranging in age from infancy to elderly; a large portion of the day spent procuring and preparing food, shelter, and other necessities of daily life; consumption of foods that provide all of the necessary macro- and micronutrients and that typically prevent nutrient deficiencies;

knowledge derived largely from one's own accumulated experiences and from direct conversation with members of a small community; and sleep patterns that are fairly consistent and that adequately if not optimally meet sleep requirements. Despite a tight integration within the social ecology, there is a great deal of freedom and personal autonomy in how daily activities are arranged and choices are made. And there is no clear separation of the self from the group or the environment.

The Western industrial society, on the other hand, is characterized as being highly linear—living in carpentered environments, with much less range of motion when sitting, standing, or walking; with a majority of time spent indoors protected from fluctuations in temperature, light, activity, and sensory input (resulting in a large degree of homogeneity in these stimuli); with a large portion of the day spent in proscribed activity, either school (children and adolescents) or work (adults); long periods of stasis (e.g., sitting at a desk job or at school); and daily connections to a much smaller social group, often of a restricted age range, while at the same time encountering brief, superficial connections to a much wider number of individuals. Furthermore, in modern households, parents are burdened with arranging schedules, feeding, entertainment, and even social activities for their pre-adult children—which poses tight time constraints that curtail the freedom of parents to engage in autonomous and social activities, such as cultivating meaningful relationships with same-sex friends. Thus, unlike the ancestral state (as evidenced from contemporary hunter-gatherer societies) fathers rarely have the freedom to spend time with their male friends, and mothers are rarely connected throughout the day to a peer group of other female friends that would offer social support and share in daily living tasks, including childcare.

What follows is a discussion of the various ways that contemporary humans live in evolutionary mismatch, and its consequences for physical and mental health (Blaisdell, Pottenger, & Torday, 2013). In discussing these cases of mismatch, I try to identify solutions that might allow us to thrive in modern society without giving up our modern-day advances and comforts.

1. Food

We diverged from the African ape niche about 6-7 million years ago when hominids moved out of dense jungle and into open woodland and savannah environments (Klein, 2009). About 2.5 million years ago, we began making and using tools to open up new feeding niches centered on carnivory. In the past 1 million years, the taming of fire allowed us to include tough, starchy tubers (underground storage organs) into our diet through cooking (Wrangham, 2009). These developments have shaped our physiology, including our digestive system and brains (Aiello & Wheeler, 1995).

Beginning about 10,000 years ago, human societies in some areas began domesticating plants and animals for use as dependable sources of food. The most important domesticated food sources were the cereal grains, which became the staples of ever larger societies. The dawn of agriculture also brought about changes in our health, but these changes tended to be for the worse (Cordain, 1999). Wherever grain-based agriculture was adopted, a number of diseases related to nutrient deficiencies arose (Cohen & Crane-Kramer, 2007).

The Industrial Revolution introduced food-processing techniques that were an economic boon for a stable modern society. Rather than having to rely on perishable goods that needed large quantities of salt or fermentation for their preservation, refining processes allowed for the development of white sugar, white flour, and seed oils that were shelf stable at room temperature for months, if not years. This resulted in an increased storage capacity and transportability that allowed these new commodities to be transported to all corners of the globe. Nevertheless, such advancements also incurred health costs. Wherever these refined foods were introduced, developmental, infectious, and chronic diseases flourished (Price, 1939).

The Western diet, high in refined sugar, flour, and oils, and often containing chemical preservatives and artificial flavors and colors, has become synonymous with an unhealthy diet. It is the standard diet against which to compare the health value of other diets, such as vegetarianism, Mediterranean, or Paleo. Consumption of a refined diet is linked to obesity, diabetes, Alzheimer's disease, cardiovascular disease, cancer, and other degenerative diseases (Lindeberg, 2010). More recently, studies using nonhuman animals have revealed the causal role of a refined or Western diet in poor mental health (Andre, Dinel, Ferreira, Laye, & Castanon, 2014) and cognitive impairments (Blaisdell et al., 2017).

Implementing diets that mimic our ancestral eating patterns holds promise for the prevention and treatment of many modern chronic diseases, often labeled as diseases of civilization (Kuipers, Joordens, & Muskiet, 2012). Following a Paleolithic style of eating that removes grains, legumes, sugar, dairy, and seed oils from the diet has been gaining traction due to its many health benefits. The good news is that the science is becoming increasingly consistent in determining what kinds of foods most people thrive on versus what kinds of foods may be problematic for sensitive individuals or even for most people. As a rule of thumb, most people seem to do well on whole foods-based diets that minimize the amount of industrial processing (e.g., refining, using chemical solvents, using ingredients produced in a laboratory, etc.), on the one hand, and poorly on highly processed and refined flour, sugar, and seed oils, on the other.

As an aside, it is interesting to consider that the convention of comparing the health effects of an experimental lifestyle intervention, such as nutritional, exercise, or sleep, against a control group that does not receive the treatment, could be reframed as the no-treatment control group actually reflecting the modern experimental condition of evolutionary mismatch! Viewed in this light, the experimental group actually corrects for the mismatch, returning one facet of modern lifestyle back to the ancestral control condition.

2. Activity

Similar to diet and nutrition, activity patterns of modern humans in Westernized societies have departed dramatically from the ancestral template. There has been an increase in sedentary behavior (e.g., sitting), especially with the advent of the Industrial Revolution, which produced machines that can do a majority of former human labor. Attempts to remedy the health ravages of sedentary behavior have often introduced problems of their own. Repetitive movement exercises, such as jogging or those involving fixed-movement machines (e.g., rowing, elliptical, cycling) may be the source of many chronic use injuries and enhanced tissue degeneration. For example, endurance athletes are at a greater risk than the general population for developing atrial fibrillations (Abdulla & Nielsen, 2009), whereas older adults who engage in light exercise are at reduced risk (Mozaffarian, Furberg, Psaty, & Siscovick, 2008). Marathon runners also have been found to have more calcified plaque in their arteries compared to a sedentary control group (Kröger et al., 2011). Furthermore, endurance training induces oxidative stress (Duca et al., 2006) and increases the stress hormone cortisol (Skoluda, Dettenborn, Stalder, & Kirschbaum, 2012). Moreover, when adults do exercise in groups, such as in a pickup game of basketball or soccer, they can experience a high rate of injury because the long periods of sedentary behavior induced by the modern lifestyle does not allow a sufficient baseline level of physical fitness and physiological robustness to protect the body from the large peak forces involved.

While high-volume endurance exercise does provide some health benefits, low-volume but high-intensity exercises such as sprinting and high-intensity interval training (HIIT) offer the same benefits at a fraction of the time investment and risk of injury and tissue degeneration (Gillen et al., 2016). Contemporary hunter-gatherers engage in a much higher volume of low-level movement, such as walking (often over uneven surfaces in bare feet), bending, squatting, climbing, and carrying, interspersed by infrequent bouts of high-intensity activity, such as sprinting, jumping, throwing, and heavy lifting (O'Keefe, Vogel, Lavie, & Cordain, 2011). This pattern of non-linearly distributed activity entrains a high level of cardiovascular fitness and a body that is robust and resilient to a wide range of forces encountered in a heterogeneous natural environment. Mimicking more ancestral types of movement patterns appears to be conducive to greater health and fitness without the costs of novel repetitive movement exercises (Gillen et al., 2014). Thus, an ideal ancestral pattern of exercise for a modern person should include a lot of low-volume, slow-paced movement such as walking, gardening, house work, etc. on an almost daily basis, and short-duration, intermittent bursts of high-intensity movements, such as sprinting, weight-bearing movements, climbing, jumping, etc., once or twice per week. This pattern, which is close to that observed among contemporary hunter-gatherer and foraging societies, may be more sustainable for a wider segment of modern society, including the frail and elderly.

3. Surroundings

In the ancestral environment, individuals spent a vast majority of their time outside, embedded in nature. Only with the development of carpentered home environments, and especially with the advent of glass windows and air-conditioning systems, have individuals started to spend a majority of their time indoors, isolated from the outside environment. What are the consequences of this dramatic change?

Inside air quality is often poor due to off-gassing of chemicals in furniture, paint, carpeting, etc. (Jones, 1999). Likewise, spending long hours indoors incurs a dramatic reduction in exposure to sunlight. As a result, we are not exposed to the intense bright light or full spectrum of light that we receive outside. Sunlight provides a stimulus for many important functions, including the production of calcitriol (active Vitamin D), nitric oxide (which regulates blood pressure and heart health (Feelisch et al., 2010)), and other key nutrients (Mead, 2008), as well as regulating circadian rhythms (van der Horst et al., 1999). Spending a majority of our time indoors has also dramatically reduced exposure to fluctuations in temperature and air movement. The drop in temperature after dusk plays as much of a role (or an even greater role) in entraining sleep onset as does the drop in illumination intensity (especially in the blue range) (Van Someren, 2000). Furthermore, indoors, we experience a reduced exposure to ambient sounds and odors typical of outside settings.

Individuals who spend more time outdoors, and especially in nature, report improvements in mental health and cognition (Mantler & Logan, 2015; Pearson & Craig, 2014). Kids choose to play in natural settings more than in more artificial, carpentered, planned settings when access to each is equally available (Lucas

& Dyment, 2010), and more natural play spaces have higher play value than do more rigidly constructed play spaces with fewer manipulable features (Woolley & Lowe, 2012). This suggests that we should prioritize more exposure to the outside throughout the day to receive important health benefits for adults, in addition to benefits to children's health and physical, cognitive, and mental development (Kemple, Oh, Kenney, & Smith-Bonahue, 2016).

4. Light Exposure

Related to the change in exposure to sunlight is the change in the type and distribution of indoor and artificial light to which we are exposed. Indoor lighting is typically much less intense than sunlight. Fluorescent lighting contains blue light, which is dramatically reduced in morning and evening sun, and thus evening and nighttime exposure to fluorescent lighting and the lighting from TV, computer, tablet, and phone screens can dysregulate circadian rhythms (Chang, Aeschbach, Duffy, & Czeisler, 2014). Dysregulated circadian rhythms are linked to increased probability and length of illness, dysregulated sleep, depressed mood, and so on. (Roenneberg & Merrow, 2016).

Some methods to reduce or prevent the disturbances of blue light include programs and settings on computer and phone screens, such as *f.lux*, and wearing blue-light blocking glasses after sunset. Light bulbs are now available on the market that do not give off blue light and thus can be used at nighttime for indoor lighting.

5. Education

Children are adapted to learn and shape intelligence naturally through play and exploration (Blaisdell, 2015). Yet 13% to 20% of children living in the United States experience a mental disorder in a given year, and these trends have been increasing over the past couple of decades (Centers for Disease Control and Prevention, 2013). This illustrates another case of evolutionary mismatch. A typical preschool and kindergarten setting, for example, is highly scheduled and structured, with many of the activities directed by the teacher rather than motivated by the children's own interests. The entire classroom of children spends proscribed periods of time on specific activities arranged by the teacher, and the children are moved from one scheduled activity to the next according to an external cue, such as clock time. Throughout their entire career in prekindergarten through twelfth-grade education, children rarely have the freedom to engage in playful exploration of whichever topic or activity they would want, for the amount of time that they want, and at their own pace. Instead, the modern educational setting requires and places value on the child sitting still for long periods of time, listening to directed instruction or performing instructed operations (e.g., reading, writing, calculations). This system is at odds with children's disposition to move, laugh, play, daydream, and socialize.

This contrasts dramatically with the way children learn and develop in hunter-gatherer societies, where kids move about freely in mixed-age groups, sometimes working together with others and other times on their own (Gray, 2015). The mismatch between modern work-based educational settings on the one hand, and the play- and freedom-based ancestral context of child cognitive, emotional, and social development on the other, may be a leading contributor to the chronic mental, psychological, and perhaps physical health issues that face children in modern society (Gray, 2011). The increase in the number of progressive schools, such as Montessori and unconventional charter schools, home schooling, and unschooling in modern American society is a symptom of a malfunctioning educational system. Unfortunately, most lower- and middle-class families do not have the time and/or financial resources to commit to such alternatives. What is really needed is an overhaul of the modern educational system to align it closely to the biological needs of child development, incorporating freedom, play and humor, cooperation, and democracy as core tenets.

6. Social Life

For most of our evolutionary history, humans lived in small band societies, consisting of on average 40 to 60 individuals in the primary social group. There were occasional gatherings among networks of bands, but these were likely rare and seasonal events. As a result, people directly knew everyone in their society, and developed close, lifelong friendships within this small community. In such a society, everyone assists with and contributes to different activities that serve the group, ranging from foraging, building shelter, constructing and mending tools and clothing, to early childcare, storytelling, and maintaining and participating in cultural activities (Sober & Wilson, 1999). Despite some division of labor— women tended more toward early childcare, while men tended toward exploration and hunting (and in some societies warfare)—band society is egalitarian, with everyone sharing in the same freedoms and autonomy on the one hand, and contributing significantly to the group and community on the other. Perhaps most important of all, belief systems, decision-making structures, and community relations were handled with a playful approach and attitude (Gray, 2009).

In contrast, modern society consists of millions of individuals, organized into multiple, partly overlapping social groups that are connected via incredibly complex networks, and enforced through coercion and policing. We are a citizen of a neighborhood, city, state, and nation. Most individuals work for a company or other organization. We are members of political parties, book clubs, parent-teacher groups, sports teams, gyms, social media groups, and so forth. A typical modern day is organized around a fairly restrictive schedule of commuting to and from work at proscribed times set by the standard work day, chauffeuring kids to and from school, preparing (or ordering) meals at specified times of breakfast, lunch, and dinner, and going to bed often much later than our ancestors in small band societies typically did. Moreover, at the end of the work or school day, we may spend a little while in extracurricular activities with friends and acquaintances, but most of our remaining time is spent in our homes with our nuclear family, but not with our friends. This restrictive and regimented daily schedule, with so little time spent socializing with a large network of close friends, significantly injures our mental health and psychological well-being. Spending time among friends helps reduce stress and improves physiological health (Uchino, Cacioppo, & Kiecolt-Glaser, 1996). Restricted access to friends, especially on a daily basis, leaves us disconnected from the support group that would naturally surround us in band society. As a result, we suffer a greater degree of anxiety, which adversely affects our mental and even physical health. Social disconnection is associated with increased systemic inflammation, a major contributor to chronic disease (Steptoe, Owen, Kunz-Ebrecht, & Brydon, 2004) and cognitive decline (Cacioppo & Hawkley, 2009).

7. Parenting

We've already mentioned how the role parents play in a contemporary nuclear family in a middle-class, dual-income household contrasts drastically with parenting roles in traditional, ancestral communities. As parents, we must feed, bathe, and clothe our children, and provide them with safety and shelter. Moreover, we must wash their clothing and chauffeur them to and from school, extracurricular activities, and "play dates." And beyond this, we are their primary source of both entertainment and socialization. Thus, contemporary parenting is highly restrictive of the parents' freedom and is laborious and tedious.

In ancestral communities, by contrast, parenting roles can be quite different. While the mother is heavily invested in childcare during her child's infancy, other adults and older children provide plenty of alloparenting. Furthermore, the ability to carry and feed the infant wherever she goes means that the mother is not as restricted in movement or daily schedules as is a mother in Western society. There is no separation of a new mother from her peer group and social network. Fathers also play a strong, supporting role in ancestral communities, but are not burdened by the labors of maintaining a household. Thus, fathers, too, have similar freedom after the birth of a child as they enjoyed beforehand. Children rapidly gain independence in the first few years of life. In fact, by the time a child is 5 or 6 years old, he or she typically spends very little time around the nuclear family and instead spends most of his or her time with the other children in the band. In most contemporary hunter-gatherer groups, children from ages 5 to 18 typically form their own separate, somewhat independent camp adjacent to the adult's camp (Gray, 2015).

It is difficult to underestimate the stress level induced by the burdens of child rearing in modern society, and it is likely to become increasingly recognized as an important evolutionary mismatch that needs to be addressed. In fact, the rise in home schooling and even unschooling movements across the United States and other modern societies reflects this growing awareness and is an attempt to restore balance to both the child and parental sides of the equation.

8. Social Structure and Governance

Hunter-gatherer band society is largely democratic and egalitarian. Cognitive biases evolved to adapt individuals to band society (Haselton & Nettle, 2006). Human behavior is naturally both altruistic and greedy (cooperative and competitive), but in hunter-gatherer society there are social constraints, and especially severe consequences for antisocial behaviors (Boehm, 1999). Antisocial and selfish individuals can be ostracized or even ejected from a group. Cultural practices also evolved to prevent individuals from mistreating or abusing others (especially women and children), from hoarding resources, and from curtailing the freedom of other individuals. Such individuals would lose out on mating opportunities, would suffer a loss of reputation, group status, and reciprocity, and might ultimately be shunned and cast out from the group.

In modern society, anonymity, population size, and the economic systems of the modern world insulate individuals from the effects of their greed or other antisocial behaviors. Unconstrained greed is now actually glorified (large mansions, conspicuous consumption, purposeful waste, etc.). In the United States, the wealthiest 160,000 families own as much wealth as the poorest 145 million families (Saez & Zucman, 2016). Anonymity in such a large society allows for behaviors that normally would be kept in check in a small band society, such as mistreatment of women, child abuse, unequal distribution of resources, and so on.

While we retain built-in biases that adapted us to the ancestral world, such as mistaking a stick for a snake (Johnson, Blumstein, Fowler, & Haselton, 2013), there has not been sufficient evolutionary time to acquire biases for evolutionarily novel immediate threats, such as cars or drugs, or for long-term dangers, such as failure to plan for retirement or to make choices that benefit our environment and address climate change (Kahneman, 2011). What are the implications of the lack of biases for dealing with long-term problems? Their absence can lead to greater disparity between the diminishing wealthy and the growing poor, increased stagnation at the level of governments and organizations in addressing global issues such as overpopulation, food production, climate change, pollution, health care, and the burden of chronic disease, to name a few (Diggs, 2017). We need to apply evolutionary thinking to re-engineer our society, governance, policy, infrastructure, and resource management and allocation, and to realign our ancestral psychology with our modern technology and systems, in order to promote a healthy population and environment (Wilson, 2011).

9. Cognitive Grounding

Humankind has been defined by its technology, from the origin of *Homo habilis* or "handy man" and associated Oldowan tool assemblages, to the more sophisticated Acheulean stone tools and control of fire evidenced at *Homo erectus* archeological sites, to the progression of ever more advanced tools constructed by *Homo sapiens* from the stone age to the bronze age to the iron age; from the steam engine to the combustion engine to the airplane to the rocket ship; from the control of electricity to the microprocessor to the wireless Internet of a globally connected world.

This last technological revolution of a continuously connected Internet that links devices such as smart phones, computers, and tablets distributed across all corners of the globe has happened so swiftly (over the past 15 or so years), and transformed our daily lives so dramatically that it is surprising how little we have noticed these changes. While this recent advancement is the true birth of the information age, and all of the wonderful advantages it brings, there is also a dark side to this new technology. It is difficult to step out of one's door these days and not observe a large segment of the population engaged with their smartphones— while walking, sitting in a café or restaurant, or even while driving a car. Aside from potential accidents due to distracted walking or driving, the ubiquitous overuse of smartphones may dysregulate normal neurofunctional connectivity and end up lowering productivity (Gazzaley & Rosen, 2016). Even ten years ago, Internet overuse and cell phone overuse were being recognized as emerging behavioral addictions (Jenaro, Flores, Gómez-Vela, González-Gil, & Caballo, 2007). The compulsive usage of smartphones and related types of "technostress" (excessive use of technology) positively correlate with negative psychological effects, such as lower perceived locus of control, increased social interaction anxiety, exaggerated materialism, and an excessive need for touch (Lee, Chang, Lin, & Cheng, 2014).

The ramifications of this major technology-driven cultural shift are clear. If left unchecked, technostress-induced distraction, disconnection from the physical, emotional, tangible world, loss of boredom and daydreaming, and the dysregulation of sleep circadian rhythms, in combination can overwhelm our coping mechanisms and induce clinical levels of anxiety, stress, and depression, as well as contribute to systemic inflammation and dysregulation of the HPA axis, and catalyze the development of chronic illness.

At the same time, the incredible flexibility and rapid development and evolution of information technology could help us to find answers to these emerging problems. When combined with evolutionary thinking, technology can be shaped to optimize human performance and health. Despite the pessimistic view presented in this chapter, there is actually much to be optimistic about. Already, there is a growing wave of new apps, devices, and programs that harness technology for human betterment. Now one can wear a daily step counter, connect to a sleep tracker at night, and download a nutrition app on the smart phone, all to optimize these health variables. The open source, embed nature of the emerging sharing economy holds promise for support from social groups and community, and even a new ecology of choices of increasing efficiency and convenience distributed across the globe. Airbnb and Uber are two examples of information-technology-driven companies that have integrated themselves into the fabric of local and international travel. Academic organizations such as the Ancestral Health Society and the Evolution Institute, academic journals such as the *Journal of Evolution and Health* and *Evolution, Medicine, and Public Health*, and the plethora of personal and commercial websites devoted to topics grounded in evolutionary mismatch (often identified by the words "Ancestral," "Paleo," or "Primal") are all testaments to the growing use of information technology and social institutions to leverage health and wellness.

References

Abdulla, J., & Nielsen, J. R. (2009). Is the risk of atrial fibrillation higher in athletes than in the general population? A systematic review and meta-analysis. *Europace, 11,* 1156–1159. doi:10.1093/europace/eup197

Aiello, L. C., & Wheeler, P. (1995). The expensive-tissue hypothesis: The brain and the digestive system in human and primate evolution. *Current Anthropology, 36*, 199–221.

Andre, C., Dinel, A. L., Ferreira, G., Laye, S., & Castanon, N. (2014). Diet-induced obesity progressively alters cognition, anxiety-like behavior and lipopolysaccharide-induced depressive-like behavior: Focus on brain indoleamine 2,3-dioxygenase activation. *Brain, Behavior, and Immunity, 41*, 10–21. doi:10.1016/j.bbi.2014.03.012

Blaisdell, A. P. (2015). Play as the foundation of human intelligence: The illuminating role of human brain evolution and development and implications for education and child development. *Journal of Evolution and Health, 1*, 1–54.

Blaisdell, A. P., Biedermann, T., Sosa, E., Abuchaei, A., Neveen, Y., & Bradesi, S. (2017). An obesogenic refined low-fat diet disrupts attentional and behavioral control processes in a vigilance task in rats. *Behavioural Processes, 138*, 142–151.

Blaisdell, A. P., Pottenger, B., & Torday, J. S. (2013). From heart beats to health recipes: The role of fractal physiology in the Ancestral Health movement. *Journal of Evolution and Health, 1*, 1–17.

Boehm, C. (1999). The natural selection of altruistic traits. *Human Nature, 10*, 205–252. doi:10.1007/s12110–999–1003-z

Cacioppo, J. T., & Hawkley, L. C. (2009). Perceived social isolation and cognition. *Trends in Cognitive Sciences, 13*, 447–454. doi:10.1016/j.tics.2009.06.005

Centers for Disease Control and Prevention. (2013). Mental health surveillance among children—United States, 2005–2011. Retrieved from www.cdc.gov/mmwr/preview/mmwrhtml/su6202a1.htm

Chang, A.-M., Aeschbach, D., Duffy, J. F., & Czeisler, C. A. (2014). Evening use of light-emitting eReaders negatively affects sleep, circadian timing, and next-morning alertness. *Proceedings of the National Academy of Sciences, 112*, 1232–1237. 201418490. doi:10.1073/pnas.1418490112

Cohen, M. N., & Crane-Kramer, G. M. M. (2007). *Ancient health: Skeletal indicators of agricultural and economic intensification.* Gainesville, FL: University Press of Florida.

Cordain, L. (1999). Cereal grains: Humanity's double-edged sword. *World Review of Nutrition and Dietetics, 84*, 19–73.

Diggs, G. (2017). Evolutionary mismatch: Implications far beyond diet and exercise. *Journal of Evolution and Health, 2*, 3.

Duca, L., Da Ponte, A., Cozzi, M., Carbone, A., Pomati, M., Nava, I., ... Fiorelli, G. (2006). Changes in erythropoiesis, iron metabolism and oxidative stress after half-marathon. *Internal and Emergency Medicine, 1*, 30–34. doi:10.1007/BF02934717

Feelisch, M., Kolb-Bachofen, V., Liu, D., Lundberg, J. O., Revelo, L. P., Suschek, C. V., & Weller, R. B. (2010). Is sunlight good for our heart? *European Heart Journal, 31*, 1041–1045. doi:10.1093/eurheartj/ehq069

Gazzaley, A., & Rosen, L. D. (2016). *The distracted mind: Ancient brains in a high-tech world.* Cambridge, MA: MIT Press.

Gillen, J. B., Martin, B. J., MacInnis, M. J., Skelly, L. E., Tarnopolsky, M. A., & Gibala, M. J. (2016). Twelve weeks of sprint interval training improves indices of cardiometabolic health similar to traditional endurance training despite a five-fold lower exercise volume and time commitment. *PLoS ONE, 11*. doi:10.1371/journal.pone.0154075

Gillen, J. B., Percival, M. E., Skelly, L. E., Martin, B. J., Tan, R. B., Tarnopolsky, M. A., & Gibala, M. J. (2014). Three minutes of all-out intermittent exercise per week increases skeletal muscle oxidative capacity and improves cardiometabolic health. *PLoS ONE, 9.* doi:10.1371/journal.pone.0111489

Gray, P. (2009). Play as a foundation for hunter-gatherer social existence. *American Journal of Play, 1,* 476–522.

Gray, P. (2011). The decline of play and the rise of psychopathology in children and adolescents. *American Journal of Play, 3,* 443–463.

Gray, P. (2015). *Free to learn: Why unleashing the instinct to play will make our children happier, more self-reliant, and better students for life.* New York: Basic Books.

Haselton, M. G., & Nettle, D. (2006). The paranoid optimist: An integrative evolutionary model of cognitive biases. *Personality and Social Psychology Review, 10,* 47–66. doi:10.1207/s15327957pspr1001_3

Jenaro, C., Flores, N., Gómez-Vela, M., González-Gil, F., & Caballo, C. (2007). Problematic internet and cell-phone use: Psychological, behavioral, and health correlates. *Addiction Research and Theory, 15,* 309–320. doi:10.1080/16066350701350247

Johnson, D. D., Blumstein, D. T., Fowler, J. H., & Haselton, M. G. (2013). The evolution of error: Error management, cognitive constraints, and adaptive decision-making biases. *Trends in Ecology and Evolution, 28,* 474–481.

Jones, A. P. (1999). Indoor air quality and health. *Atmospheric Environment, 33,* 4535–4564. doi:10.1016/S1352–2310(99)00272–1

Kahneman, D. (2011). *Thinking, fast and slow.* New York: Farrar, Straus & Giroux.

Kemple, K. M., Oh, J., Kenney, E., & Smith-Bonahue, T. (2016). The power of outdoor play and play in natural environments. *Childhood Education, 92,* 446–454.

Klein, R. G. (2009). The human career: Human biological and cultural origins. Chicago: University of Chicago Press.

Kröger, K., Lehmann, N. Rappaport, L., Perrey, M. Sorokin, A., Budde, T., … Möhlenkamp, S. 2011). Carotid and peripheral atherosclerosis in male marathon runners. *Medicine and Science in Sports and Exercise, 43,* 1142–1147.

Kuipers, R. S., Joordens, J. C. A., & Muskiet, F. A. J. (2012). A multidisciplinary reconstruction of Palaeolithic nutrition that holds promise for the prevention and treatment of diseases of civilisation. *Nutrition Research Reviews, 25,* 96–129. doi:10.1017/S0954422412000017

Lee, Y. K., Chang, C. T., Lin, Y., & Cheng, Z. H. (2014). The dark side of smartphone usage: Psychological traits, compulsive behavior and technostress. *Computers in Human Behavior, 31,* 373–383. doi:10.1016/j.chb.2013.10.047

Lindeberg, S. (2010). *Food and western disease: Health and nutrition from an evolutionary perspective.* West Sussex, UK: Wiley-Blackwell.

Lucas, A. J., & Dyment, J. E., 2010. Where do children choose to play on the school ground? The influence of green design. *Education 3-13, 38,* 177–189. doi:10.1080/03004270903130812

Mantler, A., & Logan, A. C. (2015). Natural environments and mental health. *Advances in Integrative Medicine, 2,* 5–12. doi:10.1016/j.aimed.2015.03.002

Mead, M. N. (2008). Benefits of sunlight: A bright spot for human health. *Environmental Health Perspectives, 116,* A160–A167. doi:10.1289/ehp.116-a160

Mozaffarian, D., Furberg, C. D., Psaty, B. M., & Siscovick, D. (2008). Physical activity and incidence of atrial fibrillation in older adults: The cardiovascular health study. *Circulation, 118,* 800–807. doi:10.1161/CIRCULATIONAHA.108.785626

Odling-Smee, J. (2009). Niche construction in evolution, ecosystems and developmental biology. In A. Barberousse, M. Morange, & T. Pradeu (Eds.), *Mapping the future of biology*. Dordrecht, Germany: Springer. doi:10.1007/978–1–4020–9636–5

O'Keefe, J. H., Vogel, R., Lavie, C. J., & Cordain, L. (2011). Exercise like a hunter-gatherer: A prescription for organic physical fitness. *Progress in Cardiovascular Diseases, 53*, 471–479. doi:10.1016/j.pcad.2011.03.009

Pearson, D. G., & Craig, T. (2014). The great outdoors: Exploring the mental health benefits of natural environments. *Frontiers in Psychology, 5*, 1178. doi:10.3389/fpsyg.2014.01178

Price, W. A. (1939). *Nutrition and physical degeneration*. New York: Paul B. Hoeber.

Roenneberg, T., & Merrow, M. (2016). The circadian clock and human health. *Current Biology, 26*, R432–R443. doi:10.1016/j.cub.2016.04.011

Rosell, F., Bozsér, O., Collen, P., & Parker, H. (2005). Ecological impact of beavers castor fiber and castor canadensis and their ability to modify ecosystems. *Mammal Review, 35*, 248–276. doi:10.1111/j.1365–2907.2005.00067.x

Saez, E., & Zucman, G. (2016). Wealth inequality in the United States since 1913: Evidence from capitalized income tax data. *Quarterly Journal of Economics, 131*, 519–578. doi:10.1093/qje/qjw004

Skoluda, N., Dettenborn, L., Stalder, T., & Kirschbaum, C. (2012). Elevated hair cortisol concentrations in endurance athletes. *Psychoneuroendocrinology 37*, 611–617. doi:10.1016/j.psyneuen.2011.09.001

Sober, E., & Wilson, D. S. (1999). Unto others: The evolution and psychology of unselfish behavior. Cambridge, MA: Harvard University Press.

Steptoe, A., Owen, N., Kunz-Ebrecht, S. R., & Brydon, L. (2004). Loneliness and neuro-endocrine, cardiovascular, and inflammatory stress responses in middle-aged men and women. *Psychoneuroendocrinology 29*, 593–611. doi:10.1016/S0306–4530(03)00086–6

Stutz, A. J. (2014). Embodied niche construction in the hominin lineage: Semiotic structure and sustained attention in human embodied cognition. *Frontiers in Psychology, 5*, 1–19. doi:10.3389/fpsyg.2014.00834

Uchino, B. N., Cacioppo, J. T., & Kiecolt-Glaser, J. K. (1996). The relationship between social support and physiological processes: A review with emphasis on underlying mechanisms and implications for health. *Psychological Bulletin, 119*, 488–531. doi:10.1037/0033–2909.119.3.488

van der Horst, G. T., Muijtjens, M., Kobayashi, K., Takano, R., Kanno, S., Takao, M., … Yasui, A. (1999). Mammalian Cry1 and Cry2 are essential for maintenance of circadian rhythms. *Nature, 398*, 627–30. doi:10.1038/19323

Van Someren, E. J. W. (2000). More than a marker: Interaction between the circadian regulation of temperature and sleep, age-related changes, and treatment possibilities. *Chronobiology International, 17*, 313–354. doi:10.1081/CBI-100101050

Wilson, D. S. (2011). The neighborhood project: Using evolution to improve my city, one block at a time. New York: Little, Brown.

Woolley, H., & Lowe, A. (2012). Exploring the relationship between design approach and play value of outdoor play spaces. *Landscape Research, 38*, 53–74. doi:10.1080/01426397.2011.640432

Wrangham, R. (2009). Catching fire: How cooking made us human. New York: Basic Books.

CHAPTER 13

Living Well as the First Medicine: Health and Wellness in the Modern World

Kelly G. Wilson
University of Mississippi

During the winter of 2009, at a training for clinical psychologists in Eugene, Oregon, David Sloan Wilson asked me if my understanding of psychology was informed by evolution science. That question, when it really sunk in, completely upended my career and my life. David's question was kin to Dobzhanski's assertion that nothing makes sense in biology except in light of evolution. I currently believe that nothing makes sense in psychology except in light of evolution. And what began as a project that winter to sort the evolutionary roots of modern mental health epidemics led to what I currently see as the roots of the entirety of modern healthcare epidemics.

Hundreds of articles, chapters, and books later, evolution science has combined with my own knowledge of contextual behavioral science and altered the way I treat each client I see and each therapist I train. And it has fundamentally changed how I treat myself. I have come to know what sort of creatures we are, what we most deeply need, and what are our greatest strengths and vulnerabilities. Humans have conquered most of what killed us on the savannah and have become wealthier and more successful as result of our industry and efficiency. But our science of kindness has lagged. It is time we fixed that.

Human Health and Well-Being: I Have Good News and Bad News

The Good News

Two hundred years ago, the human lifespan was half of what it is today. We died from infections and injuries and complications in birth that would never kill in the modern world. Over the past 200 years, global infant mortality has dropped from 43 to 4.3 percent and maternal mortality from 1000 per 100,000 live births to fewer than 10—orders of magnitude in change! Even in developing sub-Saharan Africa, infant and maternal mortality is roughly one-half—and lifespan is double—what it was in the wealthiest countries in Europe just 200 years ago (Roser, 2017).

The Bad News

As these historic killers have receded, they have left in their wake new and entirely modern sources of death and disability. With many new forms of illness, we are better at preventing death than we are at preventing disability. Years Lived with Disability (YLD) is a measure from public health of impairment that results from illness and injury. Although mortality is falling, YLDs are growing and becoming the dominant source of the global burden of disease. Some top sources of disability—low-back and neck pain, major depression, dysthymia, obesity, bipolar disorder, anxiety, schizophrenia, and substance-use disorders—have seen 40-50 percent increases in YLDs between 1990 and 2013. YLDs for diabetes increased 136 percent over that same twenty-three-year period (Global Burden of Disease Collaborators, 2015).

These increases are not unrelated. For example, an individual with depression has a 58-percent increased risk of becoming obese, while a person who is obese has a 53-percent increased risk of becoming depressed (Gatineau & Dent, 2011). Evidence overlap between mental and physical illness is also visible in healthcare trends. Medco is a prescription-management company that serves approximately 65 million people in the United States. Medco's sales region, comprising of Kentucky, Tennessee, Mississippi, and Alabama, is known as "the diabetes belt" because it has the highest level of diabetes prescription treatment in the United States. This same region also has the highest use of psychiatric medications, with more than 23 percent of covered individuals taking at least one psychiatric medication (Medco Health Solutions, Inc., 2011).

My reading of the literature on modern disease burden has convinced me that rising rates of the most prevalent forms of both physical and psychological problems are two faces of the same epidemic. The epidemic is occurring as a result of the intersection of the behavior of human beings in and with contexts that are enormously different from the contexts in which humans (and these patterns of behavior) evolved. For millennia, humans died from what the world did to us. It injured us and infected us with parasites and microorganisms. It assaulted us with inclement weather, insufficient food, and predators that were happy to make lunch out of us. In many regards, we evolved in a world that was too hard. Humans have conquered an astonishing array of these hardships. But the conquests have come at a cost.

The Risk Factors for Modern Illness

The risk factors for modern illnesses are twofold. The first is exposure to our engineered world itself. Our solutions to the world's hardships have had unintended consequences. Engineered foods that are conveniently fast to prepare are simultaneously more palatable and less nutritious than unprocessed, whole foods. Urban planning has given us greater living space than ever while it has removed incentives for walking and bicycling as a part of daily life. Fossil fuels and nuclear energy are behind the miracle of electrification, but they have created significant environmental problems. That said, many modern-day risks can be moderated through behavior change.

But this brings us to our second risk factor: modern world behavior change has been made much, much harder by easily accessible but unhealthy reinforcers. Reinforcers are things that are themselves desirable (a cookie), or things that make undesirable things go away (a Tylenol tablet). We will work for a cookie, because cookies taste good. We will work for a Tylenol because it makes our headache go away—which is good. The *matching law,* developed by behavior analyst Richard Herrnstein (1961), describes the lawful relationship between the ways organisms distribute behavior and the availability of reinforcement. Organisms distribute behavior in a way that "matches" the availability of reinforcement. Sometimes a single response option produces a steady stream of reinforcement. But in a more complex world, with many response options, organisms vary their behavior over time to produce maximum reinforcement using minimal energy. The most sophisticated iterations of the matching law show a remarkable ability to predict response patterns across many different species and forms of behavior, and across a wide variety of reinforcers.

The matching law makes the modern world a very treacherous place. The savannah favored our ancestors who chose the easiest path up the hill, the ones who preferred the energy-dense foods, and the ones who moved away from pain or threat of pain. We have engineered our way out of most of the problems that killed us throughout our evolutionary history, but the tendency to park in the nearest space, choose the donut, and take a pill at the slightest pain has remained.

Psychological Flexibility and Lifestyle Change

Understanding the matching law and its interface with a modern world makes it clear why change is hard. We have not birthed a generation of lazy, overindulgent humans. We have engineered a world filled with convenience and calories, a world in which convenience is marketed as intrinsically good. But these reinforcers are not static things, they are reinforcers in context. And we possess a science of behavioral context that can help us to let go of the donut and move toward healthier, richer lives.

Acceptance and commitment therapy (ACT) is a contextual behavioral model that has been tested broadly on both psychological and physical illnesses, including depression, anxiety, psychosis, chronic pain, obesity, diabetes, sedentary behavior, substance abuse and dependence, workplace stress, and smoking cessation, among others. The treatment goal of ACT is to increase *psychological flexibility*. Psychological flexibility is the ability to be aware and engaged in the present moment as a conscious human being and to persist in or change behavior based on one's most deeply held values. In the service of growing this capacity, we teach six core practices (Hayes, Strosahl, & Wilson, 2011).

Practice acceptance. In ACT, teaching acceptance is a primary activity. The counter to acceptance is avoidance. If there is a single well-founded fact in empirical clinical psychology, it is that avoidance is the kryptonite of good psychological functioning. Changes in acceptance have been shown to mediate critical lifestyle change. For example, Bricker and colleagues more than doubled the quit rates in an online ACT smoking-cessation program compared to the National Cancer Institute's online program. Fully 80 percent of the variance in quit rates could be accounted for with knowledge of change in overall acceptance of difficult physical sensations, cognitions, and emotions (Bricker, Wyszynski, Comstock, & Heffner, 2013). Acceptance can be understood as a skill, and like any skill, it can be actively practiced. Sustainable lifestyle change is supported by well-practiced acceptance. Letting go of enticing immediate reinforcers often comes with significant discomfort, which can either be accepted or avoided.

Practice defusion. Humans do not just live in the world, they live in a storied-up version of the world. When we talk to people about lifestyle change, among reactions are stories about lack of time, lack of persistence, or whether people are too tired, too late, too sick, too weak, too stupid, or too lazy to change. Or maybe there are stories about "I don't like it" or "I don't want to" or "I just don't feel like it."

These thoughts are only the enemy if we put them in charge of our lives. But they need not be the enemy. They are simply our evolutionary history, echoing up and urging us to burn less energy and consume more—a very safe strategy up until very recently. We can act without the permission of these troubling thoughts and without making them go away.

There is a growing body of evidence that suggests that attempts to suppress unwanted thoughts will backfire. As the Buddhists say: that which you resist persists. In ACT, we teach people to use a variety of exercises to grow the capacity to notice a thought, as a thought, without refuting it or acting upon it. Just noticing is a skill that has been taught with the most frightening and compelling of difficult thoughts from addiction to psychosis to chronic pain.

Change in behavior is difficult. A donut is easy. We live in a world built to entice. We can learn to notice the rise and fall of cognitive, emotional, and behavioral resistance when we turn away from the donut and move toward the gym. We can practice noticing and opening up to momentary discomfort, and in doing so, to the possibility of change.

Practice present-moment awareness. The modern world is full of distractions. For every moment we save with our efficiencies, we seem to have two new things to fill that moment. There truly is no peace in the modern world except the peace we create. The evidence for the benefits of cultivating mindful awareness is rapidly expanding and shows impacts from the behavioral to the molecular level. Meta-analyses show substantial benefits to the many common clinical syndromes, especially anxiety, depression, and stress (Khoury et al., 2013). Mindfulness-meditation practices influence genetic expression associated with energy metabolism, mitochondrial function, insulin secretion, telomere maintenance, pro-inflammatory processes, and other stress response systems. These changes are present even in novice mindfulness practitioners and greatly enhanced among experienced meditators (Bhasin et al., 2013).

Practice self-as-process. Some of the stories we inhabit are stories about ourselves, including stories about what is possible for us. In the world of lifestyle change, we can become wedded to our habitual ways of living as a sort of identity. Here, we teach flexible and fluid perspective taking. For example, I have dug through some old photos of myself from when I was small—perhaps three or four years old. I have spent some time looking into the face of that small boy and then into the mirror, to see if I can get a glimpse of him in my own eyes. I notice how much easier it is to neglect or postpone my own care than it would be to postpone care for the little boy in that old photo. I wonder: when did it become okay to neglect me? When I go for a run, sometimes I imagine that little boy with me. What might he think of this old man taking him for a run? Noticing the shifts in your thoughts about yourself, over time, can help generate a sense of self as ongoing process, apart from the predominant content of the day.

Practice cultivating connection to values. It is easy to get lost in the troubles of the day and to lose track of what is most important. But mindful and deliberate connection to what is most valued in our lives is the best protection against the hazards of the modern world. Our evolved default mode is to distribute behavior to maximize calories and minimize energy output, to move toward the reinforcers and away from pain. But a close connection with values can circumvent and even change the function of immediate reinforcers.

For example, we know that humans are deeply social mammals. Social mammals thrive, in part, by doing what the ones around them do, eating what they eat, moving as they move. My daughters are important to me, and I know that the likeliness of them eating healthy foods, getting enough high-quality sleep, and getting regular exercise is influenced by the people around them. I want to be a part of that influence toward a healthy life. As a father, I know that remonstrating them for unhealthy activity has little impact. In fact, our children and others around us rarely want to be told what to do. They most often tune out uninvited advice. But the slow, steady presence of healthy behavior around us can create change in us. We can actively choose what sort of influence we will be in this world. The fruits are slow and small, yet they are present in each act of self-care. When I run or eat a healthy meal or ensure a quiet house after ten o'clock at night, I am not just practicing self-care. I am also being a husband and a father. When I talk about these things publicly, in my classes, among my students and friends—about sustaining change, and about the meaning and purpose of that change—I extend that influence into the world around me. Typically, lifestyle change is linked to weight loss or a relatively generic version of "better health," but we can add to these reinforcers by making clear connections between self-care and deeply held values.

Practice committed action. The final and critical element of ACT practice is the construction of small specific acts which bring a person into alignment with a valued pattern of living. Commitment in ACT is not a promise about future action. Committed action is a moment-by-moment practice. Any pattern of deliberate lifestyle change in a world persistently tempting us with the easy path to donut and disability will include lapses in those patterns. In these moments, committed action is about taking the very next step, however small, that returns us to alignment with self-care and, also, importantly, with the values to which that care is connected.

Dying from Our Solutions: Some Places to Begin to Practice

#DetoxYourLife. The world used to be filled with bugs and weeds and germs. We have created a host of toxins to combat them. But many of the toxins we use to kill weeds and pests and murderous microbes have turned against us. Endocrine disruptors like BPA are ubiquitous, and bacteria are becoming increasingly resistant to our antibiotics. Many of our other medicines cure our momentary ills, but have disabling long-term effects. We need to minimize patterns of exposure to toxins, including obvious toxins like tobacco, excessive alcohol, and other drug use, but also the casual and habitual use of medications. Many modern remedies, used chronically, will themselves produce illness. Headaches resulting from medication overuse have increased 130 percent in recent years and risen to become the 18th leading cause of Years Lost to Disability (Global Burden of Disease, 2015). Fifteen percent of US adults and 40 percent of those over sixty-five are taking five or more medications each day (Kantor, Rehm, Haas, Chan, & Giovannucci, 2015). There are simply no clinical trials on the long-term impact of these medication cocktails, especially on people with extremely compromised health. You cannot eliminate all toxins, but anyone can make small steps to reduce their overall toxic load. Improvements in lifestyle can reduce and sometimes eliminate the need for medication. Look for patterns of toxins in your own life and disrupt them.

#MoveYourBody. On the savannah, providing ourselves with basics like food, water, and shelter was extremely labor intensive. We have made a world where little movement is needed, including both the duration and variability of movement. Most people reading these words make a living sitting. Sedentary behavior predicts back pain, metabolic illness, and depression, among many other modern forms of illness. Movement is good medicine for virtually every modern illness. If you want a sense of what we were evolved to do, take some six-year-olds to a well-appointed playground and watch them sprint and tumble and twirl and hang and balance and roll in the grass. Watch the range of motion expressed in their spine and arms and shoulders and knees and ankles and hips. Now follow yourself through a day and just count the basic movement patterns and how much time you devote to each. Notice the limits in your own range of motion. As we age, and sometimes quite early on, we lose that range of motion. Movement is increasingly recognized as good medicine. The Cancer Counsel of Western Australia in Perth has patients doing vigorous exercise immediately before and after chemotherapies to prevent muscle and bone loss. Exercise alone is an effective treatment for depression and analysis for publication bias suggests that its benefit has been underestimated (Schuch et al., 2016). Metabolic illness resulting from aging can be reversed or slowed with physical activity (Mora, 2013). Long-term studies of vigorous exercise like running routinely show 45-percent reductions in cardiac mortality and 30-percent reductions in death from all causes. The size of the risk reduction is even higher among some chronically ill subgroups (Lee et al., 2014).

#GetSomeSleep. In our evolutionary history, our sleep was set by the rise and fall of light and temperature. Today, both are adjustable features of any modern home. Lights and screens are everywhere. Unlike fifty years ago, broadcast television runs on a 24-hour cycle. Everybody reading these words likely carries a device in their pocket that can connect them with virtually any form of entertainment at any moment of any day. This nonstop world is tough on sleep. Nearly 25 percent of the adult US population reports insomnia complaints (Irwin, 2015). We are just learning about the purpose of sleep, but it appears that sleep has multiple vital functions. Even hibernating animals come out of hibernation to sleep—it is that essential. During deep sleep, the newly discovered glymphatic system dramatically increases the flow of cerebrospinal fluid, greatly accelerating passive clearance of cellular waste. Clearance of beta amyloids implicated in Alzheimers, as well as other proteinopathies, may yield to a better understanding of these waste-clearance processes (Warren & Clark, 2017). Sleep is a highly modifiable risk factor and is linked to a wide array of modern-world morbidity and mortality, including obesity, diabetes, cardiovascular disease, high blood pressure, kidney disease, depression, anxiety, attention deficit disorder, and alcohol use disorders, among many others (Institute of Medicine, 2006). One night of terrible sleep and you can see for yourself the profound impact sleep has on mood, attention, motivation, judgment, problem solving, and interpersonal effectiveness. Though sleep dysregulation may not play a directly causal role in many diseases, it is a clear stressor to physical and psychological systems. When you do not sleep, wastes do not clear and you carry a toxic debt into the next day. What is the impact of that toxic load on the developing brains of children or the vulnerable brains of the elderly? We simply do not know. But for my own part, I am willing to take the risks of getting regular, adequate sleep opportunity while the science is being sorted.

#EatRealFood. Nearly three of every five calories in the US diet consist of ultra-processed foods loaded with ingredients with no nutritional value (Steele et al., 2016). These products are engineered for overconsumption with optimized mouth-feel, sweetness, and saltiness. Among the outcomes of this onslaught of manufactured foods is a dramatic increase in added sugar. Consumption of added sugar rose from about four pounds per year in colonial America to 120 pounds per person per year in 1994 (Bray & Popkin, 2014). Nearly 90 percent of the added sugar in our diet comes from these ultraprocessed foods (Steele et al., 2016). Likewise, high omega-6 seed oils are in virtually every ultraprocessed food. Omega-6 to omega-3 oils occur at a ratio of 15:1 in a contemporary diet after millions of years of a 1:1 ratio. These omega-6 oils, while essential, in high doses are known drivers of inflammatory processes seen in obesity, diabetes, and many psychiatric disorders (Simopoulos, 2006). Nothing in our evolutionary history could prepare us for this onslaught of energy-rich, nutritionally empty "food." Although there is little agreement about the "ideal" diet, Paleo, low-fat, raw, vegan, and so on, it is safe to say no one—no one at all—is advocating a diet with most calories coming from ultraprocessed food. Michael Pollan seems moderate and sensible: Eat real food, not too much, mostly plants.

#SocialNetworksMatter. Humans spent most of our evolutionary history living in small groups where we knew everyone and everyone knew us. We slept together, ate together, hunted and foraged together. We mourned and danced and sang together. And, yes, we fought. Certainly between groups, savagely at times, but in more limited ways within. Mostly, however, humans won on the savannah by cooperation. We do not have the teeth, claws, muscles, speed, or stealth that many of our competitors possessed. But if you put a couple of us together, we can track, trap, trick, herd, devour, or put to service any of them (for good and ill). We needed one another on the savannah for survival. Noted experimental psychologist Harry Harlow said it best: A lone monkey is a dead monkey. But our problem-solving skill has created an unfortunate side effect: It is possible to survive in the modern world in complete isolation. However, just because something is possible does not make it healthy. When I was a child, I slept in the same bed as my little brother David until I was twelve years old—skin against skin, right up next to one another. I shared touch, sight, smell, and social connection with him each and every day. It is worth pausing to recognize that someone reading these words right now will go through this day, perhaps this week, perhaps longer, without ever knowing the kind touch of another human being. More people live alone today than at any time in history. If you are among the fortunate, consider putting the book down for a second and offering a touch in kindness to someone nearby. We are made to be together.

Social connection affects us all the way down to the molecular level. Social isolation and social hostility upregulate genes coding for stress hormones and pro-inflammatory processes, and downregulate insulin sensitivity and viral immunity, among other effects. Over a short timespan, these challenges are fine and likely even build the robustness and responsiveness of the underlying systems. But chronic exposure to hostility and/or isolation is toxic to social mammals. And humans are deeply social mammals. Besides metabolic havoc, both social isolation and social hostility can be used to model depression and anxiety with laboratory animals. In epidemiological studies, both predict virtually every modern illness that besets us. Even though our immediate health does not depend on social connection, our long-term health most certainly does.

Taking the smallest bit of time each day to maintain social connections and to deepen them is an investment in both your well-being and that of the people with whom you connect. Work by Harvard sociologist and physician Nicholas Christakis suggests that all manner of things, both good and ill, propagate through social networks. If you take care of yourself, and do so in a public way and are connected with others, you will make a contribution to the health of your social network. As these small influences propagate around you, your network will also help you to persist in your own self-care. The influence is reciprocal. You cannot care for yourself or neglect yourself without sharing this influence.

Living Well in the World We Made: Making Practice Humane

It is entirely possible to live well, richly, with purpose and vitality in the world we have made—perhaps more so than in any time in human history. But handing people a list of evolutionary mismatches about diet, exercise, and sleep will most likely add to an already too long to-do list. If we understand behavioral principles and what kind of creature we are, we can do better.

#SmallThingsMatter. The scientific literature is rife with examples of small things precipitating important change. This is true in physical health: a little bit of folate can prevent spinal defects in babies if given to pregnant moms. Recent studies of high-impact interval training, involving only seconds of high-effort exercise interspersed with a few minutes of low-effort movement, can produce impressive benefits, even for individuals with severely compromised health (Gibala, Little, MacDonald, & Hawley, 2012). The same is true in psychological treatments. In behavioral activation, a first-line treatment for depression, it is not the magnitude but rather the direction of activity that matters. Moving just a tiny amount can allow a person to be caught back up in the stream of life. Likewise, the best exposure-based treatment for anxiety often starts with very small anxiety-provoking encounters. When people contemplate lifestyle change, they often try to take on too much. Starvation diets and overlong, over-strong exercise programs are common examples.

#EverythingInteracts. Interactions are ubiquitous, from the molecular level to the level of individual behavior change. While it may not be true in a literal sense that everything interacts, it is true enough in lifestyle illness. There is simply nothing you can do that optimizes sleep that will not affect dozens of metabolic processes that impact appetite hormones, glucose management, stress hormones and their cleanup, emotion regulation, attention, and on. But the same could be said of improved social interactions, like reducing one's exposure to social hostility or social isolation. The answer to the question of where to intervene on lifestyle is: everywhere. But practically speaking: anywhere. Because of the deep interactivity of these systems, anywhere you help yourself is likely to have important downstream effects on other areas. And remember, small things matter.

#TurninginKindness. When encouraging lifestyle change, it is worth considering whether you would rather have a to-do list or a gift offered in kindness. Gifts offered in kindness do not need to be big or elaborate or expensive. The child's gift made with glue and colored paper and popsicle sticks, made with love, and offered in kindness, is sufficient. It is more than sufficient. Certainly better than any to-do list. Here is a simple ACT perspective-taking strategy: Imagine the face of someone you love with all your heart. Pause. See if you can recall a time when that person caught you looking at him or her. Pause. Look into that face. Spend just a few moments looking into those eyes. And then imagine a harder thing, that you were someone you loved like that, with that open heart. If you could offer, in kindness, some small healthy change in lifestyle, a walk, a wholesome meal, a movie with an old friend, what kindness would you offer to you? Positive reinforcement beats coercion. Kindness beats a to-do list. Before you eat the next meal, before you decide whether or not you will go to the gym, before you watch one more episode of *Law and Order*, consider what the kind thing would be, pause, and turn in kindness toward your own life.

#WeAreNotThatKindofMonkey. "I should be able to do this on my own!" is one of the singularly false and unhelpful stories my clients bring to therapy. We are simply not that kind of monkey. I am convinced of David Sloan Wilson's contention that the crown jewel in human evolution is human cooperation. More so than any other primate, humans cooperate across time and space, and in recent millennia across an increasingly interconnected globe. Acting in cooperation, humans eliminated death from smallpox in an all-out international effort between 1967 and 1977. No individual could have done it, not even an individual nation.

We can do things together that we could not do by ourselves. Lifestyle change at the individual level is no exception. Find a friend to walk with, or a stranger. Talk about your dietary change and trade recipes with people you know. Tell people when you go to yoga and ask them if they want to come along. Go a little early. Stretch and chat with the others who come early. Most people you speak to won't join you in this endeavor, but a few will, and a few of their friends, too. We are made to be together.

#PatternsofPractice. Animal models of illness typically involve patterns of exposure to stressors: social, chemical, and physical insults. I like to urge people to find patterns in their own life and to disrupt toxic patterns. This can involve very, very simple things. At work, disrupt sedentary time by taking a two-minute walk or by climbing a single flight of stairs. Cultivate healthy patterns of practice. Again, these can be very small practices. Eat dinner from salad plates instead of dinner plates. Evidence is clear that this small change will reduce your caloric intake. If you know dinner will be unhealthy, eat an apple before dinner. It will spoil your appetite just like mom said! One of the things I love about both yoga and mindfulness meditation is the concept of practice. No matter how many years people practice, they never arrive. All the way to practice is practice.

#FindYourPractice. One of the hazards of the modern world is the ready availability of negative social comparison. Competition is baked into the culture as a virtue. Competition is fine, but in some situations, it is unhelpful. If your lifestyle is in deep need of a tune-up, the last thing you need is to compare your speed on a run to someone who has run regularly for years. Virtually every fitness app presses people to set goals. When I was school age, I was incredibly fit. I grew up in a place where we spent a lot of time in the woods walking, running, climbing trees, swimming and wading in the rivers of western Washington. I had no fitness goal. I encourage people to use an image of a child at play as a model. I like to trail run. When people tell me that they hate running, that after thirty seconds they feel like they are going to die, I ask them why are they running for so long! Next time you are out walking somewhere, run for five seconds, just five, and then stop and walk again. Do this just once each day for a week. What will happen is that you will notice an urge to run for ten seconds … and so it begins. People often run too fast, too far, too soon, and hurt themselves. I love a good trail run, but it is not necessary. The data on running even just a short distance, even slowly, even infrequently, is astonishingly good. Find your own practice. The practice that suits you at this very moment. Let go of comparison. Make a start, no matter how small.

#PracticeFalling. When I first began my own yoga practice, my teacher, Stevi Self, would say, "What if falling were part of the pose?" This was some comfort, since falling was very often part of my yoga poses. After eight years of near daily practice, I still almost inevitably fall in tree pose. You do not need to be perfect. It is the pattern of practice that matters. Acute insults are rarely good models for illness. We are incredibly robust for acute insults. One night of lost sleep. One fast-food meal. One day sedentary. One stumble in tree pose. None of these create illness. When you fall from your practice, as in a breathing meditation, your job it to simply return to practice. The only people who do not fall are those who do not practice. Practice falling with grace and kindness.

Don't Drive a Car Down a River; Don't Take a Boat on the Freeway

Evolution is a profound process, and it sometimes happens very "fast." But even the "fast" evolutionary time scale dwarfs our human lifespan. Giraffes didn't get long necks in a weekend. At some point in our species' history, something very strange happened: we started talking to one another. It took some time to get from grunts and hoots to Twitter, but relatively "quickly," the ultimate adaptation of language made us the uncontested apex predators in the earth's ecosystem. With this incredible advantage, human cooperation ramped up to astonishing levels. We've solved a staggering number of environmental problems that complicated and shorted our lives, with decidedly good, but sometimes mixed results.

Yet even with smartphones, transcontinental air travel, and the microprocessor, we can't do all that much to accelerate the adaptations of our physical bodies to the environment we've created for ourselves. The staggering pace of our technological and economic progress in the last five hundred years has created a world to which we are simply not yet adapted.

If you drive your car into the river, you're not going to get where you want to go, and you're going to get wet. Take a boat onto the freeway, and you'll roll to one side and stay right where you started. Your vehicles won't be suited to their surroundings, and that can only lead you to problems. One day, we may well evolve to thrive on a diet of canola oil and regular Coke. We may come to thrive living alone, never straying from our apartments, sleeping four hours a night. But today is not that day.

As we've seen, though, the harmful behaviors we've taken to lately *can* be changed or eliminated. It may not be easy or convenient, but it *is* possible, and we know *how* to do it. What's left is a matter of intention, persistence, and the dissemination of the insights that emerge from labs around the world every day.

One thing seems clear: the narrative that chalks up every human problem to genes and brains appears to not fit the world all that well. Genes and brains aren't broken as much as they are not well matched to the world we have engineered. Another way to express this is that the environment we have created is toxic to genes and brains. We are not broken human beings. We're human beings in need of different ways of living that are more congenial to our evolutionary nature. We will not prescribe our way out of these troubles. I propose a science of healing, not a science that makes living with social and environmental toxins tolerable. Living well should be our first medicine.

References

Bhasin, M. K., Dusek J. A., Chang B., Joseph, M. G., Denninger J. W., Fricchione, G. L., … Libermann, T. A. (2013). Relaxation response induces temporal transcriptome changes in energy metabolism, insulin secretion and inflammatory pathways. *PLoS ONE, 8.* doi:10.1371/journal.pone.0062817

Bray, G. A., & Popkin, B. M. (2014). Dietary sugar and body weight: Have we reached a crisis in the epidemic of obesity and diabetes? *Diabetes Care, 37,* 950–956. doi:10.2337/dc13–2085

Bricker, J., Wyszynski, C., Comstock, B., & Heffner, J. L. (2013). Pilot randomized controlled trial of web-based acceptance and commitment therapy for smoking cessation. *Nicotine and Tobacco Research, 15,* 1756–1764. doi:10.1093/ntr/ntt056

Gatineau, M., & Dent, M. (2011). *Obesity and mental health.* Oxford: National Obesity Observatory.

Gibala, M. J., Little, J. P., MacDonald, M. J., & Hawley, J. A. (2012). Physiological adaptations to low-volume, high-intensity interval training in health and disease. *Journal of Physiology, 590,* 1077–1084. doi:10.1113/jphysiol.2011.224725

Global Burden of Disease Collaborators (2015). Global, regional, and national incidence, prevalence, and years lived with disability for 301 acute and chronic diseases and injuries in 188 countries, 1990–2013: A systematic analysis for the Global Burden of Disease Study 2013. *Lancet, 386,* 743–800. doi:10.1016/ S0140–6736(15)60692–4

Hayes, S. C., Strosahl, K., & Wilson, K. G. (2011). *Acceptance and commitment therapy: The process and practice of mindful change* (2nd ed.). New York: Guilford.

Herrnstein R. J. (1961). Relative and absolute strength of response as a function of frequency of reinforcement. *Journal of the Experimental Analysis of Behavior, 4,* 267–272.

Institute of Medicine (2006). *Sleep disorders and sleep deprivation: An unmet public health problem.* Washington, DC: National Academies Press. doi:10.17226/11617

Irwin, M. R. (2015). Why sleep is important for health: A psychoneuroimmunology perspective. *Annual Review of Psychology, 66,* 143–172. doi:10.1146/annurev-psych-010213–115205

Kantor, E. D., Rehm, C. D., Haas, J. S., Chan, A. T., & Giovannucci, E. L. (2015). Trends in prescription drug use among adults in the United States from 1999–2012. *Journal of the American Medical Association, 314,* 1818–1831. doi:10.1001/jama.2015.13766

Khoury, B., Lecomte, T., Fortin, G., Masse, M., Therien, P., Bouchard, V., … Hofmann, S. (2013). Mindfulness-based therapy: A comprehensive meta-analysis. *Clinical Psychology Review, 33*, 763–771. doi:10.1016/j.cpr.2013.05.005

Lee, D., Pate, R. R., Lavie, C. J., Sui, X., Church, T. S., & Blair, S. N. (2014). Leisure-time running reduces all-cause and cardiovascular mortality risk. *Journal of the American College of Cardiology, 64*, 472–481. doi:10.1016/j.jacc.2014.04.058

Medco Health Solutions, Inc. (2011). America's state of mind. http://apps.who.int/medicined-ocs/en/d/Js19032en/

Mora, F. (2013). Successful brain aging: Plasticity, environmental enrichment, and lifestyle. *Dialogues in Clinical Neuroscience, 15*, 45–52. PMCID:PMC3622468

Roser, M. (2017). Our world in data. Retrieved May 24, 2017 from www.ourworldindata.org

Schuch, F. B., Vancampfort, D., Richards, J., Rosenbaum, S., Ward, P. B., Stubbs, B. (2016). Exercise as a treatment for depression: A meta-analysis adjusting for publication bias. *Journal of Psychiatric Research, 77*, 42–51. doi:10.1016/j.jpsychires.2016.02.023

Simopoulos, A. P. (2006). Evolutionary aspects of diet, the omega-6/omega-3 ratio and genetic variation: Nutritional implications for chronic diseases. *Biomedicine and Pharmacotherapy, 60*, 502–507. doi:10.1016/j.biopha.2006.07.080

Steele, E. M., Baraldi, L. G., Louzada, M., Moubarac, J., Mozaffarian, D., Monteiro, C. A. (2016). Ultra-processed foods and added sugars in the US diet: Evidence from a nationally representative cross-sectional study. *British Medical Journal Open, 6*, e009892. doi:10.1136/bmjopen-2015–009892

Warren, J. D., & Clark, C. N. (2017). A new hypnic paradigm of neurodegenerative proteinopathies. *Sleep Medicine, 32*, 282–283. doi:10.1016/j.sleep.2016.12.006

Dialogue on Behavioral and Physical Health

Participants: Aaron P. Blaisdell, David Sloan Wilson, Kelly G. Wilson

David Sloan Wilson: Both of you have used your science to change health practices in your own lives. I think it would be interesting to explore that a bit.

Aaron P. Blaisdell: Sure. In 2008–2009, when I came across all these paleo diet blogs, I started tinkering in my own life. Like, what would happen if I took out most of the processed foods that I was eating, all grains, and breads, and pastas, and sugars? I went low carb because I was also reading Gary Taubes's *Good Calories, Bad Calories*, and he was championing that idea. I really slimmed down, and I found I had a lot more energy than I used to. But the most profound change is that I have this genetic disorder—erythropoietic protoporphyria—that made me highly sensitive to sunlight. Any area of the skin that was exposed to the sunlight would gradually get very inflamed and burning. About a year into being pretty religiously on this diet, I spent an afternoon with my wife out in the sun. I was thinking, *Oh man, it's going to be one of those rough evenings and nights*, but—nothing. I didn't get any reaction at all. So I started experimenting—how much sun can I handle before I start feeling that reaction? I would never get the reaction.

When my older daughter was about five or six years old, you'd get invited to all the obligatory birthday parties with all the other five- and six-year-olds in their school. I started snacking at the parties: a little bit of cake, have some chips and Doritos. My porphyria symptoms came back when I would be out in the sun. So I just cut all that stuff back out again and within a few weeks it was in remission again.

It is about lifestyle and evolutionary mismatch. So I try and put all of these ideas into practice in my own life.

Another example—blue light is what will suppress melatonin production, which is what you need at night for the regulation of sleep and falling asleep. I've learned to be smart about it with programs like f.lux, which automatically knows when the sun is setting and filters out the blue, so I'm not being exposed to the melatonin-suppressing part of the spectrum. I've actually replaced some of the lightbulbs, especially my reading lamp next to my bed, with orange-spectrum light that doesn't emit any blue spectrum.

I've started a whole society dedicated to this idea of understanding these mismatches and trying to correct them.

David: The Ancestral Health Society (http://www.ancestralhealth.org).

Kelly G. Wilson: I had a conversation with you, David, I think in the winter of '09, which got me to think of my own discipline through the lens of evolution science. Later that summer I was doing some work on values and self-care and it clicked. I said to an audience "I'm in the H. L. Mencken school of self-care: Whenever I get the urge to do strong phys-ical exercise, I lie down until it passes." Everyone in the room laughed, and I laughed. But then when I was doing this values exercise and I was imagining the face of one of my daughters—suddenly my joke was not so funny anymore. I had this moment of clarity: "When did it become okay for me to neglect myself, and when did it become okay for me to joke about that neglect?" Then I came back to evolution science with a sense of personal connection. And I started to ask this question—If you understand us as evolved beings, what would we need?

I came home from that workshop and started a yoga practice. It was the first thing that I did. I could barely do anything, after decades of being incredibly careless about my physical well-being. I spent months and months in gentle yoga for special needs. I was in there with the surgery-recovering patients and elderly, and my favorite yoga teacher likes to joke that I came to the prenatal classes. People always laugh and she's like, "No, no really!" The next box that got ticked was diet. My diet was not *completely* awful, because my wife was paying attention to what the kids were eating and stuff like that. Nevertheless, I was prob-ably getting 90% of my calories from ultraprocessed foods. I started reading about the bitter truth of sugar and I was just sort of gobsmacked. You can see the health impact of the astonishing changes that have happened in the world in our diet and exercise just in the last fifty years. A retreat in mortality levels. White males especially, where growth in health has actually plateaued. I took all of that stuff out of my diet. Now what I eat is *food*. If it has a label, it's probably not great news. Then I started running; and then within three or four months, I started barefoot running. Now I'm a barefoot running maniac. Put "running" in quotes, because I don't know if any runner would claim it.

Aaron: It's amazing how much variability is due to ontogenetic factors.

David: Things like barefoot running or the paleo diet are often accused of being junk science. Maybe that's justified, or perhaps not. But what works and might even be critical for one individual might not matter for another; their ancestry might have a big part to do with it; environmental factors, and so on. How do we do good science, and how do we protect against veering in a faddish direction?

Kelly: There's some unbelievable, wacky stuff out there. But when you start going into "good" sources, you have to settle for "good enough." For example, an enormous amount of the population is taking five or more medications. There are no randomized clinical trials on that and there never will be, because no institutional review board in their right mind would approve it.

David: On top of that, if you do a drug trial and you find that there are short-term benefits, that says nothing about long-term effects. Those can actually be reversed.

Kelly: That is true in many cases. For example, I was on proton pump inhibiters after I had head and neck cancer; I had not enough saliva, which leads to chronic and gastrointestinal vulnerability. So I took these proton pump inhibiters for a dozen years. Well, what I know now is you can induce gastrointestinal symptoms in healthy adults by as little as eight weeks of exposure to proton pump inhibitors. The label always says "up to fourteen days" or something, but that's not what the commercials say. The commercials say it's un-American to not eat your chili dog.

David: So what's the way to proceed?

Kelly: Well, you could avoid eating things that have no nutritional value. So, look at "natural ingredients"—sometimes that is the way food companies hide food ingredients that they don't want to tell you about. Take something called castoreum. It's in vanilla- and berry-flavored things, and it comes from the anal glands of beavers. It is natural. What is more natural than a beaver's ass? And do I think it's dangerous? Probably not. But is it necessary? No.

David: That's not getting at the science. How do we actually demonstrate that something is or is not good for you, especially when it's so contextual?

Kelly: Well, here's the problem. The doubt factory that is run by Big Food says "There isn't evidence for X." And so my rule is: is there a downside to leaving this out? I can't figure out any downside to not eating foods that have ingredients that are not food.

Aaron: I agree with Kelly: "Guilty until proven innocent." That is usually applied in medicine and should be applied on a policy-wide spectrum, for any new ingredients—or even things like BPA lining and things invented by industry and then utilized in products that affect not just our health, but our environment.

I think it's unwise to make policy on observational science—which is really good at generating hypotheses that can then be, and should be, tested scientifically with more laboratory, empirical-based work. The media gets caught up with this: a headline says "Red meat can do this, it's bad," then the next week, "Red meat is actually a source of this, it's really good for you," then the week after that, "Red meat is bad," and this ping pong keeps going back and forth.

Kelly: We can have both. Take the sugar evidence. There is epidemiological evidence, but also basic laboratory evidence. For example, you can expose baby rats to free access to a 25% sugar solution, and it will alter the regulation of genes having to do with the stress response. It'll alter genes that will impact neurological flexibility.

David: How does the concept of model organism survive the mismatch concept? If things are good or bad depending on your ancestral environment, the ancestral environments of the mouse or pigeon are very different from each other and from us!

Kelly: Some models hold and some don't. If you take something like movement, you can model depression and anxiety in rats by restricting their physical activity. You can also predict psychological health in humans by knowing something about their movement patterns; you can also cure psychological problems by enhancing movement patterns. When you see a convergence of these data, that's what I'm looking for.

Aaron: I agree with Kelly. Mice and rats—they're granivores. A dietary staple for them is seeds and grains. Humans only started eating grains about 10,000 years ago. In my own lab, where I've shown that a diet high in sugar will dysregulate the cognition and motivation in a rat, their control group was eating a diet that had a fair amount of grain, including wheat in it. For a rat, that would be a more ancestral-type diet. Maybe not for a human.

Knowing something about the evolutionary history and the ecology of the organism that you're picking—it doesn't take a lot of effort to do that, and to understand whether or not that would be a good animal model.

David: It seems to me, with so many people interested in this, that there should be a way to organize what amounts to randomized controlled trials or at least things that are like clinical studies—except they're done on a self-organizing basis, with a large population of people that really need to be organized. Human DNA research such as *23andme* is a bit like that.

Aaron: We need more individualized therapies and behavioral techniques that we can offer people in our daily lives. Small steps to building up your skills that work for you and rewriting your own narrative. I think teaching people to do these kinds of things is going to have the most bang for the buck and give people the most help.

David: That actually segues into the final big topic that I want to discuss with you, which has to do with ACT training in the context of our native learning abilities. We're getting a lot of mileage from thinking about Skinner's selection by consequences—operant conditioning as an evolutionary process in its own right. With relational frame theory, we're providing an overlay of language and symbolic thought that goes beyond Skinner. One of the insights from this is that if each person is an evolutionary process, then evolution often takes you where you don't want to go. What's adaptive in the evolutionary sense of the word is not always normatively good and adaptive in the everyday sense of the word. A large fraction of social and personal problems are adaptive in the evolutionary sense of the word—for example, through a reinforcement process that led in a dysfunctional direction, such as families that have reinforced each other for obnoxious behavior. What ACT does is manage the process of personal evolution. And so there's a sense in which it's a form of niche construction. We're actually creating an environment, a kind of cognitive environment or social environment or something like that, which is then steering the evolutionary process in a more helpful direction, both personally and socially.

Kelly: If you look at the distinction between short-term contingencies and long-term contingencies, there's a thing called the matching law. It's a way to quantitively predict how animals will distribute their behavior. What the matching law basically says is that animals will distribute their behavior in a way that matches the availability of reinforcers. And the thing we know about reinforcers is immediate reinforcement is good, and big reinforcers are better than small reinforcers. The problem is, we have this plethora of amazing energy-rich foods that our ancestors couldn't have hoped for.

David: A lot of the mismatches we've been talking about are actually mediated through the matching law.

Kelly: Yeah. But the problem is we're just doing what evolution said: take the easy path up the hill, prefer the energy-rich foods. The problem with small, incremental, and delayed reinforcers is that they just don't work that well. ACT, I think—in the values work in ACT, in particular—allows us to bring these small, incremental reinforcers into this moment right here. For example, I get people who say, "Well, self-care, that just doesn't interest me," and I'm like, "Oh, really? What does interest you?" Very often people say, "My kids, my family." And I'm like, "Okay, so if they watch you making sure everyone else's needs are met but you neglect yourself and joke about your own self-care, is this cool?" What values work delivers for you is, at that moment where I put myself in my truck and start to head out to the trail, in that moment I'm being the dad I want to be, I'm being the mentor I want to be, I'm being the member of the community I want to be, in that moment. I don't have to wait until later because I'm in the pattern. And values work in ACT in patterns of activity. It's not about some big outcome, it's about the pattern of activity.

For laboratory animals, you can establish preference by establishing a little bit of deprivation, but with freely moving adult humans, the things that have the biggest reinforcement value for you are almost all things that involve those really extended patterns over time: being a husband, being a teacher, being a dad. I create a niche around myself where talking about what is most valuable is a daily occurrence.

David: ACT as mental niche construction.

Kelly: I love that.

Aaron: I've basically done the same thing—constructing my own niche and realizing that I had to prioritize myself, especially now that I have kids. I really had to pull back and realize that, okay, what's important to me is not how many publications I get per year. I think we need to be able to package this kind of approach and bring it more to the blue-collar areas. People who are living in food deserts, where they don't have access to anything but convenience marts and it's hard for them to make decisions when they are working a couple of jobs, they have children, and they're trying to meet rent every month, month to month. What we want to see happen is for evolutionary scientists to help influence the way that our environments are built at a societal level.

David: I am convinced that ACT is doing something for me. As I soak this stuff up, it is becoming second nature for me. I'm confident—not that I can prove it—that ACT training has caused me to function better as a person. Take the studies on ACT bibliotherapy. Assign *Get Out of Your Mind and Into Your Life* to challenged populations of people—highly stressed Japanese students studying in America, K-12 teachers suffering from burnout. The numbers go up in the terms of increased well-being, just on the basis of reading the book. That is getting into your more blue-collar population, Aaron. It's a way of solving mismatch, but you can also think of it as a form of niche construction.

Kelly: That's what we're shooting for: the most general principles.

CHAPTER 14

Small Groups as Fundamental Units of Social Organization

David Sloan Wilson
Binghamton University

For human well-being to exist at all, it needs to exist at a variety of scales, including individuals, relatively small groups, large-scale societies, and ultimately the whole planet. In today's world, attention tends to focus on the smallest and largest scales. We yearn to thrive as individuals and yet the global calamities looming over us—climatic, environmental, political, and economic—are all too clear. These concerns often obscure the importance of well-being at the scale of small groups and its role in fostering well-being at other scales.

The dominant trends in intellectual thought during the last half-century also obscure the importance of small groups. Reductionism claims that higher-level entities can only be understood by taking them apart and studying the interactions among their parts. Methodological individualism in the social sciences is a form of reductionism that privileges the individual human as a fundamental unit and claims that all things social can be understood in terms of individual-level interactions. Rational choice theory in economics is a form of methodological individualism that treats the individual utility-maximizer (*Homo economicus*) as the fundamental unit, with utility usually operationalized as monetary wealth. Small groups are almost invisible against the background of these reductionistic intellectual traditions.

In contrast, within the context of evolutionary theory, small groups emerge as a fundamental unit of human social organization, essential for human well-being at both smaller and larger scales. Before proceeding to the details, it is important to dwell a bit longer on why the choice of theoretical background makes such a difference.

Reductionism, Holism, and Evolutionary Theory

Niko Tinbergen (1963), who pioneered the study of behavior (ethology) as a branch of biology, famously noted that four questions need to be asked for all products of evolution concerning their function, history, mechanism, and development. Tinbergen's four questions, as they are often called, map onto a two-fold distinction between ultimate and proximate causation made by the evolutionary biologist Ernst Mayr (1961), with "ultimate" encompassing Tinbergen's "function" and "history" questions and "proximate" encompassing his "mechanism" and "development" questions. These distinctions are essential to acquiring evolutionary theory's conceptual toolkit (see for example D. S. Wilson & Gowdy, 2013, for a discussion of Tinbergen's four questions oriented toward economics and public policy).

"Mechanism" in this context refers to the physical basis of an evolved trait, and "development" concerns how the trait comes into being during the lifetime of the organism. Asking these questions can be usefully treated as a reductionistic enterprise, all the way down to the molecular level. But the "function" and "history" questions are profoundly holistic. Holism is the view that the whole can't be fully explained in terms of the parts; that somehow the parts *permit*, but do not *cause*, the whole to have its properties. This concept is not mystical woo-woo—it is necessary in order to understand adaptations that evolve by natural selection (D. S. Wilson, 1988).

When I lecture on this topic, I show a slide of a typical desert with light brown sand, followed by a slide of a more unusual desert with pure white sand. Such deserts are rare but exist in regions where the underlying rock substrate is white, such as chalk or gypsum. Then I ask the audience to guess the color of most of the animals that live in white deserts. Invariably, someone suggests the answer "white."

They are right, of course, but what's interesting is to reflect upon the chain of reasoning that led them to their conclusion. If they weren't creationists, they were likely thinking that animal populations in a white desert vary in their coloration, that nonwhite prey are removed by predators (or conversely, nonwhite predators are spotted and avoided by their prey), and that offspring resemble parents in their coloration. Taken together, these thoughts lead to the prediction that animals in a white desert will also be white.

Notice that this prediction can be made without any knowledge of the physical makeup of the organisms. It can be made for all taxonomic groups, such as insects, snails, spiders, reptiles, amphibians, birds, and mammals, even though these groups differ in their genes and physical exteriors. The only assumption required is that the physical makeup of organisms produces heritable variation, which turns organisms into a kind of malleable clay that is molded into various shapes by differential survival and reproduction. Just as clay permits but does not cause the shaping influence of the sculptor, the physical makeup of an organism permits but does not cause the shaping influence of the environment. Donald Campbell (1990), who pioneered modern evolutionary thinking in the social sciences and was fiercely critical of methodological individualism, called this "downward causation."

Units of Selection as Privileged Units of Analysis

Reductionism privileges the lowest units of analysis, in principle down to molecules, atoms, and subatomic particles. Methodological individualism in the social sciences privileges individuals as the unit of analysis, but does not have a strong theoretical reason for doing so. That's why it is called "methodological." Evolutionary theory provides strong theoretical reasons to privilege units of analysis, but the privileged unit is only sometimes an individual.

The way I get this point across in a lecture is by showing a slide of a solitary bee. Not all bee species are social; in thousands of species, a single female does the work of raising her brood without any assistance. The subunits of a solitary bee, such as its organs, cells, genes, and molecules, can be studied. Likewise, we can study solitary bees at the levels of the populations, species, and the multispecies ecosystems that contain them. Nevertheless, there is something privileged about the individual bee as a unit of analysis, because it is the primary unit that natural selection has acted upon. Everything below the level of the individual is organized to help the individual survive and reproduce as a unit. And these adaptations need to be known to understand how individual bees interact in populations and ecosystems. Thus, while a solitary bee species can and should be studied at many levels, inquiry needs to be anchored at the level of the individual organism.

This much is easily understood by most audience members, just as they easily predict that most animals living in white deserts should be white. Next, I show them a slide of a honeybee colony, where individual bees work together as a cooperative unit, including reproductive division of labor similar to the gonads and somatic cells of a multicellular organism. In this case, the privileged unit of analysis is the colony and not the individual bee, because most traits in honeybees evolve by virtue of colonies surviving and reproducing better than other colonies, not bees surviving and reproducing better than other bees within colonies— although this does occur to a degree, just as cancers occur in multicellular organisms. Generalizing, we can say that the privileged unit of analysis depends upon the dominant level of selection: the individual level for solitary bees, the group level for honeybees, and the cellular level for cancers.

247

Even when a given unit of analysis is worth emphasis, it is important to consider other units and levels as well. For example, in some contexts, multispecies ecosystems can be considered units of selection and therefore assume a privileged level of analysis. In fact, however, every multicellular organism serves as a planet for an ecosystem of microbes consisting of many thousands of species and many trillions of cells, rivaling and even exceeding the number of the host's own cells. Microbiomes vary among individuals, influence phenotypic traits, and survive and reproduce along with their hosts. Thus, what is typically regarded as individual-level selection has in another sense been an example of ecosystem-level selection all along!

Human Groups as Units of Selection

Most audience members are able to follow me this far, although thinking of themselves as planets inhabited by microbiomes can take a little getting used to. Certainly, everything that I have said so far would be regarded as mainstream by most evolutionary biologists. What comes next challenges the imagination of almost everyone—that small groups were potent units of selection during human genetic evolution and that larger groups were potent units of selection during human cultural evolution over the past 10,000 years. This means that for many purposes, groups should be the privileged unit of analysis in our species. It is as wrong to focus primarily on individual humans as it is for a honeybee biologist to focus primarily on individual bees. This is a paradigmatic shift away from reductionism and methodological individualism and toward a more systemic and multilevel evolutionary conceptualization.

The scientists who have been developing this thesis number in the dozens or even hundreds, which is enough to have made substantial progress, but they are still a tiny fraction of the worldwide scientific community, not to speak of the general public. They include John Maynard Smith and Eors Szathmary (1995, 1999), Christopher Boehm (1993, 1999, 2011), Terrence Deacon (1998), Eva Jablonka and Marion Lamb (2006), Peter Richerson and Robert Boyd (2005), E. O. Wilson (2012), Peter Turchin (2015), Joseph Henrich (2015), Michael Tomasello (2009), and myself (D. S. Wilson, 2015). Here is the emerging story in brief.

In most primate species, between-group selection operates to a degree but is substantially opposed by disruptive forms of within-group selection, favoring traits that would be regarded as selfish and despotic in human terms. Even cooperation often takes the form of coalitions competing disruptively against other coalitions within the same group.

Our ancestors found ways to suppress disruptive forms of competition within groups, so that the group became the primary unit of selection. Mutual policing among group members was required, since the temptation to benefit oneself at the expense of others was ever-present. Christopher Boehm calls this reverse dominance—the group dominating the would-be alpha individual, rather than the reverse. According to Boehm (1993), "The data do leave us with some ambiguities, but I believe that as of 40,000 years ago, with the advent of anatomically modern humans who continued to live in small groups and had not yet domesticated plants and animals, it is very likely that all human societies practiced egalitarian behavior and that most of the time they did so very successfully" (p. 236).

This is enough time for guarded egalitarianism at the scale of small groups to have become deeply embedded in our innate social psychology. Teamwork—succeeding as a group rather than at the expense of others within one's group—became the signature adaptation of our species. Teamwork takes familiar physical forms such as hunting, gathering, childcare, predator defense, and offense and defense against other human groups. Teamwork also takes mental forms, such as a greatly increased capacity for symbolic thought and the ability to transmit large amounts of learned information across generations. In this fashion, almost everything that is distinctive about our species can be understood as following from a single event—a shift in the balance between levels of selection with between-group selection becoming the dominating force.

Our greatly enhanced ability to transmit learned information across generations became a rapid evolutionary process in its own right, allowing human groups to adapt to their environments much more quickly than by genetic evolution alone. This enabled our ancestors to spread over the globe, occupying all climatic zones and dozens of ecological niches, eating everything from seeds to whales (Pagel & Mace, 2004). Then the advent of agriculture created a positive feedback loop between group size and the production of food, leading to the mega-societies of today (Turchin, 2015).

Large human groups are more difficult to regulate than small groups for two reasons. First, psychological mechanisms that evolved at the scale of small groups don't necessarily work at the scale of large groups. Second, large groups are intrinsically more difficult to regulate than small groups. This led to a despotic phase in human history, in which societies were structured for the benefit of a small group of elites, ironically more like primate societies than small-scale human societies. Cultural evolution is a multilevel process no less than genetic evolution, however, and between-group selection, fueled largely but not entirely by warfare, has led to the impressively cooperative mega-societies of today. Of course, despotic societies also exist in the present, and cooperative societies are in constant danger of converting to despotism and other forms of elitism (e.g., economic despotism caused by extreme wealth inequality). The sociopolitical events swirling all around us reflect cultural evolution operating at different levels with different degrees of intensity. The cultural evolution of large-scale human society has not been a smooth upward curve. It has been more like a saw tooth curve with many reversals along the way.

These insights about human genetic and cultural evolution didn't begin to develop until the late twentieth century and are therefore new to most of the human-related disciplines, both basic and applied. I will briefly discuss implications for the basic academic study of humans before turning to implications for contextual behavioral science.

Implications for the Basic Academic Study of Humans

Even though reductionism and methodological individualism have dominated intellectual thought during the last half-century, other intellectual traditions in the human social sciences and humanities are much friendlier to the concept of human society as an organism in its own right. Modern multilevel selection (MLS) theory provides a new foundation for reviving and updating these traditions.

The French social theorist Alexis de Tocqueville, who wrote perceptively about America during the 1830s, also observed that small groups are a fundamental unit of human social organization in this passage: "The village or township is the only association which is so perfectly natural that, wherever a number of men are collected, it seems to constitute itself" (1835/1990, p. 60).

Daniel Wegner, a contemporary psychologist who conducted a fascinating series of experiments on memory as a group-level process, had this to say about the founding fathers of psychology (Wegner, 1986, p. 185):

> Social commentators once found it very useful to analyze the behavior of groups by the same expedient used in analyzing the behavior of individuals. The group, like the person, was assumed to be sentient, to have a form of mental activity that guides action. Rousseau and Hegel were the early architects of this form of analysis, and it became so widely used in the nineteenth and early twentieth century that almost every social theorist we now recognize as a contributor to modern social psychology held a similar view.

This was also true for the discipline of social psychology during the first half of the twentieth century, before it was eclipsed by methodological individualism.

Many of the classic studies, such as the Robber's Cave experiment (Sherif, Harvey, White, Hood, & Sherif, 1954) and minimal group experiments (Tajfel, 1981), make perfect sense from a modern multilevel evolutionary perspective.

Among the early social theorists was Emile Durkheim (Durkheim & Fields, 1912/1995), who famously defined religion as "a unified system of beliefs and practices relative to sacred things … which unite into one single moral community called a Church, all those who adhere to them" (p. 44). Religious believers themselves describe their communities as like a single body or a beehive, as in this passage from the seventeenth century (Ehrenpreis & Felbinger, 1650/1978): "True love means growth for the whole organism, whose members are all interdependent and serve each other. That is the outward form of the inner working of the Spirit, the Organism of the Body governed by Christ. We see the same thing working among the bees, who all work with equal zeal gathering honey" (p. 11). Durkheim initiated the tradition of functionalism, which was widely accepted among anthropologists and sociologists during the first half of the twentieth century before being eclipsed by methodological individualism. Functionalism had its own problems, including being axiomatic in its assumptions about groups as functional units, but MLS theory is not axiomatic on that point and can place the functionalist tradition on a contemporary scientific foundation.

Political scientists can't avoid talking about supraindividual entities such as institutions and states because they are part of the subject matter of their discipline. Most of the theoretical frameworks used by political scientists are largely confined to their discipline and adjacent disciplines such as economics and sociology. MLS theory places the study of human institutions on a much broader foundation that can be related to a greater diversity of other disciplines. I will provide an example in the next section involving the work of Elinor Ostrom, a political scientist by training who received the Nobel Prize in Economics in 2009.

Even a highly reductionistic discipline such as neuroscience can benefit from a multilevel evolutionary perspective. According to Coan and Sbarra (2015), the human brain is designed to "expect" the presence of supportive others. Deviations from this default condition trigger various types of stress reactions. This is the default condition of the brain because it was the default social environment for the vast majority of our ancestors for most of our evolutionary history. You can't understand the structure of the brain without understanding what it evolved to do in a given set of ancestral environments.

The intellectual traditions of reductionism and methodological individualism have produced many advances in the study of proximate mechanisms, but this knowledge is necessarily incomplete. In the future, we can look forward to a more fully rounded approach that is both holistic and reductionist by paying equal attention to all four of Tinbergen's questions, or both proximate and ultimate causation, as Mayr would put it. In this more fully rounded approach, the small group will be duly recognized as a fundamental unit of human social organization.

Ostrom's Design Principles and Contextual Behavioral Science

Unlike most basic scientific research, contextual behavioral science (CBS; Zettle, Hayes, Barnes-Holmes, & Biglan, 2016) is action oriented. It expects to predict *and influence* behavior from the very beginning, rather than treating basic and applied research as separate enterprises. Among other things, this focus means CBS needs to focus on the manipulable contextual variables that determine action, and few such factors are more important than the role of small groups in human behavior. The concept of small groups as fundamental units of human social organization has rich implications for CBS, not only at the scale of small groups, but for individuals and large-scale societies as well.

Elinor Ostrom was a key figure in establishing the relevance of small groups in modern life. A political scientist by training, she specialized in studying groups that attempt to manage common-pool resources such as forests, fields, fisheries, and irrigation systems. These groups are vulnerable to the tragedy of overuse, dubbed "The Tragedy of the Commons" by the ecologist Garrett Hardin (1968), because every member of a common-pool resource group has a temptation to use more than his or her fair share. Conventional wisdom held that the only solutions to the tragedy of the commons were to privatize the resource (if possible) or to regulate it in a top-down fashion. Ostrom showed through the creation of a worldwide database that groups are capable of managing common-pool resource groups on their own, but only if they possess eight Core Design Principles (CDPs), which are shown in Table 1, at the end of this chapter. This was so contrary to orthodox economic wisdom that Ostrom was awarded the Nobel Prize in Economics in 2009 for her achievement.

Ostrom originally framed her research in political science, but she incorporated evolutionary thinking throughout her career. Her most important book was titled *Governing the Commons: The Evolution of Institutions for Collective Action* (Ostrom, 1990; see also Ostrom, 2010). I was privileged to work with Lin (as she preferred to be called) and her postdoctoral associate Michael Cox for three years prior to her death in 2012, resulting in an article titled "Generalizing the Core Design Principles for the Efficacy of Groups" (D. S. Wilson, Ostrom, & Cox, 2013).

Even if this book chapter is your first introduction to MLS theory, the correspondence of MLS theory to Ostrom's CDPs should be clear. First and foremost, in a group that strongly implements CDP1, members will recognize themselves as a group and will have a clear understanding of its purpose. CDP2 prevents the unfair distribution of costs and benefits, or disruptive within-group selection in the language of MLS theory. CDP3-6 also discourage disruptive within-group selection in contexts such as decision making and conflict resolution. CDP7-8 concern relations among groups in multigroup societies, which need to reflect the same principles as relations among individuals within groups, a point to which I will return below.

Very simply, in a group that strongly implements the CDPs, it is difficult for group members to benefit themselves at the expense of other members, leaving teamwork as the only remaining path to success. The CDPs foster a social environment in a modern-day group similar to the social environment that existed throughout our genetic evolution, which caused us to be such a cooperative species in the first place.

The connection between the CDPs and MLS theory allows Ostrom's work to be generalized in two respects. First, the CDPs are theoretically general. They follow from the evolutionary dynamics of cooperation in all species and our own evolutionary history as a highly cooperative species. Second, for this same reason, they should apply to all small human groups, not just common-pool resource groups. Cooperation is itself a common-pool resource, vulnerable to the same tragedy of exploitation from within that Hardin called attention to regarding the exploitation of common-pool natural resources.

From a CBS perspective that is focused on influencing action, generalizing the CDPs is eminently useful by providing a practical blueprint for the design of small human groups and their assembly into larger multigroup societies (called "polycentric governance"; McGinnis, 1999; Ostrom, 2010). However, it is crucial to distinguish between a functional design principle and its implementation. As an example, all groups need to monitor agreed-upon behaviors (CDP4), but *how they do it* can be highly variable and can depend upon local circumstances. This means that the CDPs cannot be implemented in a cookie-cutter fashion. Group members must tinker with their arrangements until they get the CDPs right for themselves. The many-to-one relationship between a functional design principle and its implementation maps nicely onto the evolutionary distinction between ultimate and proximate causation.

In addition, the CDPs are only necessary and not sufficient for a group to function well. There are also Auxiliary Design Principles (ADPs) that are needed by some groups but not others to achieve their particular objectives. As one example, some groups experience much higher turnover in their members than other groups (e.g., a college fraternity). These groups must be designed with turnover in mind if they are to persist. For any given group, the ADPs are as important as the CDPs. Between the challenges of identifying the ADPs and the challenges of finding the best implementation for all the CDPs, creating an ideal social environment for a given group is not necessarily easy—but it is possible, and having the right functional blueprint is better than having the wrong blueprint or no blueprint at all.

Prosocial: A Reunification Success Story

Despite Ostrom's fame as a Nobel laureate, her legacy is still largely confined to the study of common-pool resource groups and there have been few efforts to develop practical change methods based on the CDPs. A notable exception is Prosocial, a project headed by the author with numerous members of the CBS community, including Steven C. Hayes, the coeditor of this volume, and Paul Atkins, who has written the other chapter on small groups in this chapter pair. Prosocial therefore represents an example of the reunification between CBS and evolutionary science that this volume seeks to foster for a diversity of topic areas.

Prosocial is inspired in part by the empirical database that Ostrom assembled for common-pool resource groups, which led her to derive the CDPs in the first place. The database was compiled mostly from books and articles written about the groups, not directly from the groups, and many of the accounts were descriptive, requiring coding procedures to conduct quantitative analyses. The database was crude but sufficient for Ostrom to achieve her insights. In our case, we are working directly with groups and are in a much better position to capture quantitative data. Thus, Prosocial will be a scientific database in addition to a practical change method, which is very much in the spirit of CBS.

We predict that the CDPs are needed by virtually all groups whose members must work together to achieve common goals. Further, we predict that in any particular sector (e.g., businesses, schools, churches, nonprofit organizations, neighborhoods) groups will *vary* in how well they implement the CDPs, and this variation will correlate with the ability of the groups to accomplish their given objectives. This is what Ostrom observed for common-pool resource groups, and other kinds of groups should be no different.

One test of these predictions is a survey of intentional communities, which are residential communities designed with a high degree of social cohesion and teamwork in mind (Grinde, Nes, MacDonald, & Wilson, 2017). Intentional communities have existed throughout history (see Sosis, 2000, for a review of nineteenth-century communes) and thousands exist in the present day, networked by consortia such as the Federation of Intentional Communities (FIC). We collaborated with the FIC to conduct a survey of over 100 intentional communities that measured implementation of the CDPs and various performance metrics such as Life Satisfaction.

Despite the fact that all intentional communities are committed to teamwork, they still varied in their implementation of the CDPs, which in turn correlated with Life Satisfaction, as shown in Figure 1. The two lines represent groups that are high and low in identity fusion, which represents the degree to which individuals' personal identity has merged with their group's identity. Identity fusion can be regarded as an exceptionally strong implementation of CDP1, and it has a very powerful correlation with Life Satisfaction.

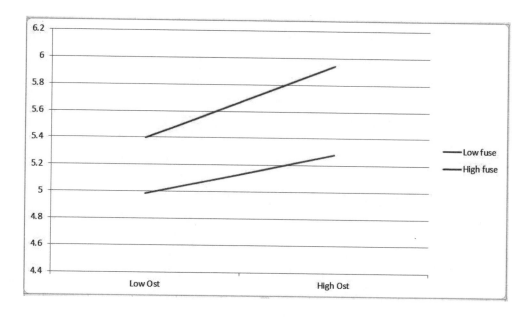

Figure 1. Life Satisfaction (y-axis) as a function of implementation of the Core Design Principles (x-axis) in a survey of intentional communities. The two lines represent communities that score high and low in the degree to which the identity of the individual is fused with the identity of their group. (From Grinde et al., 2017.)

For groups that are in need of improvement, adopting the CDPs requires change. Change can be difficult for individuals and it can be even more difficult for groups. Hence, when Prosocial facilitators begin working with groups, they begin not with the CDPs but with an exercise from acceptance and commitment training (ACT; Hayes, Strosahl, & Wilson, 2012) designed to increase psychological flexibility and the selection of actions that move the group toward its valued goals. The particular method we use is called the "matrix" (Polk & Schoendorff,

2014), but the more important point to make is that working with a group is a matter of managing a cultural evolutionary process, thereby touching upon numerous other chapters and topic areas in this book. All of this is being packaged into a practical method for working with groups, including a website and a worldwide network of facilitators (visit http://www.prosocial.world for more information).

From Group-Care to Self-Care and Earth-Care

Not only are small groups a fundamental unit of human social organization, but they may well be required to achieve well-being at both smaller and larger scales. Individuals thrive when they are members of groups engaged in meaningful activities, where they can give and receive social support and be known for their actions as prosocial individuals (Biglan, 2015). Small groups can also establish and enforce norms for prosocial action at a higher scale, such as working toward global sustainability, more effectively than an individual on his or her own.

Both of these predictions are supported by our study of intentional communities (MacDonald et al., in press), where individuals report a very high level of satisfaction and sense of meaning in life and groups succeed at collective action goals such as reducing their ecological footprint and serving as wise stewards of natural areas. In the business world, corporations that embrace social responsibility and implement the Core Design Principles report high levels of employee satisfaction, are successful at achieving their social responsibility goals, and even thrive as businesses thanks to cooperative relations with like-minded suppliers, customers, and other groups in the multigroup human ecosystem (Wilson, Kelly, Philip, & Chen, 2016). Combining a functional blueprint for the efficacy of groups with a blueprint for cooperative interactions between groups provides a strong ray of hope in a world where little else seems to be working.

Table 1. The Core Design Principles
1) Clearly defined boundaries. The identity of the group and its rights to the common resource must be clearly delineated.
2) Proportional equivalence between benefits and costs. Members of the group must negotiate a system that rewards members for their contributions. High status and other disproportionate benefits must be earned.

3) Collective-choice arrangements. Group members must be able to create their own rules and make their own decisions by consensus. People hate being told what to do but will work hard for group goals that they have agreed upon.

4) Monitoring. Managing a commons is inherently vulnerable to free-riding and active exploitation. Unless these locally advantageous strategies can be detected at relatively low cost, the tragedy of the commons will occur.

5) Graduated sanctions. Transgressions need not require heavy-handed punishment, at least initially. Often gossip or a gentle reminder is sufficient, but more severe forms of punishment must also be waiting in the wings for use when necessary.

6) Conflict resolution mechanisms. It must be possible to resolve conflicts quickly and in ways that are perceived as fair by members of the group.

7) Some recognition of rights to organize. Groups must have the authority to manage their own affairs. Externally imposed rules are unlikely to be adapted to local circumstances and violate CDP3.

8) Polycentric governance. For groups that are part of larger social systems, there must be nested enterprises. The previous principles work best in relatively small groups. Society at a larger scale must be polycentric, with groups interacting with groups, often in multiple layers.

References

Biglan, A. (2015). *The nurture effect: How the science of human behavior can improve our lives and our world.* Oakland, CA: New Harbinger Publications.

Boehm, C. (1993). Egalitarian society and reverse dominance hierarchy. *Current Anthropology, 34,* 227–254.

Boehm, C. (1999). *Hierarchy in the forest: Egalitarianism and the evolution of human altruism.* Cambridge, MA: Harvard University Press.

Boehm, C. (2011). *Moral origins: The evolution of virtue, altruism, and shame.* New York: Basic Books.

Campbell, D. T. (1990). Levels of organization, downward causation, and the selection-theory approach to evolutionary epistemology. In G. Greenberg & E. Tobach (Eds.), *Theories of the evolution of knowing.* Hillsdale, NJ: Lawrence Erlbaum.

Coan, J. A., & Sbarra, J. A. (2015). Social baseline theory: The social regulation of risk and effort. *Current Opinion in Psychology, 1,* 87–91.

Deacon, T. W. (1998). *The symbolic species.* New York: Norton.

Durkheim, E., & Fields, K. E. (1912/1995). *The elementary forms of religious life*. New York: Free Press.

Ehrenpreis, A., & Felbinger, C. (1978). An epistle on brotherly community as the highest command of love. In R. Friedmann (Ed.), *Brotherly community: The highest command of love*. Rifton, NY: Plough Publishing. (Original work published in 1650.)

Grinde, B., Nes, R. B., MacDonald, I. F., & Wilson, D. S. (2017). Quality of life in intentional communities. *Social Indicators Research*, 1–16. https://doi.org/10.1007/s11205–017–1615–3

Hardin, G. (1968). The tragedy of common sense. *Science, 162*, 1243–1248.

Hayes, S. C., Strosahl, K., & Wilson, K. G. (2012). *Acceptance and commitment therapy: The process and practice of mindful change* (2nd ed.). New York: Guilford.

Henrich, J. (2015). *The secret of our success: How culture is driving human evolution, domesticating our species, and making us smarter*. Princeton, NJ: Princeton University Press.

Jablonka, E., & Lamb, M. (2006). *Evolution in four dimensions: Genetic, epigenetic, behavioral, and symbolic variation in the history of life*. Cambridge, MA: MIT Press.

Maynard Smith, J., & Szathmary, E. (1995). *The major transitions in evolution*. New York: W. H. Freeman.

Maynard Smith, J., & Szathmary, E. (1999). *The origins of life: From the birth of life to the origin of language*. Oxford: Oxford University Press.

Mayr, E. (1961). Cause and effect in biology. *Science, 134*, 1501–1506.

McGinnis, M. D. (Ed.). (1999). *Polycentric governance and development: Readings from the workshop in political theory and policy analysis*. Ann Arbor: University of Michigan Press.

Ostrom, E. (1990). *Governing the commons: The evolution of institutions for collective action*. Cambridge, UK: Cambridge University Press.

Ostrom, E. (2010). Beyond markets and states: Polycentric governance of complex economic systems. *American Economic Review, 100*, 1–33.

Pagel, M., & Mace, R. (2004). The cultural wealth of nations. *Nature, 428*, 275–278.

Polk, K. L., & Schoendorff, B. (2014). *The ACT matrix: A new approach to building psychological flexibility across settings and populations*. Oakland, CA: Context Press.

Richerson, P. J., & Boyd, R. (2005). *Not by genes alone: How culture transformed human evolution*. Chicago: University of Chicago Press.

Sherif, M., Harvey, L. J., White, B. J., Hood, W. R., & Sherif, C. W. (1954). *The robber's cave experiment: Intergroup conflict and cooperation*. Middletown, CT: Wesleyan University Press.

Smith, J. M. *See* Maynard Smith, J.

Sosis, R. (2000). Religion and intragroup cooperation: Preliminary results of a comparative analysis of utopian communities. *Cross-Cultural Research, 34*, 70–87.

Tajfel, H. (1981). *Human groups and social categories: Studies in social psychology*. Cambridge, UK: Cambridge University Press.

Tinbergen, N. (1963). On aims and methods of ethology. *Zeitschrift für Tierpsychologie [Journal of Animal Psychology]*, 20, 410–433.

Tocqueville, A. de (1990). *Democracy in America*. New York: Penguin Classics. (Original work published in 1835.)

Tomasello, M. (2009). *Why We Cooperate*. Boston: MIT Press.

Turchin, P. (2015). *Ultrasociety: How 10,000 years of war made humans the greatest cooperators on earth*. Storrs, CT: Baresta Books.

Wegner, D. M. (1986). Transactive memory: A contemporary analysis of the group mind. In B. Mullen & G. R. Goethals (Eds.), *Theories of group behavior*. New York: Springer-Verlag.

Wilson, D. S. (1988). Holism and reductionism in evolutionary biology. *Oikos, 53*, 269–273.

Wilson, D. S. (2015). *Does altruism exist? Culture, genes, and the welfare of others*. New Haven, CT: Yale University Press.

Wilson, D. S., & Gowdy, J. M. (2013). Evolution as a general theoretical framework for economics and public policy. *Journal of Economic Behavior and Organization, 90*, S3–S10. doi.org/10.1016/j.jebo.2012.12.008

Wilson, D. S., Kelly, T. F, Philip, M. M., & Chen, X. (2016). Doing well by doing good: An Evolution Institute report on socially responsible businesses. https://evolution-institute. org/wp-content/uploads/2016/01/EI-Report-Doing-Well-By-Doing-Good.pdf

Wilson, D. S., Ostrom, E., & Cox, M. E. (2013). Generalizing the core design principles for the efficacy of groups. *Journal of Economic Behavior and Organization, 90*, S21–S32. doi:10.1016/j.jebo.2012.12.010

Wilson, E. O. (2012). *The social conquest of earth*. New York: Norton.

Zettle, R. D., Hayes, S. C., Barnes-Holmes, D., & Biglan, T. (Eds.). (2016). *The Wiley handbook of contextual behavioral science*. West Sussex, UK: Wiley-Blackwell.

CHAPTER 15

Prosocial: Using CBS to Build Flexible, Healthy Relationships

Paul W. B. Atkins

Institute for Positive Psychology and Education, Australian Catholic University

"All life is interrelated. Somehow we are tied in a single garment of destiny, caught in an inescapable network of mutuality, where what affects one directly affects all indirectly. ... I can never be what I ought to be until you are what you ought to be, and you can never be what you ought to be until I am what I ought to be. This is the way the world is made."

—Martin Luther King, Jr. (1960/2005)

We live on the cusp between being overcome by the physical realities of our impacts upon the earth and the possibility of global cooperation. Technology has enabled tremendous acceleration of both self-interest and sharing. The dominant economic narrative of self-interest, and the patterns of transactional, "I-It" (Buber, 1958), "what's in it for me" relationships that appear to be celebrated in popular culture cannot be sustained. We need more cooperative behavior that values individual autonomy at the same time as acknowledging the perspectives and values of others. We need an approach that is clear enough to help many people to live more peacefully, meaningfully, and respectfully with one another but that does not rely upon simplistic and moralistic admonishments to be kinder or more helpful to one another.

Prosocial addresses these challenges using both of the intellectual traditions upon which this volume is based. Evolutionary theory has explored why those who cooperate might survive, thrive, and reproduce more effectively than those who act out of self-interest. And contextual behavioral science (CBS) has explored both (a) how human language simultaneously enables tremendous feats of problem solving and cooperation, while also enabling new forms of suffering and misunderstanding that undermine cooperation, and (b) strategies for behavior change that reach far beyond psychoeducation and transform individual and collective motivation, sense-making, and valuing.

The principles and process of Prosocial from an evolutionary perspective were described in the previous chapter. In brief, the Prosocial process involves six steps:

1. Initial engagement and measurement of current state of team interaction.

2. Individual matrix (Polk, Schoendorff, Webster, & Olaz, 2016).

3. Group matrix.

4. Consideration of team interaction in terms of eight core design principles (Wilson, Ostrom, & Cox, 2013).

5. Action planning.

6. Closing and final measurement of state of team interaction.

My aim in this chapter is to supplement the previous chapter with insights from CBS regarding behaviors needed to most effectively implement the principles. Training and practice in perspective taking and psychological flexibility are the central focuses of the early phases of Prosocial, involving the individual and group matrix procedures. I will argue that enhanced perspective taking and psychological flexibility transforms implementation of the core design principles from potentially rigid constraints upon behavior to powerful and effective guides for enhancing cooperation, managing diversity, and acting to create systemic change.

I begin by exploring the implication of contextual behavioral science for thinking about cooperation. I then outline two key behavioral repertoires that are the focus of this chapter: perspective taking and psychological flexibility. I conclude by describing how these two repertoires transform the core design principles.

Forms of Learning and Their Impact in Groups

Contextual behavioral science is based in an evolutionary perspective upon behavior. Behavior varies and is selected for continuation or repetition according to its functionality in serving the organism's goals. Variation in behavior occurs through learning. Humans are unique in their capacity to make use of arbitrary symbols as cues to transform how they respond to events. However, before focusing most of this chapter on our symbolic capabilities, I wish to briefly mention the more ancient processes of respondent and operant learning. Such automatic bodily and emotional learning processes are crucial parts of behavior in groups that are often ignored when we focus on more rational, verbal processes.

Much of our emotional responding to other people has its foundations in respondent learning. For example, a child with a caring mother with brown eyes may learn to associate brown eyes with feelings of warmth and safety and may consequently trust people with brown eyes more. Conversely, a child who grows up in a household where arguments often escalate into violence may learn that conflict predicts violence and therefore become fearful in the presence of even mild disagreements. Just as Pavlov's dogs learned to salivate at the sound of a bell, for this child, raised voices become a conditioned stimulus for a conditioned fear response, leading to avoidance of even productive forms of conflict. Such automatic, emotional reactions are pervasive in groups, sometimes have great impact, and are relatively difficult to control.

Operant learning is based upon the consequences of behavior. A person with performance anxiety might notice that spending a lot of time preparing for a presentation sometimes increases the likelihood of the presentation going well. If that contingency (preparation leading to the presentation going well) increases the frequency of the behavior (a greater likelihood of preparation), operant learning is occurring. Alternatively, the person might learn that avoiding presentations (behavior) decreases her anxiety (removal of a negative consequence) and thus may become increasingly likely to avoid presentations: another form of operant learning. These sorts of nonverbal processes can be found everywhere in groups. For example, the vast majority of managers I have coached have an unexamined history of avoiding conflict because this avoidance behavior is often immediately reinforced with reductions in discomfort and anxiety.

Both humans and nonhuman animals can learn respondently or operantly. But humans can also learn to relate events symbolically. Relational frame theory (RFT; Hayes, Barnes-Holmes, & Roche, 2001) is perhaps the best-supported current behavioral approach to human language and higher cognition, with an impressive list of empirical predictions that have subsequently been confirmed (Hayes, Gifford, & Ruckstuhl, 1996; see chapter 4 in this volume). According to RFT, the core of human language is the ability to derive mutual and combinatorial relations among events based in part on social cues. This process of derivation then alters the behavioral impact of these events.

These symbolic relations dramatically enhance the human capacity to cooperate (Hayes & Sanford, 2014). Although human groups engage in nonverbal processes such as co-location and shared toil in physical tasks, the vast majority of their behavior is symbolic in nature. Symbolic learning can sometimes undermine cooperation, as when group members derive unhelpful interpretations of the motives of others. But in the main, symbolic learning massively enhances the potential for cooperation. Using language, humans in groups discuss their aims and objectives, difficulties they might encounter, and how to overcome them. We plan for the future and remember what worked and what did not in the past. And we track behavior and seek to change it using language processes.

Symbolic learning also enables humans to cooperate more effectively by making and following instructions. Within CBS, rule-governed behavior is behavior under the control of a "contingency specifying stimulus" such as an "if-then" rule. So, for example, one group member might instruct another that "when we have a meeting [antecedent or context], you should write down what we discuss in the form of minutes [behavior], so that people can better understand and remember the action points from the meeting [consequence of the behavior]." Rules can also be descriptive rather than prescriptive. For example, a person's behavior may be influenced by a personally held "self-rule" that "people who openly disagree [behavior] in meetings [antecedent context] are likely to be seen as troublemakers and poor team-players [consequence]."

Bringing behavior under the stimulus control of rules can be enormously helpful in groups. Group members can be quickly instructed about how to perform complex behaviors efficiently and effectively without any need for trial-and-error learning. And groups can formulate helpful social norms, such as the Prosocial principles outlined in the previous chapter and below.

But behavior under the control of rules can also be problematic. Rule-governed behavior tends to be insensitive to changing contexts (Hayes, Brownstein, Haas, & Greenway, 1986; Hayes, Brownstein, Zettle, Rosenfarb, & Korn, 1986). So, for example, policies and procedures evolved to solve past problems might be followed long after they have become unhelpful. Second, social and self-rules for guiding behavior can often be relatively arbitrary and unhelpful. So, for example, a leader may hold a self-rule that, when a subordinate comes to him with a problem, he should seek to fix the problem in order to be seen as a good leader. This self-rule, while apparently benign, is often unhelpful in modern organizations seeking to support and empower staff to find solutions for themselves.

In summary, our capacity to cooperate is influenced both by primitive non-symbolic, emotional, and experiential forms of learning, and by more phylogenetically recent forms of symbolic or verbal learning. Symbolic learning is relational and transforms the effects of stimuli so that a person can respond to a given situation more effectively (or not) as a result of instruction rather than direct experience. Rules as understood in this way specify the consequences of particular behaviors in particular contexts; and they can be explicit social agreements or more tacit self-rules that regulate behavior. In the next two sections, I describe two behavioral repertoires that enable humans to cooperate in vastly more complex ways than other animals, but which can also be enhanced within a person's life span to improve cooperation: perspective taking and psychological flexibility.

Skills for Cooperation in Groups

Perspective Taking

Our capacity to take the perspective of another allows us to coordinate our behavior with the anticipated behavior of others. Perspective taking is thus foundational to human relationships (Mead, 1934; Piaget, 1969), and when it is impaired, it leads to serious difficulties in social engagement (Doherty, 2012). From a CBS/RFT point of view, perspective taking is enabled by symbolic relations such as I/YOU, HERE/THERE, and NOW/THEN. These relations are acquired through multiple exemplar training during language acquisition. The classic "object permanence" experiments, where a child assumes that because "I" know that an object has been moved "YOU" also will know, even though you did not see it being moved, amply illustrate that appropriately using the I/YOU distinction is a complex skill that is slowly learned over time (McHugh, Barnes-Holmes, & Barnes-Holmes, 2004).

The acquisition of skilled deictic framing behavior can be accelerated in children (Weil, Hayes, & Capurro, 2011), and there is abundant evidence that it is possible to increase the behavioral relevance of others' emotional and cognitive states simply by instructions like "please try to imagine how your partner [thinks or] feels about what is happening" (Leong, Cano, Wurm, Lumley, & Corley, 2015, p. 1177). Such instructions to take the perspective of others in groups can improve the quality of negotiation (Galinsky, Maddux, Gilin, & White, 2008) and cooperation. This is important in the light of evidence that even adults adept in perspective-taking skills often spontaneously fail to take the perspective of others (Keysar, Lin, & Barr, 2003).

A distinction is sometimes drawn between cognitive and affective perspective taking. In the former, the person attends to the other's thoughts and reasoning, while in the latter, the person attends to the other's feelings. Oswald (1996) demonstrated that affective perspective taking leads to more empathic arousal as well as more helping and cooperative behavior in the perspective taker. Affective perspective taking makes the targets of the perspective taking feel more validated and reduces their subjective experience of pain (Leong et al., 2015). Affective perspective taking also appears to enhance cognitive perspective taking, but the reverse is not true. It is possible to attend to the other's thinking while ignoring emotional cues. While such "cold" perspective taking may be adaptive under certain circumstances such as competitive/distributive negotiations (Galinsky et al., 2008), more cooperative/integrative negotiation situations require accurate understanding of the other's feelings (Fisher & Shapiro, 2006).

The value of perspective taking is, like all behaviors, contextually dependent. In competitive contexts, perspective taking can *increase* competition and *decrease* cooperation if there is an expectation that others will also behave competitively (Epley, Caruso, & Bazerman, 2006). In this situation, attending to the behavior of others appears to create a reactive form of egocentrism.

Human empathic concern involves the transformation of the stimulus functions that can arise when we take the perspective of another, but it is sensitive to relational framing regarding both the self and other. Atkins and Parker (2012), for example, explored how empathy can be interfered with by appraisals that the other is irrelevant to oneself or undeserving of empathy (Goetz, Keltner, & Simon-Thomas, 2010). Similarly, if people fear that they cannot cope with feeling the other's pain, they may divert attention to minimizing their personal distress rather than empathizing with the other (Roseman & Smith, 2001).

A critical determinant of whether we feel empathy or personal distress is whether the self is in a frame of coordination with the other—that is, "if you are upset then I too must be upset" (Atkins, 2013). People who are afraid of conflict due to conditioned emotional responses may feel intense upset when putting themselves in the shoes of others facing conflict. The development of adaptively complex self-other (I/YOU) differentiation appears to involve a dialectic of first separating from the other but then reintegrating through caring for the other's welfare (Kegan, 1994). This process has been manipulated experimentally. When taking the perspective of a person who is suffering, instructions to imagine how the perspective taker would feel if she too were in the same situation can lead to personal distress as well as empathy (Batson, Early, & Salvarani, 1997).

Overall, the evidence suggests that efforts to prompt "warm" perspective taking and empathic concern will be useful in groups, provided that group members are adequately skilled at differentiating their own experience from that of others. Understanding the feelings of others, as well as their thoughts, reasoning, and purposes, is a critical precursor to cooperative behavior.

Psychological Flexibility in the Context of Groups

Psychological flexibility is the ability to connect with the present moment and pursue personally chosen values and goals even when experiencing unwanted psychological events (e.g., difficult thoughts, feelings, images, and so on). It is now well established that psychological flexibility enhances individual well-being (see especially chapters 6, 10, 12, and 16) but it also impacts group-related outcomes such as performance, engagement, and relationships. Cross-sectionally, psychological flexibility is associated with general mental health, job satisfaction, fewer absences from work, work engagement, and task performance (Bond, Lloyd, & Guenole, 2013).

Longitudinally, Bond and Bunce (2000) showed that psychological flexibility mediated improvements in employee well-being and propensity to innovate when compared to a problem-focused intervention. Bond and Flaxman (2006) showed that psychological flexibility predicted general mental health, but also fostered job performance and learning new skills. In this latter study, psychological flexibility also interacted with job control, suggesting that when the situation affords choice, psychological flexibility is even more important. This is an important finding for Prosocial, which is built on the idea of enhancing inclusivity and control within groups. For a review of the effects of psychological flexibility in work settings, see Bond, Lloyd, Flaxman, and Archer (2015).

The effects of psychological flexibility upon quality of relationships have been somewhat less studied. Atkins and Parker (2012) provided a comprehensive model of the ways that psychological flexibility can enhance noticing of suffering in another, reduced reactivity to appraisals that inhibit compassion, enhanced feelings of empathic concern, and committed action in the direction of prosociality. Perspective taking and empathic concern both correlate negatively, while experiential avoidance correlates positively with social anhedonia (Vilardaga, Estévez, Levin, & Hayes, 2012). The same pattern is true for generalized prejudice (Levin et al., 2016). There is some evidence that the mindfulness aspects of psychological flexibility can enhance forgiveness of others who have transgressed. Johns, Allen, and Gordon (2015) studied couples where one partner had been unfaithful and found that, even controlling for general empathy and perspective taking, mindfulness predicted greater forgiveness. Thus, the emotional regulation capabilities associated with mindfulness act over and above empathy and perspective taking.

Psychological flexibility can be understood in terms of three subprocesses: openness, awareness, and engagement. Being willing to have difficult experiences in the service of individual and collective values and goals is critical to effective group functioning. Without openness to experience, group members may not speak up about issues with which they disagree because they fear conflict, or take on too much work because they are afraid of social disapproval, or refuse challenges because they are afraid of failing, or storm out in anger rather than finding ways of staying constructively engaged with the group. In all these cases, people act to alleviate short-term pain and discomfort even at the expense of their long-term interests.

Flexible sense of self is another critical aspect of psychological flexibility relevant to groups. For example, a person might get angry if someone disagrees with her point of view—she may be fearful that the disagreement might reflect badly on her reputation in the eyes of others. Such a reaction is an attempt to defend a particular conceptualization of the self in the eyes of others, such as being competent or kind. From a CBS perspective, this is about having a view of self as a set of category labels that need to be defended rather than as an observer of experience. Building psychological flexibility prior to team development allows people to defuse from thoughts, beliefs, and assumptions about the self, other individuals, and the team, while also enhancing a capacity to accept the inevitable psychological difficulties that arise when we cooperate.

The last subprocess, engagement, involves the clarification and commitment to behaving in line with values. Prosocial values are associated with more prosocial responding (Caprara & Steca, 2007) as well as greater empathic concern and perspective taking (Silfver, Helkama, Lonnqvist, & Verkasalo, 2008). Atkins and Parker (2012) discussed how we are less likely to extend compassion to another if we do not see the other as relevant to our own goals, or if we see the other as undeserving of compassion. For example, Batson, Eklund, Chermok, Hoyt, and Ortiz (2007) showed that people were more likely to attempt to take the perspective of another, extend empathic concern, and ultimately help someone who had acted kindly than of someone who had acted unkindly toward another. The extent and nature of our valuing of others depend upon whether we even seek to cooperate.

Individuals with more prosocial values are more motivated to pursue prosocial goals and, indeed, appear to experience positive affect if they are able to fulfill their prosocial values (Schwartz, 2010). Batson, Ahmad, and Tsang (2002) discuss how individuals may be motivated to help others from values of egoism (helping others in order to accrue personal benefits), altruism (helping specific individuals we care for), collectivism (helping groups we care for), or principlism (upholding moral principles of universal and impartial good). Each of these value sets has strengths and weaknesses, and many people are likely motivated by all of them at different times. What is critical is engendering a discussion of the value of cooperation that helps people clarify and extend their understanding of sources of reinforcement available in cooperating with others.

How Perspective Taking and Psychological Flexibility Transform Ostrom's Design Principles

In this final section, I explore some examples of how perspective taking and psychological flexibility transform the functions of the core design principles of Prosocial. Note that in order to better link the Core Design Principles (CDPs) to the purposes of Prosocial, some minor rewordings of the CDPs are used.

The first CDP is *shared identity and common purpose.* As the group members use the matrix (Polk et al., 2016) to develop both their own and a shared group map of what they care about and what gets in the way, they are operating at two levels. On the surface, this is about sharing information—finding out what we each care about and how those interests and values overlap as a group, and finding out what trips us up separately and when we are together. But on a deeper level, if done well, this process of sharing our hopes and fears builds perspective taking, empathy, safety, trust, and a sense of "we" as a group.

During this process, group members become more vulnerable and human. Moreover, this encourages others to take risks when they may otherwise have let fear inhibit them from speaking up. Group members see how others have the same difficulties, and they learn that most of the labels we use to describe ourselves and others (I am smart, he is stupid, I am introverted, she is extroverted, etc.) create illusory separation and unnecessary suffering. Instead of judging, criticizing, and defending, group members learn to notice what is happening and to explore actions that might help them move toward individual and collective goals, thereby enhancing shared identity and common purpose.

CDP2 is about *equitable distribution of costs and benefits.* Costs and benefits are not the same for everyone. One person might see group leadership as a benefit while another might see it as a cost; others might value or deplore spending time on other tasks. Once we better understand the perspective of others, we can also better understand what they see as a cost or a benefit, and if our values move us in the direction of caring for others, we can work more efficiently to maximize benefits and minimize costs on an individual basis. Furthermore, psychological flexibility can help us accept the pain of momentary or perceived inequity, and still act in the direction of what matters for the long-term aims of the group. The Prosocial process transforms the meaning of relationships to be less about transactions and more about shared effort and co-creation.

CDP3, *inclusive decision making,* is important in Prosocial because it helps build motivation for collective action, at the same time as protecting against individual interests dominating over shared interests. Inclusive decision making can take many forms, including consultation, voting, or consensus, for important but nonurgent decisions. An efficient form of consensus, used widely in organizational and community groups, is to ask if anyone "objects" to a decision, rather than whether it is everyone's first preference. Groups can converge more quickly on final decisions with an emphasis upon canvassing objections, and everybody still feels involved. This principle is at the heart of Sociocracy, Holocracy, and other formalized systems for group collaboration.

Whatever approach is used, enhanced perspective taking and psychological flexibility help group members navigate the turbulent and sometimes emotionally challenging waters of sharing and advocating for different ideas and options. Flexible perspective taking enhances group members' capacity to listen to and value the concerns of others, while considering the consequences of decisions from the perspective of others. Psychological flexibility can help group members to sit with, rather than react to, the discomfort of disagreements, and to exert more job control (Bond & Bunce, 2003). Psychological flexibility also reduces impulsive or risk-averse decision making—when people are more psychologically flexible they are less likely to gamble or to discount the benefits of long-term goals. Psychological flexibility shifts people from being under the control of short-term interests to responding to their long-term and more deeply held interests.

If the group has a leader, psychological flexibility can help that leader handle the stress of making decisions in the face of complexity and uncertainty. Another common dynamic is that leaders make decisions to help them look good, rather than in the interests of the group. As leaders learn that they no longer need to defend their egos, they can free up their attention to listen to others and base their decisions on better information. It is well known that more transformational leaders tend to be those who are better able to act in the face of difficult emotions, such as uncertainty and doubt (Heifetz, Grashow, & Linsky, 2009).

Thus far, the design principles have been primarily about creating optimal conditions for cooperation. The next two CDPs, *tracking agreed-upon behaviors* (CDP4) and *graduated responses to transgressions* (CDP5), are key to managing self-interested behaviors. CDP4 is enhanced by better perspective taking, which increases the likelihood of noticing actual as opposed to expected behavior in others. Furthermore, the matrix generates information about *what* needs to be tracked—valued individual and group behaviors and defensive behaviors that are likely to interfere with effective cooperation. For example, if members of a group know that, when the leader feels insecure, he or she gets more dictatorial or avoidant or even excessively friendly, then they can act to help the leader and the group to act more effectively when these behaviors are noticed. Similarly, if members of a group know that when they get stressed, they are more likely to argue, then they can put in place processes for noticing when that is happening and ameliorate the effects. Tracking is not just about selfish or unhelpful behavior, it can also help align well-meaning, cooperative efforts that are pulling in different directions. Ostrom (1990) found that tracking behavior was most effective if conducted by the group as a whole. Clarity of purpose can help build a shared sense of responsibility and care within the group for performance improvement.

CDP5, *graduated responses to transgressions*, is also transformed by perspective taking and psychological flexibility. Perspective taking will improve understanding of the reasons for transgressions and it can also transform the "feel" of being sanctioned. For example, being dismissed from an organization (i.e., exiled by the group) can be done in a way that is harsh and punitive, or it can be done in a way that maximizes the opportunities for the person leaving the group to learn and grow from the experience. Psychological flexibility is associated with increased willingness to tackle difficult conversations without avoidance, while also noticing and mitigating tendencies to act judgmentally or punitively (out of self-righteousness, for example). Finally, it can help increase forgiveness where appropriate (Johns et al., 2015). Psychological flexibility helps people use their power kindly.

CDP6 is about *fast and empathetic conflict resolution*. Conflict is an inevitable part of cooperation: Whenever people are empowered to bring their whole selves to a cooperative venture, their interests will inevitably diverge at some points. Perspective taking increases the likelihood that different perspectives will be sought and listened to. Psychological flexibility can help to manage the virtually ubiquitous fear of conflict by increasing coping self-efficacy, reducing ego involvement and verbal defensiveness (Heppner et al., 2008), and reducing aggression (Lakey, Kernis, Heppner, & Lance, 2008). Learning to take the perspective of others, to empathize with them and tolerate the pain of conflict, helps transform the meaning and value of conflict from something to be avoided into something to be used for learning and growth. This shift from a punishment orientation to a learning orientation is a profound cultural shift for most groups.

The last two principles of Prosocial are about how the group engages with other groups: *authority to self-govern* (CDP7) and *appropriate relations with other groups* (CDP8). Just as the first six principles afford the possibility of healthy, supportive relationships between individuals, the last two CDPs highlight the need for perspective taking and psychological flexibility at the group level. By clarifying the importance and value of the broader ecology within which the group sits, groups are more likely to care about and notice the perspectives of other groups. And psychological flexibility allows *groups* to move in the direction of what they care about even in the presence of difficult experiences stemming both from inside and outside their groups.

In conclusion, this chapter has explored how contextual behavioral science transforms and deepens our efforts to implement the core design principles of Prosocial. While it is perfectly possible for groups to attempt to improve their performance through adopting practices consistent with the core design principles alone, I have argued that perspective taking and psychological flexibility dramatically increase the likelihood of effective implementation of the principles. While the core design principles tell us broadly *what* we need to do to enhance cooperation, contextual behavioral science tells us *how* we need to do it to optimize both our effectiveness and our humanity.

References

Atkins, P. W. B. (2013). Empathy, self-other differentiation and mindfulness. In K. Pavlovich & K. Krahnke (Eds.), *Organizing through empathy*. New York: Routledge.

Atkins, P. W. B., & Parker, S. K. (2012). Understanding individual compassion in organizations: The role of appraisals and psychological flexibility. *Academy of Management Review, 37*, 524–546.

Batson, C. D., Ahmad, N., & Tsang, J. A. (2002). Four motives for community involvement. *Journal of Social Issues, 58*, 429–445.

Batson, C. D., Early, S., & Salvarani, G. (1997). Perspective taking: Imagining how another feels versus imagining how you would feel. *Personality and Social Psychology Bulletin, 23*, 751–758.

Batson, C. D., Eklund, J. H., Chermok, V. L., Hoyt, J. L., & Ortiz, B. G. (2007). An additional antecedent of empathic concern: Valuing the welfare of the person in need. *Journal of Personality and Social Psychology, 93*, 65–74.

Bond, F. W., & Bunce, D. (2000). Mediators of change in emotion-focused and problem-focused worksite stress management interventions. *Journal of Occupational Health Psychology, 5*, 156–163.

Bond, F. W., & Bunce, D. (2003). The role of acceptance and job control in mental health, job satisfaction, and work performance. *Journal of Applied Psychology, 88*, 1057.

Bond, F. W., & Flaxman, P. E. (2006). The ability of psychological flexibility and job control to predict learning, job performance, and mental health. *Journal of Organizational Behavior Management, 26*, 113–130.

Bond, F. W., Lloyd, J., Flaxman, P. E., & Archer, R. (2015). Psychological flexibility and ACT at work. In R. D. Zettle, S. C. Hayes, D. Barnes-Holmes, & A. Biglan (Eds.), *The Wiley handbook of contextual behavioral science*. West Sussex, UK: Wiley-Blackwell.

Bond, F. W., Lloyd, J., & Guenole, N. (2013). The work-related acceptance and action questionnaire: Initial psychometric findings and their implications for measuring psychological flexibility in specific contexts. *Journal of Occupational and Organizational Psychology, 86*, 331–347.

Buber, M. (1958). *I and Thou* (R. G. Smith, Trans.). New York: Scribner.

Caprara, G. V., & Steca, P. (2007). Prosocial agency: The contribution of values and self-efficacy beliefs to prosocial behavior across ages. *Journal of Social and Clinical Psychology, 26*, 218–239.

Doherty, M. (2012). Theory of mind. In L. McHugh & I. Stewart (Eds.), *The self and perspective taking: Contributions and applications from modern behavioral science.* Oakland, CA: Context Press.

Epley, N., Caruso, E. M., & Bazerman, M. H. (2006). When perspective taking increases taking: Reactive egoism in social interaction. *Journal of Personality and Social Psychology, 91*, 872–889.

Fisher, R., & Shapiro, D. (2006). *Beyond reason: Using emotions as you negotiate.* New York: Penguin.

Galinsky, A. D., Maddux, W. W., Gilin, D., & White, J. B. (2008). Why it pays to get inside the head of your opponent: The differential effects of perspective taking and empathy in negotiations. *Psychological Science, 19*, 378–384.

Goetz, J., Keltner, D., & Simon-Thomas, E. (2010). Compassion: An evolutionary analysis and empirical review. *Psychological Bulletin, 136*, 351–374.

Hayes, S. C., Barnes-Holmes, D., & Roche, B. (2001). *Relational frame theory: A post-Skinnerian account of human language and cognition.* New York: Plenum.

Hayes, S. C., Brownstein, A. J., Haas, J. R., & Greenway, D. E. (1986). Instructions, multiple schedules, and extinction: Distinguishing rule-governed from schedule-controlled behavior. *Journal of the Experimental Analysis of Behavior, 46*, 137–147.

Hayes, S. C., Brownstein, A. J., Zettle, R. D., Rosenfarb, I., & Korn, A. P. (1986). Rule-governed behavior and sensitivity to changing consequences of responding. *Journal of the Experimental Analysis of Behavior, 45*, 237–256.

Hayes, S. C., Gifford, E. V. & Ruckstuhl, L. E., Jr. (1996) Relational frame theory and executive function. In G. R. Lyon & N. A. Krasnegor (Eds.), *Attention, memory, and executive function.* Baltimore: Brookes.

Hayes, S. C., & Sanford, B. (2014). Cooperation came first: Evolution and human cognition. *Journal of the Experimental Analysis of Behavior, 101*, 112–129. doi:10.1002/jeab.64

Heifetz, R. A., Grashow, A., & Linsky, M. (2009). *The practice of adaptive leadership.* Boston: Harvard Business Press.

Heppner, W. L., Kernis, M. H., Lakey, C. E., Campbell, W., Goldman, B. M., Davis, P. J., & Cascio, E. V. (2008). Mindfulness as a means of reducing aggressive behavior: Dispositional and situational evidence. *Aggressive Behavior, 34*, 486–496.

Johns, K. N., Allen, E. S., & Gordon, K. C. (2015). The relationship between mindfulness and forgiveness of infidelity. *Mindfulness, 6*, 1462–1471.

Kegan, R. (1994). *In over our heads: The mental demands of modern life.* Cambridge, MA: Harvard University Press.

Keysar, B., Lin, S. H., & Barr, D. J. (2003). Limits on theory of mind use in adults. *Cognition, 89*, 25–41.

King, M. L., Jr. (1960/2005). The three dimensions of a complete life. Sermon delivered at the Unitarian Church of Germantown. In T. Armstrong, C. Carson, & A. Clay (Eds.), *The papers of Martin Luther King, Jr.: Threshold of a new decade, January 1959–December 1960* (Vol. 5). Berkeley, CA: University of California Press.

Lakey, C. E., Kernis, M. H., Heppner, W. L., & Lance, C. E. (2008). Individual differences in authenticity and mindfulness as predictors of verbal defensiveness. *Journal of Research in Personality, 42,* 230–238.

Leong, L. E. M., Cano, A., Wurm, L. H., Lumley, M. A., & Corley, A. M. (2015). A perspective-taking manipulation leads to greater empathy and less pain during the cold pressor task. *Journal of Pain, 16,* 1176–1185.

Levin, M. E., Luoma, J. B., Vilardaga, R., Lillis, J., Nobles, R., & Hayes, S. C. (2016). Examining the role of psychological inflexibility, perspective taking, and empathic concern in generalized prejudice. *Journal of Applied Social Psychology, 46,* 180–191.

McHugh, L., Barnes-Holmes, Y., & Barnes-Holmes, D. (2004). Perspective-taking as relational responding: A developmental profile. *Psychological Record, 54,* 115–144.

Mead, G. H. (1934). *Mind, self and society.* Chicago: University of Chicago Press.

Ostrom, E. (1990). *Governing the commons: The evolution of institutions for collective action.* Cambridge, UK: Cambridge University Press.

Oswald, P. A. (1996). The effects of cognitive and affective perspective taking on empathic concern and altruistic helping. *Journal of Social Psychology, 136,* 613–623.

Piaget, J. (1969). *The psychology of the child.* New York: Harper.

Polk, K. L., Schoendorff, B., Webster, M., & Olaz, F. O. (2016). *The essential guide to the ACT matrix.* Oakland, CA: Context Press.

Roseman, I. J., & Smith, C. A. (2001). Appraisal theory: Overview, assumptions, varieties, controversies. In K. Scherer & T. Johnstone (Eds.), *Appraisal processes in emotion: Theory, methods, research.* New York: Oxford University Press.

Schwartz, S. H. (2010). Basic values: How they motivate and inhibit prosocial behavior. In M. Mikulincer & P. R. Shaver (Eds.), *Prosocial motives, emotions, and behavior: The better angels of our nature.* Washington, DC: American Psychological Association.

Silfver, M., Helkama, K., Lonnqvist, J. E., & Verkasalo, M. (2008). The relation between value priorities and proneness to guilt, shame, and empathy. *Motivation and Emotion, 32,* 69–80.

Vilardaga, R., Estévez, A., Levin, M. E., & Hayes, S. C. (2012). Deictic relational responding, empathy and experiential avoidance as predictors of social anhedonia: Further contributions from relational frame theory. *Psychological Record, 62,* 409–432.

Weil, T. M., Hayes, S. C., & Capurro, P. (2011). Establishing a deictic relational repertoire in young children. *Psychological Record, 61,* 371–390.

Wilson, D. S., Ostrom, E., & Cox, M. E. (2013). Generalizing the core design principles for the efficacy of groups. *Journal of Economic Behavior and Organization, 90,* Supplement(0), S21–S32. doi:http://dx.doi.org/10.1016/j.jebo.2012.12.010

Dialogue on Small Groups

Participants: Paul W. B. Atkins, Steven C. Hayes, David Sloan Wilson

Steven C. Hayes: The readers need to know that all three of us have worked together on the Prosocial project for years. At this point in the book, people have read about Prosocial, so perhaps we might just start with a few random comments on things that are notable or unexpected about the Prosocial project.

Paul W. B. Atkins: What we're really about is managing the relationship between group interests and individual interests and the interests of the broader system. We're managing three sets of relationships. And I think that the way that we introduce psychological flexibility and perspective taking puts the individual in the picture in a way that allows us to then have a profound conversation about the relationship between self-interest and group interest.

David Sloan Wilson: If you're an economist, and if you're in many business groups, then you think you know it's all about self-interest. That's the only thing you ever appeal to. The design principles in Prosocial provide a different kind of functional blueprint for what a group wants to become. The first element, psychological flexibility, increases the capacity for change. Change is difficult. I've started to think about ACT—acceptance and commitment therapy—as an evolutionary process at the individual level that changes evolution at other levels.

Steve: All forms of change can be thought of from an evolutionary point of view. Psychotherapy and behavioral change have to do with variation, selection, and retention, in context at the right dimension and level. Preparing individuals for change is a matter of creating enough flexibility so that there is some variation from which to select, and noting the context so that you can do it wisely—and then focusing on what the selection criteria are, so that you build new patterns you really want. But you need to do it also at the right level. I imagine a number of readers coming in and just hearing about contextual behavioral science and evolution science would be thinking, "The biological unit is the small unit. And the psychological unit is the bigger unit." And here suddenly the individual is seen in the context of a larger group, but the larger unit is viewed from the point of view of evolution science. So that's interesting.

Maybe this would be a way to enter back into the design principles themselves. The design principles, of course, were originally developed for common-pool resources and Lin Ostrom's Nobel was won for that— not for a change method. Does that change how her work lands when you use it with groups?

David: It's totally the case that her work is studied mostly in the context of common-pool resource groups. And we generalized it in two ways. One is to show that the design principles follow very generally from multi-level selection theory, from the evolutionary dynamics of cooperation. Once you take that on, it becomes clear that these design principles should really apply to all groups. The Zeitgeist in business groups is often one of complete selfishness, social Darwinism; we expect that everyone is motivated by self-interest, and presume that a competition within groups is a good thing. The whole model of leadership is often very top-down and so in the context of some backgrounds, the core design principles, as a blueprint for a well-functioning group, are highly counterintuitive and counseled against by the prevailing worldview.

Paul: "Commoning" is an alternative to this dominant worldview, *Homo economicus*. Ostrom was really arguing against just relying on top-down regulation or competitive market forces between individuals. She was pointing to this third way, which is the possibility of creating shared agreements that coordinate activity. CBS really adds something about the power of the stories that we tell ourselves. If we say we're self-interested economic beings, then that's what we become. But if we say there's this possibility of cooperation, let's look at small groups as the unit of the analysis, then that might be more helpful for us, as a way of organizing society. So for me there's a deep link between Ostrom's work and prosociality.

David: This paradigm is a third way between total laissez-faire on one hand, and command-and-control on the other hand. It's called polycentric governance, because you have to address the larger scales in addition to the smaller scales. Those design principles—they've got teeth. I mean, it's like altruism with attitude, basically. They really dictate what you can and cannot do. The first design principle defines the group and who's in it, and if you're not in it, you don't get to share the common-pool resource. So prosociality requires not just figuring out what's prosocial within the group, but also how does a group fit into a larger, multigroup social organization that *in turn* is prosocial.

Steve: Can we look at this idea that units of selection are privileged? At each level you can't understand the higher level by understanding the ones that are still higher and lower. For example, cultural practices may be selected at a very high level, but they are tied to individual behaviors that may be selected within the lifetime of the individual. At each level, we can't just stay at that level in order to understand it. Does that change your view, David, about this unit of selection being privileged?

David: If you want some unit to be adaptive, to be sustainable, to be functional, then the selection has to be at the level of that unit. The invisible hand pretends that the maximization of lower-level self-interest robustly benefits the common good. That is profoundly not the case. You can have a group that's been selected at the group level, that functions well as a group. But when you delve into the mechanisms, you find that the group actually does not require the members to have the welfare of the group in mind. And just to give a really simple example, reputation is huge in human groups. One reason that people behave for the good of their groups is that they're passionate about their reputations. If they misbehave, their reputation will tank, and then they'll do poorly as an individual. That's a pretty self-interested motive. And yet at the same time, it's a powerful mechanism for the group to work well.

Paul: It's really important that we become more skilled at talking about the selection forces at multiple levels at the same time. The narrative is not "Let's all be more cooperative." The narrative is "How can we understand the needs and the selection pressures on multiple levels at the same time?" Individuals, small groups, large groups. When we privilege a group, we're sort of privileging relationship and love and connection. When we privilege the individual, we're tending to privilege individual power or capacity to achieve one's own ends. I think that's the narrative we're trying to create in the work that we're doing together.

David: When we think, *What kind of group do we want it to be?* we're operating in ultimate causation mode. It's in that sense that unless we think on behalf of the whole group, then the group is not going to function well.

Steve: I imagine a lot of people who are interested in these particular chapters in the book have practical reasons for being interested. It could be that they actually work with groups; they certainly all are members of groups. What could you say about somebody who came into this and said, "Well, why would I think that this will actually be more effective than other group methods that are out there?" Are these design principles—as augmented by the psychological flexibility, perspective-taking kind of things from the CBS point of view—are they a relatively coherent set?

Paul: What strikes me about the principles when I look at them is that there are different levels of generality. Monitoring graded behaviors and graded sanctions is quite a specific prescription for how to do things. But the notion of equity of distribution of costs and benefits seems to happen at a much broader level of generality. Having appropriate relations to other groups is not about the group itself but the next level up. I think that's incredibly helpful because it builds on the whole notion that we've been talking about of "groups all the way down" and the need to be able to flexibly move between different levels of organization and understand selection by consequences at each level of organization. So that's all a long way of saying I think the principles are coherent, and I think they're functional, but I also think they're by no means the last word.

David: I'd like to remind people what Ostrom's done for the common-pool resource group. She found that they varied in their efficacy. There was a bell-shaped curve. Some performed beautifully, others were train wrecks, most were in the middle. And then those core design principles basically explain the variation. Now when we generalize the core design principles, we expect to find the same things for all kinds of groups. If I were to study business groups or churches or neighborhoods or whatever, I would expect to find a bell-shaped curve. Some would work beautifully, nobody had to teach them; others would be train wrecks; most would be in the middle. So we're not saying that nothing else works. That's not the case.

Steve: One of the major reasons to believe that these are relatively complete is the consilience with evolutionary theory. And what has happened here in this project, we have two evolutionarily focused perspectives. One in the behavioral sciences, one from biological sciences.

Paul: What we're building is a set of abstract principles that are broad enough to be flexible in many, many contexts—it means that people have to do the work to apply it to their own particular context; they need to make it much more specific.

David: Ostrom made this point herself, that the core design principles are functional principles and their implementation needs to be highly contextual, back to a form of contextualism. Every group needs to monitor, but on the other hand exactly how they monitor depends, richly depends, on the particular situation. For every functional arrangement, there are many possible implementations. It happens that way a lot in evolution, and it happens that way a lot in groups.

Steve: Part of what's hard about a functional approach is that you do have to hold off on these more topographical, just-do-it-this-way kind of solutions. That's been true historically with functional approaches in the behavioral sciences. I'd like to turn to a slightly different direction. Would you agree with me that evolution science has had a relatively hard time producing a robust applied science?

David: Well, let's look at applications just within the biological sciences—you get into things like animal and plant breeding and things like that. Now if you ask the same question within the human-related sciences, you'll get a completely different answer. But I think the reason for that is this terrible, complex history of evolutionary thinking in human affairs. Evolutionary thinking was exorcised from human behavioral sciences for the most part and it's only now that it's being given a fair shot.

Paul: So I think we're starting to talk about multiple strings of evolution or dimensions of evolution, and cultural evolution. And that's entering the popular press, the popular parlance, to a much greater extent than it had been previously.

David: More and more I've been speaking in terms of evolution and complexity. When we're trying to change a complex system, then nobody's smart enough for that. There will always be unforeseen consequences. This forces a certain strategy of adaptive change, which is highly cautious, highly experimental; likely to be highly contextual, so that what works in one context is not going to work in another.

Steve: If you think of processes just as sequences of change, you can become overwhelmed by the number of processes that are involved in every given situation. But if you think of the processes that fit within an evolutionary perspective, variations that can be selected and impact on selection and retention in other parts of the system—things that can actually be changed in some way, and when they do, they impact in that way—you go from a purely descriptive, almost overwhelming level of complexity to something much simpler. Evolutionary thinking gives you a small set of analytic tools with which to evaluate that complexity and to enter into it.

All the more reason that there should be a robust applied evolution science in the behavioral domain. And this project is an attempt to sort of begin to do that by bringing people who are used to doing interventions at the psychological level into contact with those who are used to thinking about evolutionary principles at higher levels.

David: There seem to be so many ways that you can go wrong. I think a more sophisticated evolutionary approach would help solve some of those problems.

Steve: What do you think you've learned about the central topic of the book itself? Bringing reconciliation between behavioral science perspectives and evolution science.

Paul: Sure. Well, I've learned an awful lot from both of you. Mechanistic psychologists tend to be individualistic. I'd always been dissatisfied with that. I'd always had a leaning toward much more systemic, process-based perspectives. Prosocial allows us to ask how we can bring science into the discussion of alternative models of organizing that really privilege human caring, and relationships, and spirituality, and all the things that matter so much beyond the dominant economic narrative.

David: My relationship with the CBS folks is really one of the most important things that's happened to me in my life. And I'll venture the prediction that the CBS side is going to take on evolution a long time before the evolutionists are going to take on CBS. I think there's a number of barriers that my colleagues face. One is this ridiculous polarity that got set up between evolutionary psychology and the so-called standard social science model. So even those evolutionists that are thinking about psychology, for example, don't see the Skinnerian tradition or anything that followed it as part of an evolutionary perspective. So that's something that needs to be overcome. It's going to take quite a lot of time. And then another barrier which really makes me sad is the whole concept of pragmatism and contextual behavioral science—that we can do science in a real-world context, and that we could actually alleviate suffering, improve the world in various ways, and remain in contact with the highest quality of basic science. That is something that most of my colleagues don't get. The idea of doing something that's like that, it's just way beyond their comfort zone. They want to stay in the ivory tower.

Steve: Well, that's actually what I was pointing to earlier, David, when I said it's been hard to get a really robust applied evolution science going, from the standpoint of the behavioral science side. Doing so, long run, will likely have a much more important impact than anything CBS per se can produce. But for the immediate future, the CBS crew, I think, has really shown a receptivity to those concepts and they're willing to just get out there and do it. And one of the things I've said to my CBS colleagues is that this is very important for another reason that you may not be fully aware of. It isn't by accident that at the point at which the behavior analysts and behavior modifiers were stepping forward culturally, all of a sudden they were being slapped down with *Clockwork Orange* and the conflation of brain washing or psychosurgery with trying to make a positive difference in culture and families and communities. I think the same thing would have happened if people on the evolution science side had stepped forward. Cultures resist change—not just individuals. But the therapists are used to this. And the evolution scientists are going to have to walk through that territory in order to really get into producing direct cultural change. So this is an interesting partnership that we've got going. Do you have thoughts on that?

David: Well, when we approached the evolutionists with this volume, we had a very high rate of acceptance. We picked the best and the brightest, we contacted them with our project, and most of them said, "I'll write it!" So there is a sense in which I think maybe I overstated their reluctance. And I would love to think that our volume, and these webinars we're filming, will be a giant step toward that unification and moving both behavioral and evolution science forward. That's our goal.

CHAPTER 16

Variation and Selection in Psychopathology and Psychotherapy: The Example of Psychological Inflexibility

Steven C. Hayes

University of Nevada, Reno

Jean-Louis Monestès

LIP/PC2S, Grenoble Alpes University, France

Why do human being suffer amidst plenty? We can understand the suffering of those who face deprivation or war, but that hardly explains the increase of mental and behavioral health challenges in the developed world. Thirty years ago, depression was the fourth leading cause of disability and disease worldwide, after respiratory infections, diarrheal illnesses, and prenatal conditions. Twenty years ago it was the third leading cause. Ten years ago it became the second (Ferrari et al., 2013).

There are many statistics like that, and they can easily tempt us into the details. For example, we might dive into what depression is, and generate ideas about why it might be increasing. But depression is only a small part of the picture. The next strand would then have to be picked up (perhaps chronic pain, addiction, or anxiety) and treated in the same way. As behavioral causes for the many faces of poor mental and behavioral health were examined one at a time, we would almost certainly lose the big picture. We would lose the original question. Why do human beings suffer amidst plenty?

One way to proceed with a big question of that kind is to focus on well-researched processes that are known to lead to poor mental outcomes in many different areas. We can then consider why this process is toxic and what is known about how to change it. The advantage of such a strategy is that it turns the central question into one with an empirical focus, but without losing track of the big picture.

In this chapter, we will briefly examine psychological inflexibility (Hayes, Luoma, Bond, Masuda, & Lillis, 2006) and its subcomponents, and will consider why it is problematic and how it can be changed. Having established an example of knowledge in contextual behavioral science, we will then examine how the components of psychological inflexibility deal with variation and selection, and how contextual behavior science fits with modern evolution science, before turning back to our original big question.

Psychological Inflexibility

Psychological inflexibility constitutes one of the more studied psychological processes. If you enter "psychological inflexibility" and its various specific subprocesses into search engines such as "Web of Science," several hundred studies will appear. In its positive form, psychological flexibility refers generally to being open, aware, and actively engaged. In contrast, psychological inflexibility corresponds to interacting with the world in a way that is psychologically closed, mindless, disconnected, and often functionally repetitive.

Psychological inflexibility is arguably one of the more toxic psychological process known, as defined by its breadth and magnitude of impact and its tendency to mediate or moderate other psychological processes known to be problematic (Chawla & Ostafin, 2007). High-quality longitudinal and experience-sampling studies show that psychological inflexibility predicts the development and elaboration of psychological distress, over and above baseline levels (e.g., Kashdan, Barrios, Forsyth, & Steger, 2006; Spinhoven, Drost, de Rooij, van Hemert, & Penninx, 2014). Cross-sectional (e.g., Bond et al., 2011) and meditational studies (e.g., Fledderus, Bohlmeijer, & Pieterse, 2010) point in the same direction.

It is an odd empirical fact: when emotions and thoughts are avoided (Abramowitz, Tolin, & Street, 2001), or are treated as literally true (Gillanders et al., 2014), their frequency or intensity tends to increase, leading to still more escape, avoidance, and belief. Each of the six subcomponents of psychological inflexibility (experiential avoidance, cognitive fusion, inflexible attention to the past and future, dominance of the conceptualized self, deficits in valuing, and the inability to make and keep behavioral commitments) have this same basic pattern: they tend to foster a self-amplifying loop in which the output of a process is the input to that same process (see Hayes, Sanford, & Feeney, 2015).

Psychological inflexibility can, however, be changed, and when it is, these negative trajectories turn positive. Although a number of methods alter psychological inflexibility, acceptance and commitment therapy (ACT: Hayes, Strosahl, & Wilson, 2012) is an intervention method that is consciously based on psychological inflexibility as its core target for change. As an example, consider the inflexibility process of "cognitive fusion," the tendency for verbal rules to dominate over other sources of behavioral regulation. ACT targets this process using "cognitive defusion" methods that are designed to reduce the behavioral impact of difficult self-statements. One such method is word repetition (Masuda, Hayes, Sackett, & Twohig, 2004), in which challenging thoughts are distilled down to a single word, which is then repeated over and over for about 30 seconds until it loses all meaning. Another such method is silly voices (Eilers & Hayes, 2015), in which difficult thoughts are repeated aloud in the voice of cartoon characters or least-favored politicians. As another example, ACT targets weaknesses in values clarity by helping people explore chosen qualities of being and doing, an intervention process that alone changes wide varieties of action (Chase et al., 2013). Over 200 randomized trials and over 50 mediational studies show that ACT is helpful across a wide range of mental and behavioral health problems, and is helpful because it alters psychological inflexibility (an updated list of such studies can be found at http://www.contextualscience.org/state_of_the_act_evidence).

Psychopathology as a Loss in Functional Behavioral Variability

From a contextual behavioral science point of view, we can think of a psychological organism as a group populated by many different individual behaviors and "species" of behaviors, all interacting with each other and with their environment. These behaviors vary and are selected within the lifetime of the organism by means of learning processes, such as operant and classical conditioning, that allow for ontogenetic adaptation. They are passed across generations by such processes as social learning, cultural evolution, niche construction, and genetic assimilation.

Evolution writ large depends on variation and selective retention (Campbell, 1960). Intentional evolution requires the application of these evolutionary processes in the right context, on the right dimension, and at the right level of selection (Wilson, Hayes, Biglan, & Embry, 2014). Given the extensive body of evidence on psychological inflexibility as a behavioral pathogen, and on positive outcomes when inflexibility is changed, it seems worth examining these six evolutionary processes (variation, selection, retention, context, dimension, and level) when considering the development and change of behavior. That said, given the shortness of the present chapter, we will focus most heavily on variation and selection, while just mentioning the other four processes (see Hayes, Monestès, & Wilson, 2018)

Variation needs to be thought of in both a functional and a formal sense. Actions are defined functionally when we focus on how they transition the organism from a given antecedent state of affairs to a given consequential condition. For example, actions that lead to successful prey capture, successful mating, and the construction of favorable environments for child rearing are functionally distinct.

Formal variability in behavior is a matter of having many or few behavioral topographies within a given functional type. For example, if I know many routes to the store, I can get there easily by changing directions if a particular road is blocked. If I know only one route, I will not be certain to get there if the same thing happens.

Psychopathology can result from either formal or functional behavioral rigidity (or both). When change is necessary, those with both greater formal and functional diversity of behaviors are more likely to change and adapt.

Contingency learning can both increase and decrease behavior variability, depending on the type of variability being considered. Operant learning increases entire classes of formal behavioral variations, all serving a common function. For example, if a young bird sees other birds removing the tops off milk bottles, the bottle top may acquire the function of a food source, but trial-and-error learning may establish the particular forms of action that remove the top (Sherry & Galef, 1984). Learning allows the exploration of the environment, resulting in the selection of environmental niches, which in turn changes the contacted contingencies that establish new (i.e., functionally different) behaviors through reinforcement. When a behavior alters the physical, behavioral, social, or cultural niche in which it takes place, a positive behavioral loop (Monestès, 2016) can be established that further increases functional variation. Thus, learning can increase both formal and functional variation.

But learning can also, under some conditions, *narrow* behavioral variability. If one set of antecedents or consequences dominates over others, reduction in functional variability may crowd out other behavior types. For example, a person addicted to a drug may become disinterested in his family, or sex, or educational pursuits as the proportion of his day functionally linked to the drug increases. In this case, functional variation is narrowed. If access to the drug is cut off, enormous *formal* variability may be seen—doctor shopping, criminal activity, conning family members—but they are all in the service of a given functional behavioral species: the addiction itself. In this case, formal variation increased, but functional variation remained low.

Thus, each individual's behavioral repertoire can manifest itself with high or low levels of formal and functional variation. Table 1 presents the different cases of high and low formal and functional levels of variation and their common adaptive or problematic (i.e., psychopathological) outcomes.

Table 1. Outcomes of Interactions between High and Low Levels of Functional and Formal Variations

| | | FUNCTIONAL VARIATION | |
		LOW LEVEL	HIGH LEVEL
FORMAL VARIATION	**LOW LEVEL**	STEREOTYPY	CREATIVE SYMBOLISM
	HIGH LEVEL	UNPRODUCTIVE EXPLORATION (INFLEXIBILITY)	ADAPTIVE EXPLORATION

At one extreme, formal and functional variation are at a high level, which is common in the normal exploration of the environment that results in discovering adaptive ways of behaving or of maintaining adaptation to the environment. Functional and formal variations can also evolve in opposite directions. While functional variation is uncommon at a low level of formal variation, it is nonetheless possible, at least for humans, via creative symbolism, that is, via the attribution of different meanings to different occurrences of the same behavior. For example, eating an egg can be to increase access to nutrition or to curry favor with the goddess Éostre.

It is our argument that psychopathological issues correspond most closely to low levels of functional variation. One case corresponds to minimal formal variation and low functional variation, what we are calling here stereotypical behaviors, such as what can be observed in autism. Indeed, repetitive behaviors of people with autistic spectrum disorders tend to correlate with the severity of the disorder (Mirenda et al., 2010), notably because they represent obstacles to adaptive behaviors. The other case of psychological issues corresponds to high levels of formal variation with minimal functional variation, what we are calling here "inflexibility." This corresponds to an unproductive exploration of the environment in search of the same consequences. We already used the example of drug addiction, but another example would be attempting to avoid difficult emotions. Anxiety disordered clients, for example, commonly try a wide variety of formally different behaviors to reach a single desired function—getting rid of the anxiety. From the outside, the behavior looks extremely variable—such as avoiding giving a talk, drinking alcohol before social events, or running ten miles per day to downregulate arousal—but it is in fact highly invariant in a functional sense. Many forms of psychopathology appear to be characterized by engagement in myriad ways that a specific but unhelpful function can be achieved.

Language as the Main Source of Functional Variation Loss in Humans

Why would human psychology be readily dominated by relatively rigid forms of behavior even in relatively successful economic and social situations? One possibility is that characteristic forms of human language and cognition are central to both of these results. Said in another way, our greatest success evolutionarily speaking is our greatest weakness.

Human language and cognition can enormously impact the formal variation of behavior: they can increase it or decrease it. For example, the simple instruction "Never tell the same joke twice" might lead to a virtually endless string of different behavioral forms (i.e., different jokes), each with a broadly similar function (i.e., making people laugh). Conversely, with a compliant child, the instruction "Don't ever go out into the street without an adult" could virtually eliminate that topography from their repertoire, even for situations where it could be adapted (house on fire, for example). Underlying relational learning processes account for this effect. There is considerable evidence that human language and cognition are relational, not associative (Hayes, Barnes-Holmes, & Roche, 2001). The ability to derive relationships regardless of the form of related events begins with the mutual entailment of an object-name relation, such that when an object gives rise to characteristic signs, those signs will often lead to provision of the object by derivation (and vice versa). This likely first emerged as a cooperative social process (for that analysis, see Hayes & Sanford, 2014), but it rapidly extended to relational (that is, symbolic) networks based on a wide variety of different types of relationships (difference, opposition, comparison, hierarchy, causality, and so on). This view is explored in chapter 4 of the current volume.

The impact of relational learning on formal behavioral phenotypes has been shown in myriad studies (Dymond & Roche, 2013) that largely conform to predictions made decades ago when the relational nature of human language was just beginning to be understood (Hayes, Gifford, & Ruckstuhl, 1996). Language does not just alter formal behavioral variation, however. It can change other evolutionarily relevant processes (see Hayes et al., 2018, for a more comprehensive description). Language can make remote or probabilistic consequences effective, or create verbal motivational processes. At the same time, language and cognition can decrease functional variability by diminishing control over behavior by nonverbal processes. Like an invasive plant can do to other plants, human symbolic behavior rapidly crowds out other behaviors. This reduces behavioral functions in the sense that symbolically established predictive and comparative functions begin to dominate over all others. For example, when faced with difficulties, humans tend to plan, reason, and problem-solve—all symbolically established functions. This begins naturally to turn human thought, emotion, memory, or sensation from events to be experienced into problems to be solved. If someone tells you that losing love is horrifically painful, you may begin to look at your partner's loving reactions with suspicion and fear of painful disappointments to come. If something important (such as the opportunity to get a new and much desired job) produces anxiety, you may try to tamp down the fear by withdrawing from the opportunity. In other words, psychological inflexibility is in part a natural extension of human symbolic problem solving.

Human verbal behavior dominates other forms of action for several reasons. One is that it is so broadly useful, and thus tends to be overused. When verbal behaviors are selected, they diminish the ability of other behaviors to survive, or for new ones to appear, simply because time and attention are limited resources. Selection of behaviors can diminish variation in a functional sense even if it increases variation in a formal sense (Neuringer, 1986, 2002, 2004). If many moments are characterized by the participation of human thought, reasoning, or verbal communication, over time, fewer and fewer behaviors of other kinds stand unaffected.

In addition, human verbal relations tend to become dominant because they are symbolic, in the sense of being relational responses that are somewhat detached from the physical forms and contingencies. This property makes them an "all terrain" behavior. Language and cognition can potentially apply to anything, anywhere, and at any time. For example, when someone is told that he mustn't be weak, anything that can be analyzed as a sign of weakness can become painful, from a B+ at school to bad luck in games to experiencing a painful but normal emotion. This, in turn, decreases sensitivity to actual consequences of action (Monestès, Villatte, Stewart, & Loas, 2014).

The social reinforcement that comes from being right in creating verbal understanding, and the immediate sense of coherence that comes from assembling consistent verbal networks (see Quinones & Hayes, 2014), can sustain verbal networks even in the absence of other forms of behavioral effectiveness. Understanding gives humans the illusion of mastering the world merely by being able to describe it. Patients with a rash of unknown origin may be distressed because they do not know what they have, while patients who are told they have idiopathic dermatitis will likely be reassured, even though it means exactly the same thing.

As functional variability decreases and functions that can readily be combined with literal verbal events increase in an assimilated form, other functions that do not combine well with literal, evaluative, predictive verbal events may weaken. Awe, a sense of transcendence, love, or peace of mind may all wane as a single functional class of behaviors becomes more and more dominant. It is a kind of behavioral selfishness in which other types of action that could be useful as part of a cooperative group of actions get little time or attention. Elements of the verbal problem-solving repertoire selfishly get more time and attention than is their functional due. This is an example of a kind of adaptive peak, in which processes produce self-amplifying loops or self-sustaining processes that restrict further development via normal evolutionary processes.

Let's expand on an earlier example that does not require human language and cognition so that the process can be understood: drug addiction. The consumption of addictive substances can produce powerful reinforcers and gradually crowd out other forms of behavior. Each step in the "hill climbing" of an addiction can be highly appetitive, but the addiction's self-amplifying and self-sustaining qualities may eventually so reduce access to other behavioral events that there is little time for anything else. Further development via normal evolutionary processes is restricted by the highly aversive effects of withdrawal from addictive substances.

In summary, we are arguing that psychopathology is a form of symbolic problem solving that disrupts the normal interplay of the expansion and contraction of variability, thereby lessening functional variability. Verbal relations can foster this change for the reasons specified above: their general utility, their "all terrain" nature, the ubiquity of social support, and the reinforcing effects of literal coherence. Said in a colloquial way, we begin to live in our heads as psychopathology takes over because the very source of our highest achievements, human symbolic problem solving, is fostering a self-sustaining contraction of functional variability.

Successful Psychotherapy as an Increase in Flexibility

Psychotherapy and intentional behavior change can be thought of as forms of applied evolution science, with the recovering of variability selected by chosen qualities of action as the main goal. Successful psychological intervention promotes context-appropriate selection and retention of successful variations of the right dimensions at the right level of selection (Hayes & Sanford, 2015; Wilson et al., 2014).

It is the task of psychotherapy to increase context-appropriate functional flexibility in all relevant behavioral domains, including overt behavior, emotion, and cognition. This is suggested by the fact that changes in these forms of psychological flexibility are known to mediate outcomes in many forms of therapy (e.g., Arch, Wolitzky-Taylor, Eifert, & Craske, 2012).

Consider how this might be done. For one thing, psychotherapy is deliberately built on an intimate, caring relationship, but it turns out that relationships of that kind tend themselves to model psychological flexibility. Instead of experiential avoidance, caring therapists model experiential openness. Almost any feeling, memory, urge, or action can be talked about openly, and the therapist is unlikely to recoil in horror or wag a finger of shame. Judgmental attitudes are put to the side and instead, the experiences of the clients are examined with a sense of dispassionate curiosity. Rigid attention to the past, in the form of rumination, or to the future, in the form of worry, is rarely supported and instead, the focus is on what can be done now. By being willing to explore all aspects of a client's life more openly, awareness is modeled over maintaining a narrow process of self-presentation. What the client deeply cares about is given center stage, and actions are evaluated based on their benefits to a vital life. Phrased broadly as we have just done, you can see that a therapeutic relationship in psychotherapy tends to model psychological flexibility. If clients internalize that environment, their own flexibility increases.

Some studies have looked at this idea by measuring both the quality of the working alliance in therapy, and the psychological flexibility of clients (e.g., Gifford et al., 2011). Both of these processes mediate therapy outcomes, but when they are placed in a multiple mediator model, the client's flexibility reduces or eliminates the impact of the working alliance (Gifford et al., 2011). There is an easy way to put this result into words: modeling and supporting psychological flexibility in the relationship matters in outcomes, but the impact of this relationship on clients is determined by the degree to which they internalize it and then adopt this posture with themselves. What remains of the therapeutic relationship after that effect is statistically abstracted is not helpful in terms of outcome.

The goal of psychotherapy is in part to reduce the domination of literal, temporal, judgmental language that makes open contact with the inner and outer world impossible, and to establish more flexible responses that are free to maintain their contact with the current context and that can be selected based on their promotion of qualities of being and doing, that is, their promotion of human values (Ritzert, Forsyth, Berghoff, Barnes-Holmes, & Nicholson, 2015). In this way, literal, temporal, judgmental cognition can be used when it is useful (e.g., when repairing a car by following a manual), but other forms of action (e.g., observation, description, perspective taking, feeling, appreciation, nonverbal behavior) can come to the fore when they are more useful (e.g., when seeking peace of mind).

The process underlying changes of this kind fits the model we have been proposing: new functions are being established. If particular emotions have strong avoidance functions, acceptance work adds other functions to those same emotions, such as curiosity, observation, or description. If frightening or judgmental thoughts pull for argument, compliance, or resistance, defusion work adds other functions such as play, acknowledgment, humor, or self-kindness. Adding new functions leads to greater behavioral flexibility, and the consequences of action can begin to select what works.

As symbolic behavior moves away from avoidant forms of problem solving, it can instead be used to increase functional flexibility—low formal variation can be related to high functional variation (what we called "creative symbolism" in Table 1). For verbal organisms, any behavior's function can be verbally transformed without any formal change of that behavior, simply by introducing different contextual cues that bring attention to different relational networks focused on the consequences of action. The same behavior can be emitted in the same way while having different meanings for the person, provided that the behavior is linked to different relational networks. For example, suppose a person working on an assembly line is suffering from the boring repetition of movements, raising her anger against her parents who refused to pay for her higher education. But at that very same moment she might also see her commitment as a proof of her dedication to her children, or a challenge for her to find the best way to optimize the manufacturing process. In this case, the function of repetitively similar actions could be highly varied, from punishing to reinforcing, from boredom to a challenge, depending on how contextual cues point to the symbolic meaning of the action (i.e., to what this action "stands for").

Far from being a passive activity, when combined with openness and awareness, such transformations might encourage a worker to seek changes in the work environment, such as to agitate for a "take your child to work" day, or to insist on safer work conditions, so that work can actually foster family success. This is not hypothetical: it is an empirical fact that acceptance and values interventions naturally impact work environments in this way (Bond & Bunce, 2000).

The entire psychological flexibility model can be understood in terms of the six key concepts (variation, selection, retention, context, dimension, and level) necessary to intentional use of evolution science. In this chapter, we have emphasized variation and selection, but therapists also increase flexible attention to the inner and outer context of action, foster retention by practice, keep dimensions of action in focus (e.g., not allowing the symbolic dimension to crowd out other inheritance streams), and consider levels of selection (e.g., fostering positive outcomes for the whole person, not a dominant behavioral component; or fostering the success of couples and families, not just individuals). These ideas are more fully explored elsewhere (e.g., Hayes & Sanford, 2015; Hayes et al., 2018).

Conclusion

Contextual behavioral science is a form of behavioral science that is consciously located under the umbrella of evolution science. We focused in this chapter on variation and selection as they impact psychopathology and behavior change. Psychopathology involves formal or functional inflexibility, and psychotherapy helps by adding in new forms of flexibility and linking these to context and to values-based selection. By thinking of applied psychological work as applied evolution science, the range of issues is simplified without being simplistic—exactly the combination that seems likely to be practically helpful to those who work in the business of deliberate behavior change.

References

Abramowitz, J. S., Tolin, D. F., & Street, G. P. (2001). Paradoxical effects of thought suppression: A meta-analysis of controlled studies. *Clinical Psychology Review, 21,* 683–703.

Arch, J. J., Wolitzky-Taylor, K. B., Eifert, G. H., & Craske, M. G. (2012). Longitudinal treatment mediation of traditional cognitive behavioral therapy and acceptance and commitment therapy for anxiety disorders. *Behaviour Research and Therapy, 50,* 469–478.

Bond, F. W., & Bunce, D. (2000). Mediators of change in emotion-focused and problem-focused worksite stress management interventions. *Journal of Occupational Health Psychology, 5,* 156–163.

Bond, F. W., Hayes, S. C., Baer, R. A., Carpenter, K. M., Guenole, N., Orcutt, H. K., … Zettle, R. D. (2011). Preliminary psychometric properties of the Acceptance and Action Questionnaire–II: A revised measure of psychological inflexibility and experiential avoidance. *Behavior Therapy, 42,* 676–688.

Campbell, D. T. (1960). Blind variation and selective retention in creative thought and other knowledge processes. *Psychological Review, 67*, 380–400.

Chase, J. A., Houmanfar, R., Hayes, S. C., Ward, T. A., Vilardaga, J. P., & Follette, V. M. (2013). Values are not just goals: Online ACT-based values training adds to goal-setting in improving undergraduate college student performance. *Journal of Contextual Behavioral Science, 2*, 79–84.

Chawla, N., & Ostafin, B. (2007). Experiential avoidance as a functional dimensional approach to psychopathology: An empirical review. *Journal of Clinical Psychology, 63*, 871–890.

Dymond, S., & Roche, B. (2013). *Advances in relational frame theory: Research and application.* Oakland, CA: Context Press.

Eilers, H. J., & Hayes, S. C. (2015). Exposure and response prevention therapy with cognitive defusion exercises to reduce repetitive and restrictive behaviors displayed by children with autism spectrum disorder. *Research in Autism Spectrum Disorders, 19*, 18–31. doi:10.1016/j.rasd.2014.12.014

Ferrari, A. J., Charlson, F. J., Norman, R. E., Patten, S. B., Freedman, G., Murray, C. G. L., … Whiteford, H. A. (2013). Burden of depressive disorders by country, sex, age, and year: Findings from the Global Burden of Disease Study 2010. *PLoS Medicine, 10*. doi:10.1371/journal.pmed.1001547

Fledderus, M., Bohlmeijer, E. T., & Pieterse, M. E. (2010). Does experiential avoidance mediate the effects of maladaptive coping styles on psychopathology and mental health? *Behavior Modification, 34*, 503–519.

Gifford, E. V., Kohlenberg, B., Hayes, S. C., Pierson, H., Piasecki, M., Antonuccio, D., & Palm, K. (2011). Does acceptance and relationship-focused behavior therapy contribute to bupropion outcomes? A randomized controlled trial of FAP and ACT for smoking cessation. *Behavior Therapy, 42*, 700–715.

Gillanders, D. T., Bolderston, H., Bond, F. W., Dempster, M., Flaxman, P. E., Campbell, L., … Remington, R. (2014). The development and initial validation of the Cognitive Fusion Questionnaire. *Behavior Therapy, 45*, 83–101.

Hayes, S. C., Barnes-Holmes, D., & Roche, B. (2001). *Relational frame theory: A post-Skinnerian account of human language and cognition.* New York: Plenum.

Hayes, S. C., Barnes-Holmes, D., & Wilson, K. G. (2012). Contextual behavioral science: Creating a science more adequate to the challenge of the human condition. *Journal of Contextual Behavioral Science, 1*, 1–16. doi:10.1016/j.jcbs.2012.09.004

Hayes, S. C., Gifford, E. V., & Ruckstuhl, L. E., Jr. (1996). Relational frame theory and executive function. In G. R. Lyon & N. A. Krasnegor (Eds.), *Attention, memory, and executive function.* Baltimore: Brookes.

Hayes, S. C., Luoma, J., Bond, F., Masuda, A., & Lillis, J. (2006). Acceptance and commitment therapy: Model, processes, and outcomes. *Behaviour Research and Therapy, 44*, 1–25. doi:10.1016/j.brat.2005.06.006

Hayes, S. C., Monestès, J.-L., & Wilson, D. S. (2018). Evolutionary principles for applied psychology. In S. C. Hayes & S. Hofmann (Eds.), *Process-based CBT: Core clinical competencies in evidence-based treatment.* Oakland, CA: New Harbinger Publications.

Hayes, S. C., & Sanford, B. (2014). Cooperation came first: Evolution and human cognition. *Journal of the Experimental Analysis of Behavior, 101*, 112–129. doi:10.1002/jeab.64

Hayes, S. C., & Sanford, B. (2015). Modern psychotherapy as a multidimensional multilevel evolutionary process. *Current Opinion in Psychology, 2,* 16–20. doi:10.1016/j.copsyc.2015.01.009

Hayes, S. C., Sanford, B. T., & Feeney, T. (2015). Using the functional and contextual approach of modern evolution science to direct thinking about psychopathology. *Behavior Therapist, 38,* 222–227.

Hayes, S. C., Strosahl, K., & Wilson, K. G. (2012). *Acceptance and commitment therapy: The process and practice of mindful change* (2nd ed.). New York: Guilford.

Kashdan, T. B., Barrios, V., Forsyth, J. P., & Steger, M. F. (2006). Experiential avoidance as a generalized psychological vulnerability: Comparisons with coping and emotion regulation strategies. *Behaviour Research and Therapy, 9,* 1301–1320.

Masuda, A., Hayes, S. C., Sackett, C. F., & Twohig, M. P. (2004). Cognitive defusion and self-relevant negative thoughts: Examining the impact of a ninety year old technique. *Behaviour Research and Therapy, 42,* 477–485. doi:10.1016/j.brat.2003.10.008

Mirenda, P., Smith, I. M., Vaillancourt, T., Georgiades, S., Duku, E., Szatmari, P., … Zwaigenbaum, L. (2010). Validating the Repetitive Behavior Scale-Revised in young children with autism spectrum disorder. *Journal of Autism and Developmental Disorders, 40,* 1521–1530. doi:10.1007/s10803–010–1012–0

Monestès, J.-L. (2016). A functional place for language in evolution: Contextual behavior science contribution to the study of human evolution. In S. C. Hayes, D. Barnes-Holmes, R. D. Zettle, & A. Biglan (Eds.), *The Wiley handbook of contextual behavior science.* West Sussex, UK: Wiley-Blackwell.

Monestès, J.-L., Villatte, M., Stewart, I., & Loas, G. (2014). Rule-based insensitivity and delusion maintenance in schizophrenia. *Psychological Record, 64,* 329–338. doi:10.1007/s40732–014–0029–8

Neuringer, A. (1986). Can people behave "randomly"?: The role of feedback. *Journal of Experimental Psychology General, 115,* 62–75.

Neuringer, A. (2002). Operant variability: Evidence, functions, and theory. *Psychonomic Bulletin and Review, 9,* 672–705.

Neuringer, A. (2004). Reinforced variability in animals and people: Implications for adaptive action. *American Psychologist, 59,* 891–906.

Quinones, J., & Hayes, S. C. (2014). Relational coherence in ambiguous and unambiguous relational networks. *Journal of the Experimental Analysis of Behavior, 101,* 76–93. doi:10.1002/jeab.67

Ritzert, T. R., Forsyth, J. P., Berghoff, C. R., Barnes-Holmes, D., & Nicholson, E. (2015). The impact of a cognitive defusion intervention on behavioral and psychological flexibility: An experimental evaluation in a spider fearful non-clinical sample. *Journal of Contextual Behavioral Science, 4,* 112–120. doi:10.1016/j.jcbs.2015.04.001

Sherry, D. F., & Galef, B. G., Jr. (1984). Cultural transmission without imitation: Milk bottle opening by birds. *Animal Behaviour, 32,* 937–938.

Spinhoven, P., Drost, J., de Rooij, M., van Hemert, A. M., & Penninx, B. W. (2014). A longitudinal study of experiential avoidance in emotional disorders. *Behavior Therapy, 45,* 840–850.

Wilson, D. S., Hayes, S. C., Biglan, T., & Embry, D. (2014). Collaborating on evolving the future. *Behavioral and Brain Sciences, 34,* 438–460.

CHAPTER 17

Reconciling the Tension Between Behavioral Change and Stability

Renée A. Duckworth

*Department of Ecology and Evolutionary Biology,
University of Arizona*

The ability of organisms to show highly flexible behaviors in response to life-stage transitions, seasonal cues, and environmental change often seems at odds with observations of consistent differences in behavior among individuals, populations, and species. I suggest that resolving this apparent contradiction requires understanding of the proximate mechanisms underlying both behavioral change and stability. A review of neuroendocrine mechanisms suggests that stable differences in behavior among individuals is underlain by structural variation in neuroendocrine components, and this relatively inflexible scaffold is needed to enable more flexible components to function. In particular, there is evidence that patterns of investment in distinct brain circuits produce tradeoffs in neural function that account for variation in personality traits among individuals. Thus, understanding where individuals fall on this spectrum of tradeoffs may help in constructing more individualized approaches to human behavioral change that take into account differences among individuals in how they perceive the world and make decisions.

What Is Behavioral Change?

Behavior is the activity of an individual and constitutes every action (or inaction) the individual engages in over the course of its life. Because organisms are constantly changing their activities on a daily, monthly, or annual cycle, it often seems that the default state of behavior is one of change; however, this change has consistent patterns. In particular, individuals often show consistent differences in the level or intensity of expression of particular behaviors either across contexts (Sih, Bell, & Johnson, 2004) or over time within a context (Mischel & Shoda, 1995), reflecting a core stability in personality or temperament. In this chapter, I explore why such stability in expression of behavior occurs and how it might influence the ability of an individual to change its behavior over time. Answering these questions requires an understanding of the underlying physical basis for the expression of behavior. But before going into neuroendocrine mechanisms, it's important to delineate what I mean by behavioral change, since there are many different types of behavioral change and different mechanisms are linked to different types of change, including the inability to change.

In evolutionary biology, the 1980s brought a revival of interest in phenotypic plasticity (Pigliucci, 2005), yet most studies of plasticity focused on traits that were developmentally plastic, in which variable environmental conditions during ontogeny (i.e., from conception to reproductive maturity) produced drastically different adult phenotypes that were fixed for the life of the organism (Piersma & Van Gils, 2011). However, as the study of phenotypic plasticity matured, it became clear that the focus on developmental plasticity left out a number of plastic traits that changed in response to environmental variation throughout an organism's life. Filling in this gap, Piersma and Drent (2003) suggested the term "phenotypic flexibility" for types of plasticity that are reversible.

Yet even this expansion of the definition of plasticity to include more flexible traits that change within the lifetime of an organism does not fully capture the range of environmental influences on behavior. This is because flexibility is an inherent component of every behavior—all behaviors are reversible because they are only expressed in response to an internal or external stimulus; yet, at the same time, their *level* of expression may be developmentally plastic and highly consistent in adulthood, or it may be developmentally plastic and still retain some flexibility throughout life (Duckworth, 2009). Because developmental plasticity implies limits to postontogenetic change, it is often neglected as a cause of variation in behavioral traits. However, the processes that occur during ontogeny are fundamentally similar for all biological systems, including the neuroendocrine system, and thus, behaviors are no different from morphological traits in the importance of developmental plasticity to influence postontogenetic expression (Duckworth, 2015).

During development, tissues differentiate, grow, and mature, and this process requires a massive incorporation of resources to build an organism. Because of this, early ontogeny is a unique window when traits are particularly sensitive to environmental inputs, and once developmental processes come to an end, some components of traits are essentially fixed in expression for life. This process applies equally to the physical basis of behavioral traits—the nervous and endocrine systems must also go through a process of cellular differentiation and growth in which they are particularly susceptible to environmental influences (Knudsen, 2004). Thus, similar to morphological traits, this finite period of growth and tissue organization is a sensitive period for behavioral development, when environmental information is incorporated on a large scale, shaping neuroendocrine structures and essentially limiting the range of variation that is possible later in life. While well established for the phenotype in general, the role of early development in producing behavioral variation has been controversial, particularly for complex human behavior. This stems in large part from a discomfort with the notion that complex behavior can be reduced to neural physiology (Bickle, 2006; Marshall, 2009). Moreover, there seems to be a general implicit assumption in psychology that complex human behavior requires the input of a complex social environment, and therefore any processes that occur after birth override prenatal influences on variation in behavior. This is evidenced by the fact that studies of human behavioral development tend to focus on the influence of the early social and cultural environment experienced by children, rather than the physiological processes that might influence behavioral development during the prenatal environment (Buss & Plomin, 2014). Yet the most massive and profound changes in the neuroendocrine system occur during the period from conception to birth. Elements of neuroendocrine structure and organization are entrenched during this time that cannot be changed later in life. This is evident in the severe effects of early exposure to diseases, drugs, and malnutrition on brain development and function (Kolb, 1995; Riley & McGee, 2005). However, what is less clear is whether more subtle environmental variation can influence brain development during this period in ways that permanently alter the normal range of variation in the behavior of adults. Given the ubiquity of such subtle, and often adaptive, environmental influences on other morphological traits, there is no logical reason to suppose that neuroendocrine systems would be immune to such influences, and recent studies of epigenetic effects, some of which are induced even before conception, support this idea. For example, in rats, housing fathers in a more complex environment before mating influenced both brain methylation patterns and behavior of their offspring (Mychasiuk et al., 2012). This particular finding mirrors a growing body of evidence in animals for early developmental effects on the expression of natural variation in behavior (Meaney & Szyf, 2005). Moreover, studies on temperament traits in humans show early developmental effects on behavior that are evident even in the fetus (Werner et al., 2007). Thus, permanent differences among individuals in temperament or personality traits may in part be due to environmental influences that act very early in development and organize structural components of the neuroendocrine system, since these are the components that are most difficult to change postdevelopment (Duckworth, 2010, 2015).

Evidence Linking Personality Variation and Neuroendocrine Structure

Given the complexity of the neuroendocrine system, the idea that personality varies with the structure of neuroendocrine components, such as the size and composition of distinct brain regions, may seem overly simplistic; however, there is substantial evidence in both animal and human studies for a link between the two.

Variation in brain morphology is linked to affective and personality disorders in humans—for example, pituitary size has been linked to schizophrenia (Nordholm et al., 2013; Takahashi et al., 2011) and hippocampal, hypothalamic, and amygdala volumes have been linked to borderline personality disorder (Kuhlmann, Bertsch, Schmidinger, Thomann, & Herpertz, 2013; Schmahl, Vermetten, Elzinga, & Bremner, 2003). There is also evidence that structural variation in the brain is correlated with the normal range of variation in personality (DeYoung et al., 2010). For example, Bjornebekk and colleagues (2013), using the five-factor model of personality variation, found that neuroticism is negatively correlated with overall brain size, white matter microstructure, and frontotemporal surface area. Cremers and colleagues (2011) found that orbitofrontal and right amygdala volume were both positively related to extraversion, and Fuentes and colleagues (2012) showed that individual differences in anxiety-related personality traits were associated with reduced size of brain structures related to emotional control and self-consciousness. These are but a few of the now numerous studies that have found personality variation reflected in underlying structural variation in the brain (see Kennis, Rademaker, & Geuze, 2013, for review). Such correlations do not necessarily prove that structural variation in the brain is the cause of personality, but they are at least consistent with the idea.

Experimentally teasing apart cause and consequence of links between brain structural morphology and behavior is difficult in humans due to ethical considerations. However, evidence from animal studies supports a causal link between the two. For example, artificial selection on natural variation in guppy (*Poecilia reticulate*) brain size produced a correlated response in personality traits (Kotrschal et al., 2014), suggesting that changes in brain size across generations are causally linked to changes in personality traits. Another study used transgenic mice to attempt to understand the frequently observed correlation in humans between smaller hippocampal volume and anxiety-related personality disorders (Persson et al., 2014). By creating mice that expressed the CYP2C19 gene, a gene that has an enzyme metabolic function in the human brain, they showed that mice with the gene developed a smaller hippocampus compared to mice without it and had impaired adaptation to stress as adults. Because the gene is expressed only in the fetal brain, it provides evidence that morphological changes in the brain precede behavioral changes observed in adults later, suggesting a causal role for reduced hippocampal volume during ontogeny and increased stress and anxiety in adulthood.

A causal connection between structural variation in the brain and personal variation may reflect tradeoffs between relative investment in different brain regions. Because structural components of the neuroendocrine system are the most stable across an organism's life and the least likely to change on short time scales (Duckworth, 2015), their potential influence on personality variation provides a mechanistic basis for consistency of behavior across an individual's life and also suggests that personality itself provides a stable framework for the expression of behavioral flexibility and change. Moreover, if subtle differences in structural variation of the brain do reflect tradeoffs in investment, it suggests that personality variation should be influenced by variation across all regions of the brain (i.e., there is no one region that can explain personality variation) and, in turn, personality variation should influence all aspects of behavior. In this sense, personality variation essentially lays the foundation on which behavioral responses are built. Thus, personality does not preclude behavioral flexibility and may even enable it, in the same way that the relatively inflexible structure of the skeletal system in vertebrates enables an animal to walk, run, and do other flexible activities.

Do Tradeoffs in Neural Processes Underlie Personality Variation?

Behavior is the final outcome of integration between many underlying neurobiological processes and information about current context and internal organismal state. Sensory information is processed in the brain by cognitive, motivation, and emotion circuits that, in turn, influence the decisions an individual makes and the transitory state of the individual's mood. The ultimate outcome of integration among all components of this dynamic system is behavior.

There is evidence that structural variation in the brain, rather than having isolated effects on distinct behaviors, influences how information is gathered, integrated, and processed via tradeoffs in function. An example of this is in decision-making processes. Recent studies have shown that personality not only influences the process of decision making but also can influence decision-making competence (Bensi, Giusberti, Nori, & Gambetti, 2010; Dewberry, Juanchich, & Narendran, 2013; Maner et al., 2007). A potential basis for this link between personality and decision making is the speed-accuracy tradeoff, in which decisions can be made slowly with high accuracy or fast with high error rate (Chittka, Skorupski, & Raine, 2009). This tradeoff has a clear neurobiological basis (Bogacz, Wagenmakers, Forstmann, & Nieuwenhuis, 2009) and studies have shown distinct patterns of brain activity and connectivity among both individuals that preferentially prioritize speed versus accuracy and individuals that vary in their ability to flexibly adjust their level of caution (sometimes prioritizing speed, sometimes accuracy) (Forstmann et al., 2010; Perri, Berchicci, Spinelli, & Di Russo, 2014). Moreover, variation in how individuals deal with this tradeoff has been shown to relate to a variety of personality dimensions, such as risk sensitivity (Nagengast, Braun, & Wolpert, 2011), agreeableness (Bresin, Hilmert, Wilkowski, & Robinson, 2012), and neuroticism (Socan & Bucik, 1998).

Another potential neural tradeoff is between executive functions and the default mode network. Executive functions are cognitive processes that include control of attention, reasoning, problem solving, and planning, whereas the default mode network activates in the absence of external task demands and is associated with cognitive processes such as mind-wandering, future thinking, and perspective taking.

These two functions are associated with distinct regions of the brain and show an antagonistic pattern of activation—when attention to a task is required, the regions related to executive function are activated, and the default mode network is deactivated (Fox et al., 2005). Interestingly, individual variation in creative cognitive ability corresponds to structural variation in the brain regions associated with the default mode network (Jung, Mead, Carrasco, & Flores, 2013). Moreover, executive control and default mode networks are not always antagonistic in activation. Executive control prevails when study participants are required to have focused attention on an external task, and the default mode network prevails when there is no task. However, when study participants were asked to complete a task that required internal reflection, the two networks were activated simultaneously in a cooperative fashion (Beaty, Benedek, Kaufman, & Silvia, 2015). Importantly, both individual variation in executive function and structural variation in the brain underlying this network correlate with variation in personality; a strong executive function is associated with high conscientiousness, low neuroticism, and higher agreeableness (Williams, Suchy, & Rau, 2009). Moreover, individuals who showed higher impulsiveness had lower cortical thickness in areas associated with executive function (Schilling et al., 2012). Finally, there is evidence for competition between executive function, particularly in areas associated with higher-level reasoning and future planning and emotion circuits of the limbic system that are associated with making decisions that give immediate rewards (McClure, Laibson, Loewenstein, & Cohen, 2004).

Taken together, these studies suggest that there is substantial variation among individuals in decision-making processes that is related to personality, and that this variation is underlain by variation in structural components of the brain. This suggests that personality variation among individuals may reflect where they fall along the spectrum of various neurological tradeoffs in the brain. Because there are multiple interacting tradeoffs (e.g., speed versus accuracy, attentional control versus mind-wandering, and immediate versus delayed gratification), this may explain why personality variation is a multidimensional complex trait that cannot easily be summarized on a single axis.

Studies on stress-induced phenotypes and coping styles in animals suggest that variation in personality traits may be due to the interaction between genetic variation and maternally induced stress during early development (Duckworth, 2015), and the phenotypes associated with stress are consistent with the tradeoffs outlined above. Many long-term consequences of early developmental stress are caused by a resetting of fetal hypothalamic–pituitary–adrenal (HPA) axis sensitivity, which is a major cause of variation in many behavioral traits, including personality (Koolhaas et al., 1999; Meaney & Szyf, 2005; Seckl & Meaney, 2004). The HPA axis comprises the hypothalamus, anterior pituitary, and adrenal gland, along with the hormones they secrete and to which they respond. In the brain, activity in the interconnected amygdala, hippocampus, and hypothalamus activates and regulates the HPA axis (Charmandari, Tsigos, & Chrousos, 2005). Interestingly, these are the parts of the brain that seem to be most intimately tied to the tradeoffs outlined above, since the amygdala mediates value judgments about external stimuli (Janak & Tye, 2015), the hypothalamus mediates reactions to stress and can impair performance in the prefrontal cortex specifically in relation to executive functions (Phelps, Lempert, & Sokol-Hessner, 2014), and the hippocampus is part of a functional loop designed to detect novelty and to transmit behaviorally significant information into the storage of long-term memory (Lisman & Grace, 2005), thus acting as a liaison between sensory systems, information acquiring and valuation systems, and higher cognitive processes. Finally, these systems are also the most highly influenced by developmental stresses experienced in the prenatal environment (Charil, Laplante, Vaillancourt, & King, 2010).

Stress-induced behavioral phenotypes are thought to be adaptive, preparing individuals for a harsh environment (Badyaev, 2005; Korte, Koolhaas, Wingfield, & McEwen, 2005; Wells, 2003). For example, in animals, aggression is related to stress coping style, with more aggressive individuals being bolder, less exploratory, taking more risks in the face of potential dangers, and showing lower behavioral flexibility compared to less aggressive individuals. The aggressive strategy requires higher energy consumption and is thought to be advantageous in stable environments where food is abundant; whereas the nonaggressive, more behaviorally flexible type is thought to flourish in more stressful environments where resources are scarce. There is evidence from wild birds that such differences may be maintained by fluctuating selection, because food availability, which fluctuates across years, was a major determinant of survival of birds that differed in exploratory behavior (Dingemanse, Both, Drent, & Tinbergen, 2004).

Most importantly, the way that these distinct personality types deal with environmental challenges and stress is very different—proactive animals, because they rely on routines, are better at performing tasks despite minor distractions, but adapt slowly to changes in the environment; whereas reactive types are easily distracted but adapt to novel conditions faster (Coppens, de Boer, & Koolhaas, 2010). Such differences in adaptability between proactive and reactive types has been hypothesized to lead to divergent risks of disease and psychological problems in humans (Korte et al., 2005).

Implications for Human Behavioral Change

If a major cause of personality variation among individuals is stress-induced structural variation in the neuroendocrine system that cannot change substantially postdevelopment, then what are the implications of this for human behavioral change? First, it impacts our understanding of which aspects of behavior cannot be easily changed postdevelopment, and second, it implies that there may be substantial differences among individuals in how behavioral changes arise.

One of the main functions of the brain is to integrate novel information to allow the organism to respond flexibly to challenges. Yet, at the same time, the adult brain must preserve the circuitry and synaptic organization necessary to maintain continuity of behavior and long-term memories. Moreover, even though neural plasticity persists throughout the life span (Lledo, Alonso, & Grubb, 2006), it is highly constrained in the adult compared to the developing brain (Kolb, 1995). Large-scale reorganization of axons, dendrites, and myelination are limited in the adult brain because these structures provide a stable scaffold underlying neural circuits and, as a result, changes in structure are local and often short term (Bavelier, Levi, Li, Dan, & Hensch, 2010). Thus, the tension between behavioral stability and change is reflected in a similar tension between stability and change at the neural level. In general, neural rewiring is costly (Laughlin & Sejnowski,

2003), and this may lead to significant switching costs to changing behavior patterns (Wood & Runger, 2016).

If, as I suggest, personality variation reflects where individuals are on different ends of a continuum in neural tradeoffs, individuals should have different thresholds for responding to threat, engaging in social interaction, getting distracted, exploring their environment, and seeking out novelty. In animal studies, divergent ends along the personality spectrum are typically thought to reflect fitness tradeoffs, with some individuals being better adapted to more stressful or variable environments while others perform best in more benign or stable environments. However, in humans, some axes of personality variation are typically viewed as nearly universally positive or negative. For example, neuroticism is typically thought of as negative, and higher scores on this personality axis are associated with numerous psychological and health problems. Neurotic individuals are known to have a heightened sensitivity to threat (Perkins, Arnone, Smallwood, & Mobbs, 2015) and one relatively unexplored possibility is that high neuroticism is part of a stress-induced phenotype in humans that enables individuals to respond faster to potential threats. If so, then there may also be benefits to high neuroticism that perhaps are not evident in stable and benign environments, where thoughtful and reflective decision making is most highly valued. As evidence for this, individuals who were more neurotic did not experience the speed-accuracy tradeoff such that, under pressure, their accuracy increased with faster decision times (Bell, Mawn, & Poynor, 2013). Moreover, people high in neuroticism were less accurate if they slowed down on a task following error feedback (Robinson, Moeller, & Fetterman, 2010). These studies suggest that, in some contexts, there are benefits to high neuroticism, a point that has been made by others in the context of understanding the evolutionary maintenance of personality variation (Nettle, 2006) and attachment styles (Ein-Dor, Mikulincer, Doron, & Shaver, 2010). At the other end of the spectrum, extraversion is usually associated with positive health and psychological outcomes; yet at extreme levels, extraverts have a higher risk of bipolar disorder (Watson, Stasik, Ellickson-Larew, & Stanton, 2015). Thus, attributing strictly positive or negative attributes to personality dimensions irrespective of context is not warranted. Instead, problems are more apt to arise when individuals are at the extremes of any of the personality dimensions, and this raises the question of whether it is possible to moderate neural tradeoffs to allow individuals to become more centered in neural functions.

If personality variation is underlain by numerous tradeoffs in neural functioning that, in turn, are underlain by variable patterns of investment in distinct brain regions, then techniques that enable better integration across distinct cognitive, emotion, and sensory circuits may be the most powerful strategies for producing long-term behavioral change. There is evidence that mindfulness-based meditation practices, which engage sensory, emotion, and cognitive functions (Brown, Ryan, & Creswell, 2007), can change the underlying structure of the brain in multiple areas that are known to affect personality, potentially moderating some of the tradeoffs in neural function discussed above (Holzel et al., 2011). There is also evidence that physical activity can impact brain structural properties, enhance cognition and aspects of executive function, and affect mood, emotion, and anxiety (Cotman, Berchtold, & Christie, 2007; Voelcker-Rehage & Niemann, 2013). Finally, neurobiological studies of cognitive behavioral therapies for anxiety and depression suggest that these strategies work by rebalancing activity in regions of the brain associated with emotion regulation and executive function (Clark & Beck, 2010). Thus, interventions can have powerful effects on human behavior by changing the underlying activity patterns and structural organization of the brain associated with functional tradeoffs. In particular, strategies of intervention that engage multiple attentional, physiological, cognitive, or emotional circuits in the brain, such as mindfulness, physical activity, or cognitive therapy programs, may be particularly effective in rebalancing tradeoffs among different components. Moreover, one unexplored possibility is that different personality types may respond differentially to different types of therapies that touch on distinct aspects of these tradeoffs. While recent assessments of various therapy techniques show that many are effective (Hayes, Villatte, Levin, & Hildebrandt, 2011), there is always a portion of the population that is not responsive to treatment. It would be informative to investigate whether personality variation explains some of the variance in treatment efficacy.

In conclusion, I suggest that a better understanding of both stability and flexibility of the physical basis of behavior is essential to understanding behavioral change. Therapies that take into account where individuals fall on the spectrum of various neurological tradeoffs that influence their attention, level of anxiety, and decision-making strategies may be most effective in producing long-lasting behavioral change. Thus, recognizing the limits of behavioral flexibility and patterns of behavioral stability across individuals may be the most important step in understanding how lasting behavioral change occurs.

Acknowledgments

I thank Alex Badyaev for discussion and insightful feedback on this chapter and NSF DEB-1350107 for support.

References

Badyaev, A. V. (2005). Stress-induced variation in evolution: From behavioral plasticity to genetic assimilation. *Proceedings of the Royal Society of London B, 272,* 877–886.

Bavelier, D., Levi, D. M., Li, R. W., Dan, Y., & Hensch, T. K. (2010). Removing brakes on adult brain plasticity: From molecular to behavioral interventions. *Journal of Neuroscience, 30,* 14964–14971. doi:10.1523/JNEUROSCI.4812-10.2010

Beaty, R. E., Benedek, M., Kaufman, S. B., & Silvia, P. J. (2015). Default and executive network coupling supports creative idea production. *Scientific Reports, 5,* 10964. doi:10.1038/srep10964

Bell, J. J., Mawn, L., & Poynor, R. (2013). Haste makes waste, but not for all: The speed-accuracy trade-off does not apply to neurotics. *Psychology of Sport and Exercise, 14,* 860–864. doi:10.1016/j.psychsport.2013.07.001

Bensi, L., Giusberti, F., Nori, R., & Gambetti, E. (2010). Individual differences and reasoning: A study on personality traits. *British Journal of Psychology, 101,* 545–562. doi:10.1348/000712609X471030

Bickle, J. (2006). Reducing mind to molecular pathways: Explicating the reductionism implicit in current cellular and molecular neuroscience. *Synthese, 151,* 411–434. doi:10.1007/s11229-006-9015-2

Bjornebekk, A., Fjell, A. M., Walhovd, K. B., Grydeland, H., Torgersen, S., & Westlye, L. T. (2013). Neuronal correlates of the five factor model (FFM) of human personality: Multimodal imaging in a large healthy sample. *Neuroimage, 65,* 194–208. doi:10.1016/j.neuroimage.2012.10.009

Bogacz, R., Wagenmakers, E.-J., Forstmann, B. U., & Nieuwenhuis, S. (2009). The neural basis of the speed-accuracy tradeoff. *Trends in Neurosciences, 33,* 10–16.

Bresin, K., Hilmert, C. J., Wilkowski, B. M., & Robinson, M. D. (2012). Response speed as an individual difference: Its role in moderating the agreeableness-anger relationship. *Journal of Research in Personality, 46,* 79–86. doi:10.1016/j.jrp.2011.12.007

Brown, K. W., Ryan, R. M., & Creswell, J. D. (2007). Mindfulness: Theoretical foundations and evidence for its salutary effects. *Psychological Inquiry, 18,* 211–237. doi:10.1080/10478400701598298

Buss, A. H., & Plomin, R. (2014). *Temperament (PLE: Emotion): Early developing personality traits* (Vol. 3). Hove, UK: Psychology Press.

Charil, A., Laplante, D. P., Vaillancourt, C., & King, S. (2010). Prenatal stress and brain development. *Brain Research Reviews, 65,* 56–79. doi:10.1016/j.brainresrev.2010.06.002

Charmandari, E., Tsigos, C., & Chrousos, G. (2005). Endocrinology of the stress response. *Annual Review of Physiology, 67,* 259–284. doi:10.1146/annurev.physiol.67.040403.120816

Chittka, L., Skorupski, P., & Raine, N. E. (2009). Speed-accuracy tradeoffs in animal decision making. *Trends in Ecology and Evolution, 24,* 400–407.

Clark, D. A., & Beck, A. T. (2010). Cognitive theory and therapy of anxiety and depression: Convergence with neurobiological findings. *Trends in Cognitive Sciences, 14,* 418–424. doi:10.1016/j.tics.2010.06.007

Coppens, C. M., de Boer, S. F., & Koolhaas, J. M. (2010). Coping styles and behavioural flexibility: Towards underlying mechanisms. *Philosophical Transactions of the Royal Society of London B: Biological Sciences, 365,* 4021–4028. doi:10.1098/rstb.2010.0217

Cotman, C. W., Berchtold, N. C., & Christie, L. A. (2007). Exercise builds brain health: Key roles of growth factor cascades and inflammation. *Trends in Neurosciences, 30*, 464–472. doi:10.1016/j.tins.2007.06.011

Cremers, H., van Tol, M. J., Roelofs, K., Aleman, A., Zitman, F. G., van Buchem, M. A., … van der Wee, N. J. (2011). Extraversion is linked to volume of the orbitofrontal cortex and amygdala. *PLoS ONE, 6.* doi:10.1371/journal.pone.0028421

Dewberry, C., Juanchich, M., & Narendran, S. (2013). Decision-making competence in everyday life: The roles of general cognitive styles, decision-making styles and personality. *Personality and Individual Differences, 55*, 783–788. doi:10.1016/j.paid.2013.06.012

DeYoung, C. G., Hirsh, J. B., Shane, M. S., Papademetris, X., Rajeevan, N., & Gray, J. R. (2010). Testing predictions from personality neuroscience: Brain structure and the big five. *Psychological Science, 21*, 820–828. doi:10.1177/0956797610370159

Dingemanse, N. J., Both, C., Drent, P. J., & Tinbergen, J. M. (2004). Fitness consequences of avian personalities in a fluctuating environment. *Proceedings of the Royal Society B: Biological Sciences, 271*, 847–852. doi:10.1098/rspb.2004.2680

Duckworth, R. A. (2009). The role of behavior in evolution: A search for mechanism. *Evolutionary Ecology, 23*, 513–531.

Duckworth, R. A. (2010). Evolution of personality: Developmental constraints on behavioral flexibility. *Auk, 127*, 752–758.

Duckworth, R. A. (2015). Neuroendocrine mechanisms underlying behavioral stability: Implications for the evolutionary origin of personality. *Annals of the New York Academy of Sciences, 1360*, 54–74. doi:10.1111/nyas.12797

Ein-Dor, T., Mikulincer, M., Doron, G., & Shaver, P. R. (2010). The attachment paradox: How can so many of us (the insecure ones) have no adaptive advantages? *Perspectives on Psychological Science, 5*, 123–141. doi:10.1177/1745691610362349

Forstmann, B. U., Anwander, A., Schafer, A., Neumann, J., Brown, S., Wagenmakers, … Turner, R. (2010). Cortico-striatal connections predict control over speed and accuracy in perceptual decision making. *Proceedings of the National Academy of Sciences, 107*, 15916–15920.

Fox, M. D., Snyder, A. Z., Vincent, J. L., Corbetta, M., Van Essen, D. C., & Raichle, M. E. (2005). The human brain is intrinsically organized into dynamic, anticorrelated functional networks. *Proceedings of the National Academy of Sciences, 102*, 9673–9678. doi:10.1073/pnas.0504136102

Fuentes, P., Barros-Loscertales, A., Bustamante, J. C., Rosell, P., Costumero, V., & Avila, C. (2012). Individual differences in the behavioral inhibition system are associated with orbitofrontal cortex and precuneus gray matter volume. *Cognitive, Affective, and Behavioral Neuroscience, 12*, 491–498. doi:10.3758/s13415-012-0099-5

Hayes, S. C., Villatte, M., Levin, M., & Hildebrandt, M. (2011). Open, aware, and active: Contextual approaches as an emerging trend in the behavioral and cognitive therapies. *Annual Review of Clinical Psychology, 7*, 141–168. doi:10.1146/annurev-clinpsy-032 210-104449

Holzel, B. K., Lazar, S. W., Gard, T., Schuman-Olivier, Z., Vago, D. R., & Ott, U. (2011). How does mindfulness meditation work? Proposing mechanisms of action from a conceptual and neural perspective. *Perspectives on Psychological Science, 6*, 537–559. doi:10.1177/1745 691611419671

Janak, P. H., & Tye, K. M. (2015). From circuits to behaviour in the amygdala. *Nature, 517*, 284–292. doi:10.1038/nature14188

Jung, R. E., Mead, B. S., Carrasco, J., & Flores, R. A. (2013). The structure of creative cognition in the human brain. *Frontiers in Human Neuroscience, 7*, 330. doi:10.3389/fnhum.2013 .00330

Kennis, M., Rademaker, A. R., & Geuze, E. (2013). Neural correlates of personality: An integrative review. *Neuroscience and Biobehavioral Reviews, 37*, 73–95. doi:10.1016/j.neu biorev.2012.10.012

Knudsen, E. (2004). Sensitive periods in the development of the brain and behavior. *Journal of Cognitive Neuroscience, 16*, 1412–1425.

Kolb, B. (1995). *Brain plasticity and behavior.* Mahwah, NJ: Lawrence Erlbaum.

Koolhaas, J. M., Korte, S. M., De Boer, S. F., Van Der Vegt, B. J., Van Reenen, C. G., Hopster, H., … Blokhuis, H. J. (1999). Coping styles in animals: Current status in behavior and stress-physiology. *Neuroscience and Biobehavioral Reviews, 23*, 925–935.

Korte, S. M., Koolhaas, J. M., Wingfield, J. C., & McEwen, B. S. (2005). The Darwinian concept of stress: Benefits of allostasis and costs of allostatic load and the trade-offs in health and disease. *Neuroscience and Biobehavioral Reviews, 29*, 3–38.

Kotrschal, A., Lievens, E. J., Dahlbom, J., Bundsen, A., Semenova, S., Sundvik, M., … Kolm, N. (2014). Artificial selection on relative brain size reveals a positive genetic correlation between brain size and proactive personality in the guppy. *Evolution, 68*, 1139–1149. doi:10.1111/evo.12341

Kuhlmann, A., Bertsch, K., Schmidinger, I., Thomann, P. A., & Herpertz, S. C. (2013). Morphometric differences in central stress-regulating structures between women with and without borderline personality disorder. *Journal of Psychiatry and Neuroscience, 38*, 129–137. doi:10.1503/jpn.120039

Laughlin, S. B., & Sejnowski. (2003). Communication in neuronal networks. *Science, 301*, 1870–1874.

Lisman, J. E., & Grace, A. A. (2005). The hippocampal-VTA loop: Controlling the entry of information into long-term memory. *Neuron, 46*, 703–713. doi:10.1016/j.neuron.2005 .05.002

Lledo, P. M., Alonso, M., & Grubb, M. S. (2006). Adult neurogenesis and functional plasticity in neuronal circuits. *Nature Reviews Neuroscience, 7*, 179–193. doi:10.1038/nrn1867

Maner, J. K., Richey, J. A., Cromer, K., Mallott, M., Lejuez, C. W., Joiner, T. E., & Schmidt, N. B. (2007). Dispositional anxiety and risk-avoidant decision-making. *Personality and Individual Differences, 42*, 665–675. doi:10.1016/j.paid.2006.08.016

Marshall, P. J. (2009). Relating psychology and neuroscience: Taking up the challenges. *Perspectives on Psychological Science, 4*, 113–125.

McClure, S. M., Laibson, D. I., Loewenstein, G., & Cohen, J. D. (2004). Separate neural systems value immediate and delayed monetary rewards. *Science, 306*, 503–507. doi:10 .1126/science.1100907

Meaney, M. J., & Szyf, M. (2005). Environmental programming of stress responses through DNA methylation: Life at the interface between a dynamic environment and a fixed genome. *Dialogues in Clinical Neuroscience, 7*, 103–123.

Mischel, W., & Shoda, Y. (1995). A cognitive-affective system theory of personality: Reconceptualizing situations, dispositions, dynamics, and invariance in personality structure. *Psychological Review, 102*, 246–268.

Mychasiuk, R., Zahir, S., Schmold, N., Ilnytskyy, S., Kovalchuk, O., & Gibb, R. (2012). Parental enrichment and offspring development: Modifications to brain, behavior and the epigenome. *Behavioural Brain Research, 228*, 294–298. doi:10.1016/j.bbr.2011.11.036

Nagengast, A. J., Braun, D. A., & Wolpert, D. M. (2011). Risk sensitivity in a motor task with speed-accuracy trade-off. *Journal of Neurophysiology, 105*, 2668–2674. doi:10.1152/jn.00804.2010

Nettle, D. (2006). The evolution of personality variation in humans and other animals. *American Psychologist, 61*, 622–631.

Nordholm, D., Krogh, J., Mondelli, V., Dazzan, P., Pariante, C., & Nordentoft, M. (2013). Pituitary gland volume in patients with schizophrenia, subjects at ultra high-risk of developing psychosis and healthy controls: A systematic review and meta-analysis. *Psychoneuroendocrinology, 38*, 2394–2404. doi:10.1016/j.psyneuen.2013.06.030

Perkins, A. M., Arnone, D., Smallwood, J., & Mobbs, D. (2015). Thinking too much: Self-generated thought as the engine of neuroticism. *Trends in Cognitive Sciences, 19*, 492–498. doi:10.1016/j.tics.2015.07.003

Perri, R. L., Berchicci, M., Spinelli, D., & Di Russo, F. (2014). Individual differences in response speed and accuracy are associated to specific brain activities of two interacting systems. *Frontiers in Behavioral Neuroscience, 8*, 251.

Persson, A., Sim, S. C., Virding, S., Onishchenko, N., Schulte, G., & Ingelman-Sundberg, M. (2014). Decreased hippocampal volume and increased anxiety in a transgenic mouse model expressing the human CYP2C19 gene. *Molecular Psychiatry, 19*, 733–741. doi:10.1038/mp.2013.89

Phelps, E. A., Lempert, K. M., & Sokol-Hessner, P. (2014). Emotion and decision making: Multiple modulatory neural circuits. *Annual Review of Neuroscience, 37*, 263–287. doi:10.1146/annurev-neuro-071013-014119

Piersma, T., & Drent, J. (2003). Phenotypic flexibility and the evolution of organismal design. *Trends in Ecology and Evolution, 18*, 228–233. doi:10.1016/s0169-5347(03)00036-3

Piersma, T., & Van Gils, J. A. (2011). *The flexible phenotype: A body-centered integration of ecology, physiology, and behaviour.* Oxford: Oxford University Press.

Pigliucci, M. (2005). Evolution of phenotypic plasticity: Where are we going now? *Trends in Ecology and Evolution, 20*, 481–486. doi:10.1016/j.tree.2005.06.001

Riley, E. P., & McGee, C. L. (2005). Fetal alcohol spectrum disorders: An overview with emphasis on changes in brain and behavior. *Experimental Biology and Medicine, 230*, 357–365.

Robinson, M. D., Moeller, S. K., & Fetterman, A. K. (2010). Neuroticism and responsiveness to error feedback: Adaptive self-regulation versus affective reactivity. *Journal of Personality, 78*, 1469–1496. doi:10.1111/j.1467-6494.2010.00658.x

Schilling, C., Kuhn, S., Romanowski, A., Schubert, F., Kathmann, N., & Gallinat, J. (2012). Cortical thickness correlates with impulsiveness in healthy adults. *Neuroimage, 59*, 824–830. doi:10.1016/j.neuroimage.2011.07.058

Schmahl, C. G., Vermetten, E., Elzinga, B. M., & Bremner, J. D. (2003). Magnetic resonance imaging of hippocampal and amygdala volume in women with childhood abuse and borderline personality disorder. *Psychiatry Research: Neuroimaging, 122*, 193–198. doi:10.1016/s0925-4927(03)00023-4

Seckl, J. R., & Meaney, M. J. (2004). Glucocorticoid programming. *Annals of the New York Academy of Sciences, 1032*, 63–84. doi:10.1196/annals.1314.006

Sih, A., Bell, A., & Johnson, J. C. (2004). Behavioral syndromes: An ecological and evolutionary overview. *Trends in Ecology and Evolution, 19,* 372–378.

Socan, G., & Bucik, V. (1998). Relationship between speed of information-processing and two major personality dimensions—Extraversion and neuroticism. *Personality and Individual Differences, 25,* 35–48.

Takahashi, T., Zhou, S. Y., Nakamura, K., Tanino, R., Furuichi, A., Kido, M., … Suzuki, M. (2011). Longitudinal volume changes of the pituitary gland in patients with schizotypal disorder and first-episode schizophrenia. *Progress in Neuro-Psychopharmacology and Biological Psychiatry, 35,* 177–183. doi:10.1016/j.pnpbp.2010.10.023

Voelcker-Rehage, C., & Niemann, C. (2013). Structural and functional brain changes related to different types of physical activity across the life span. *Neuroscience and Biobehavioral Reviews, 37,* 2268–2295. doi:10.1016/j.neubiorev.2013.01.028

Watson, D., Stasik, S. M., Ellickson-Larew, S., & Stanton, K. (2015). Extraversion and psychopathology: A facet-level analysis. *Journal of Abnormal Psychology, 124,* 432–446. doi:10.1037/abn0000051

Wells, J. C. K. (2003). The thrifty phenotype hypothesis: Thrifty offspring or thrifty mother? *Journal of Theoretical Biology, 221,* 143–161. doi:10.1006/jtbi.2003.3183

Werner, E. A., Myers, M. M., Fifer, W. P., Cheng, B., Fang, Y., Allen, R., & Monk, C. (2007). Prenatal predictors of infant temperament. *Developmental Psychobiology, 49,* 474–484. doi:10.1002/dev.20232

Williams, P. G., Suchy, Y., & Rau, H. K. (2009). Individual differences in executive functioning: Implications for stress regulation. *Annals of Behavioral Medicine, 37,* 126–140. doi:10.1007/s12160-009-9100-0

Wood, W., & Runger, D. (2016). Psychology of habit. *Annual Review of Psychology, 67,* 289–314. doi:10.1146/annurev-psych-122414-033417

Dialogue on Psychopathology and Behavior Change

Participants: Renée Duckworth, Steven C. Hayes, Jean-Louis Monestès, David Sloan Wilson

David Sloan Wilson: I'd like to focus on an analogy that Renée made of a skeleton as something that by definition is both inflexible and designed to move. This wonderfully illustrates how flexibility requires inflexibility. How can we apply this analogy to behavioral flexibility and inflexibility, which are central concerns of acceptance and commitment therapy/training?

Renée A. Duckworth: It's important to understand where different aspects of behavioral flexibility come from. Learning, for example, is a very different kind of behavioral flexibility compared to the constant transition between behaviors that individuals go through daily and throughout their lives. Over the course of the day, an organism will go through an entire repertoire of different behaviors like foraging, sleeping, and searching for mates. But that type of flexibility has nothing to do with learning as a mechanism. Part of what I understood from Steve's and Jean-Louis's chapter is that one of the ways that psychopathology can manifest is when learning leads to an endpoint where the organism is stuck in a particular response, such as a drug addiction, which is a sort of neurological trap. The organism is getting immediate, positive feedback in the brain—pleasure from taking a drug. And that sends a signal to the organism: "Keep doing this." And so an inherently flexible mechanism of behavioral change can lead to a restriction of other types of behaviors. For me, to understand the function of behavioral flexibility, one must characterize under what category of flexibility it falls, whether it's coming from a learning mechanism, from a daily rotation through a repertoire of behaviors, or from age-related changes. These are all very different kinds of flexibility.

David: Let me weigh in with a few observations. I am struck after reading both chapters by how different they were, even though both were tasked with the same general topic. Both excellent, by the way, and both very evolutionary, and yet very different. So a lot of work is required to integrate them with each other. Another observation is that contextual behavioral science is oriented toward the need to change. It could well be the case that ninety percent of individuals function just fine; their personalities and learning mechanisms got them to a good place—so we don't think much about them! We think about the minority of people who are not functioning well. In that context, inflexibility is bad by definition and the need to change is good. That's the main space within which contextual behavioral scientists operate. My next observation and kickoff for more discussion is that contextual behavioral scientists, in their treatment both of Skinnerian processes and of the special effects of language—more or less assume a universal human nature and don't say much about personality. This is something that can be studied in nonhuman species. Once we realize that nonhuman species have profound individual differences, we can ask what it means for their learning processes. Presumably all personalities are capable of learning, but perhaps in different ways that we should be taking into account if we're going to accomplish change. To what extent do contextual behavioral scientists actually take this into account when working with people? Maybe implicitly there's an understanding that people are different in ways that are not likely to change (their inflexibility), which can inform how they can be helped to change (their flexibility).

Steven C. Hayes: The issue of a personality has a long history in psychology. One of the problems that you don't face if you're using animal models is the dominance of self-report data, such as the Big Five list of personality factors that emerge from self-reporting. Some of these differences are bound by differences in linguistic processes—we're going all the way back to Osgood's semantic differential and the relatively small number of distinctions we can make reliably. There are only so many dimensions that human language can reveal when put through the filter of population-based factor analysis, which looks only at individuals in the context of a collective. When factor analysis was first put together, there was an argument: Should we be looking at rows and consistencies within individuals across time, or columns and consistencies in the population at a given point in time? Column thinking won. That is hostile to a contextual behavioral science perspective, which is based on trying to understand the processes within individuals across time and *then* collecting them. It's not that the nomothetic level is unimportant, but we need to be careful about our levels of analysis. With the arrival of experience sampling and different analytic methods, you can now look individual by individual across time with many samples. With this approach, you reach completely different conclusions! And so the idea that we have five personality factors—I don't know that to be true.

David: That was actually on my big question list, so let me expand upon it and pass it to Renée. I am also skeptical about the Big Five as an artifact of factor analysis, exactly as you said. Factor analysis will create orthogonal dimensions out of any data cloud. Now that we're getting a mechanistic approach to individual differences, thanks to the great work of people like Renée, what is the mechanistic research telling us about the so-called Big Five? All individuals have norms of reaction, so they're all responding to the environment in that sense. What we mean by different personalities is different norms of reaction. You need to get that kind of longitudinal data on individuals in order to measure their norms of reaction. Walter Mischel championed this view for human personality psychology and was referenced by Renée in her chapter. So, Renée, what do you have to say about all this? This is very important stuff.

Renée: I was getting really excited when listening to both of your comments. I am not in the field of psychology, so I am woefully ignorant of many of the debates, but I'm getting up to speed as I read the human personality literature. For people who study animals, it's both a blessing and a curse that we don't have insight into what the animals are thinking; we can't ask them. The blessing is that we're not confused by whatever kinds of errors might occur in a self-reporting context, because I'm not convinced that self-reporting alone is the best way to assess personality. So we're not cursed with that sort of difficulty. But at the same time, that means that we can only get one dimension of information from animals. We can only observe what they do: their actual behavior. And what was interesting to me when I read the human personality literature was the lack of studies that do that for humans—that construct an ethogram to see if the personality criteria correlate with any behavior. Self-reporting and saying what you think about something are not behavior that you show on a daily basis out in the real world, but linking them would be totally fascinating. I'm not saying that nobody's doing this, because I read some fascinating studies during the course of this project, which were making this very point.

David: There are some. But not nearly as much as there needs to be.

Renée: To me, that would be the best of both worlds. To go back to the points that you were making, I'm agnostic about the Big Five and the best way to measure personality. That's the reason I have such a fascination with the neuroendocrine basis of it. I strongly believe that there are consistent differences among individuals in the way they perceive and interact with the world. I think that the Big Five probably capture some of those dimensions, because they are often repeatable and correlated with some of the neuroendocrine components I outlined in my chapter. But I think the reason I'm fascinated with trying to understand tradeoffs at the neurological level is because it can help us to identify the ways that individuals perceive the world differently. I'm not a steadfast proponent of one particular personality scheme; I think that defining personality is still an open question.

Jean-Louis Monestès: I want to recall that our goals in contextual behavioral science are to predict and influence. We think that predicting is important because there are regularities in behaviors. I don't know if we could call that "personality," but it means that in most comparable situations people are going to behave the same way. But at the same time, we think that we can influence behavior, so we've got both sides at the same moment. If I go back to your question, David, about personality, what disturbs me about this concept is the fact that it needs to be structural. I can't remember from your chapter, Renée, but are you are saying that personality builds during a window of time and after that it's difficult to change? The fact that it is structural is not a problem for us, because we are not thinking of individuals as tabula rasa. But the problem is that as soon as you define structure, you might forget that the structure is going to be in interaction with changing environments. Do you see what I mean?

Renée: I think it's about degree. The brain is a fascinating organ. Its function is to be flexible, to allow the organism to gather information, to learn, and to respond flexibly to the environment. But that flexibility shouldn't prevent us from acknowledging that brain development has a profound impact on behavior for the rest of an organism's life. This is true for any bodily system. Development is when an organism is the most open to incorporation of environmental variation in its form and function. I don't want what I wrote to come across as denial about the extensive flexibility in adult behavior in all organisms. But I do think it's important to acknowledge the limits in order to understand what kind of flexibility is possible.

David: Let me bring in some of the other chapters in this volume to give a concrete example of fast and slow life histories—with harsh environments calling for fast life history strategies and benign environments calling for slow life history strategies. If an individual experiences a harsh environment early in life and adopts a fast life-history strategy, we know from attachment theory that it can be not so easy to change. The mechanisms of phenotypic plasticity are sufficiently diverse that we could find examples of everything from prenatal effects that become fixed all the way to a chameleon-like ability to toggle back and forth between behaviors in adulthood; what's sometimes called phenotypic flexibility as opposed to plasticity. These are all among the range of possibilities. I think we all know this—that if somebody experienced very hard times, it might not be easy for them to change, which is why therapy is required. What do you guys think? Your experiences as therapists must take this kind of thing into account.

Steve: You do need to take it into account. But we come back to this problem: If you're focused on prediction and influence and you look to the literature, the contingencies that are being talked about don't tell you very much about what you need to adjust. In part, it's methodological. It's not just self-report. There's a whole literature on learning strategies linked to the Big Five. My summary of it is that the interactions are very weak, they don't tend to replicate from study to study, and it doesn't tell you very much. It's mostly been examined because of learning strategies linked to educational outcomes. The Big Five will account for between nine and thirteen percent of how you do in school. The most important one is openness to experience, because being closed off from experience makes you tend to rely on superficial learning strategies, not the deeper ones that require a bit of emotional openness, like exploring different hypotheses, holding several different alternatives in mind at the same time, or considering other points of view. These deeper learning strategies require a certain amount of tolerance of ambiguity.

But now if I bring that into my clinical work, it doesn't tell me very much about what to do, because openness to experience is what I'm targeting already. I would like a theory of personality based on what has happened to human beings over their lifetime. To me, self-report is fine. It's just another behavioral domain; be careful not to say that self-report *is* the other things. The correlation between self-report and actual behavior is typically about 0.3, but when you look within individuals across time, it's around 0.6. But our methodologies typically don't do that. Even our psychometric evaluation of self-report instruments are of the collective, whatever the heck that is. And this is not an interest in individuals and groups, I mean, this is the amorphous group then being taken to be the individual, which is just not right.

What's in our chapter is essentially the emergence of behavioral patterns or rigidities that could be seen as personality types. So, for example, if you become a heavy kind of "I'm right" person, if you get into that cul-de-sac, it will easily take you into areas that will be called a personality. I think we're seeing it right now, while watching our president perform. That's an obvious personality you're looking at there. And it's the one that's heavily dominated by "I'm right," "I'm the smart one," "I'm the brilliant one"—I mean, there are a few core central attractors there that seem to predict a lot of that behavior. That's exactly the kind of thing that we're looking at clinically. What we wrote about in our chapter is an example—it gives you the tools for how we would get a more behaviorally sensible view of personality that could be applied person by person. We're going to need more help from people like Renée, but I'm a little skeptical about going too quickly to the constancy of the brain until we put it through that filter, because the brain after all is a dependent variable, as well as an independent one. It's a plastic organ.

David: I was going to ask about that. When you do a study and you find you can trace behavioral differences to a brain and anatomy differences, what is the cause and effect there? To what extent does that reflect the outcome of a developmental process?

Renée: I'll be the first person to admit that this is just an idea right now. The correlations that I talked about in my chapter are a first attempt to build a case. If you didn't have the correlations at all, then there'd be little reason to look further for structural causes of personality variation. The other thing that we're working on in my lab right now is exploring the basis of constraints on flexibility of the brain. The question is where are the limits to flexibility—could a person even change personality type if they worked hard enough at it? Possibly, if that is the main goal. But to me it seems that would be a silly goal, because it would require so much effort and so much work. My perspective is that preexisting structural tradeoffs channel behavior in a certain way. That's not to say that change is impossible, but that the bar for change is higher in one direction than it is in another. It's just simply a fact that there's limited space in the brain and this leads to tradeoffs.

Steve: One of the problems there is you need to know the brain's reaction to context. It could be that a very unusual context can take something that previously looked difficult and show that it is actually relatively easy. And let me give you an example. Let's say that you had a stroke and you now have one hand that's quite useful and one that isn't. In the journal *Stroke*, about ten years ago, they looked at what would happen if you tied down your functional limb, just strapped it to your body so you can't use it. So to begin with, you had an insult to the brain and you had a stroke—we know exactly why you're not able to use that hand. But after tying down the other (functional) hand, within a period of about six months, the brain developed an entirely different set of neurological pathways, and sure enough, the previously inadequate limb is now quite adequate, and the brain changed to support that.

David: Neuroplasticity is something we have a lot to learn about.

CHAPTER 18

Evolution and Contextual Behavioral Science: Where Are We and Where Are We Going? A Concluding Dialogue Between the Editors

Steven C. Hayes
University of Nevada, Reno

David Sloan Wilson
Binghamton University

Steven C. Hayes: It might be a good place to start just to talk about our own history—where we were coming from, and how we came together in pursuing this topic over a period of years.

David Sloan Wilson: I've studied evolution in relation to human affairs inside the ivory tower throughout my career, but about ten years ago I decided to see how this theory could be used in the real world. And pretty quickly I came into contact with folks like you, starting out with our friend Tony Biglan, who introduced me to you and others. At the time, I knew very little about contextual behavioral science, and what I thought I knew, especially about major figures such as B. F. Skinner, was really all wrong. This volume is an effort to do for our respective fields what I think has taken place in our own minds.

Steve: I'll start with something that might startle people, but I shared it with you some time ago. When our relationship began to emerge, I literally went home and cried. I was reflecting the feelings of lots of behavioral psychologists who deeply value evolutionary thinking and have thought of their work that way, but have faced decades of disconnection, hostility, and dismissal from evolutionists. It was like coming in from the cold.

You mentioned Skinner. I am probably a neo-Skinnerian. Many of my ideas he probably wouldn't agree with if he was still alive, but he was an intellectual hero and a focus of my early training. I thought of him as an evolutionary psychologist. He thought of himself that way.

David: He did.

Steve: He wrote repeatedly about the similarity of selection within the lifetime of an individual (the focus of his work), and within cultures and genetically across the lifetime of individuals in other fields of study. But that analogy was firmly rejected by most evolutionists at the time. It was considered to be a loose metaphor—loose to the point of distortion.

The unified framework that we're exploring here is one that is profoundly important to the wing of behavioral science that thinks of itself in functional and contextual ways. It views itself as being under the umbrella of evolution science. Fortunately, evolution science has progressed so amazingly fast, due to persons like yourself and the people who are authors of the chapters in this book, that we can now seriously think of an integrated framework for understanding, predicting, and influencing behavior that is constructed by an active collaboration between these fields. I think both fields are beginning to warm to this connection.

David: Contextual behavioral science started out in an evolutionary direction within the philosophical tradition of Pragmatism. Among other things, CBS is also a philosophy of doing science. Science in the context of everyday life. This has been brought out throughout this entire book. The Skinnerian tradition, which was evolutionary in Skinner's own mind, became strongly rejected from the so-called standard social science model with the emergence of evolutionary psychology. It was a bizarre turn of events, that's only now being healed. There are all sorts of wounds that need to be healed. It's not an arid intellectual topic, is it?

Steve: It's not. People are emotional about it because they sense the importance of it. And they sense the history. I have two primary intellectual mentors in my life, John Cone and David Barlow, and one traces his academic lineage back to the Chicago school of functional psychology and scholars such as Dewey and Angell, the other goes directly to William James. Pragmatism is itself a response to Darwin, and my mentors applied Pragmatism to the behavior of scientists themselves: viewing science itself as an issue of variation and selection; of adaptation.

David: Evolutionary biology is also richly contextual. You can't understand any species except in relation to its environment. So contextualism is baked into the DNA of evolutionary biology, but mostly with respect to genetic evolution. And contextual behavioral science emphasizes context, but mostly with respect to behavioral flexibility, not genetic evolution. So they're both contextual in their own way, but at different scales.

When the authors of the present volume were brought together, those similarities and differences tended to reveal themselves. So the evolutionists were often talking about the products of genetic evolution as a sort of closed form of phenotypic plasticity. For example, people in harsh environments are programmed to develop a fast life-history strategy, as opposed to people who grow up in nurturing environments, who are programmed by genetic evolution to develop a slow life-history strategy. And then the CBS people would say, no matter who walks through our door—which is typically someone who grew up in a harsh environment; otherwise we wouldn't be seeing them—what do we do to make them more flexible?

Steve: It is a difference in terms of level of analysis. You can see both the plausibility and the possible value of an integrated framework in the chapters of this book. People need to understand: these chapters were not only written by different teams, they were written with *no specification of what authors would write about, other than the generic area.* There was no attempt to make sure that they were talking about things that would overlap! So here are these chapters that over and over again overlap in significant ways, and yet they're different.

If you were to back up and look at the state of behavioral science and evolutionary science some decades ago, I don't think we could have done this then. There have been some changes in the two fields that made it possible.

People forget that animal learning emerged from comparative psychology. Watson and Skinner cut their eye teeth in psychobiology. But there wasn't a good language for how to talk about how these different levels related 80-90 years ago. And on the evolution science side, the lack of an appreciation for the way that environment and behavior actively up- and down-regulate genes and are then involved in phenotypic selection and even in genetic accommodation over time was unappreciated. The field was just too gene-centric.

David: This book could not have been written fifteen years ago; that's how much has been happening, which on the evolution side falls under the umbrella of the extended evolutionary synthesis. In my field there's so much debate: is it extended, is it a synthesis, is it not a synthesis—there's all this angst.

You have things like that, too. One of the ways that Skinner overreached was by trying to explain language and symbolic thought as simple operant conditioning. And that turned out not to the case; there needs to be a more extended theory of language.

And then on the evolutionary side, thinking about symbolic thought from an evolutionary perspective really traces to *Evolution in Four Dimensions* by Eva Jablonka, who is part of our volume, with Marion Lamb; and then of course Terry Deacon's book, *The Symbolic Species*—he's also part of our volume, I'm proud to say.

Steve: Readers may be almost startled by how close some of these perspectives on language are. Look at Terry Deacon's chapter: it has an intensely relational basis, and his thinking is *very* friendly to RFT even though he knew little about it; you see that also in the dialogue involving Eva. Operant conditioning goes back to the Cambrian, yet what you and I are doing right now is arguably only hundreds of thousands or maybe a couple of million years old.

Once evolution science agrees that cells are systems for turning environment and behavior into biology, it has to put behavioral science at its core but in a different way. This issue of language cries out for an analysis that's systematic and careful. On the behavioral side, we need to be far more sophisticated about the multidimensional and multilevel nature of human beings. Behavior involves genetic and epigenetic and cultural and other factors, and not simply those that happen within the lifetime of an individual, and selection operates at different levels simultaneously.

David: Let's focus on some of the differences that remain. As we wrote in the *Behavioral and Brain Sciences* article "Evolving the Future," there's a lot of mileage you can get out of thinking that the behavioral system is like the immune system, with both an innate component and an adaptive component. The innate component is genetically evolved and has closed phenotypic plasticity: plan A, plan B, plan C, which are triggered by environmental stimuli. The adaptive component is open-ended, like antibodies. In many ways, the Skinnerian tradition presents the adaptive side; the evolutionary psychology tradition represents the innate side.

You see that in the dialogues. On the CBS side, everybody has flexibility or actually can obtain flexibility. On the evolutionary side, I love the metaphor that Renée Duckworth used with a skeleton: In order to be flexible, we need a skeleton; but a skeleton is inflexible. And so flexibility and inflexibility kind of go hand in hand.

Steve: It's a good point. To put it in context, though, many of the behavioral folks are applied science people, and they're trying to be of help. And there's been such a long history of misused structural assumptions harming people—you have this personality type or this level of intellect and you have to just accommodate to that fact. Then it turns out that with clever enough work it's only partially true, or sometimes not true at all. But I get your point and I think it's a good one. Within your field don't you see the same movement toward Evo-Devo in areas in which development itself was sometimes dismissed as just part of the life-cycle of genes?

These assumptions that are deeply developmental or genetic have to really be looked at carefully. If you meditate for eight weeks, about seven percent of your genes will be systematically up- or downregulated. Think about that!

David: One of the things that contextual behavioral science does, tracing back to the tradition of Pragmatism, is question the entire relationship between basic and applied science. And the standard view is that basic science is curiosity-driven and it asks the big questions and seeks understanding and prediction. It might be useful down the road, but utility is not what you keep in mind. And then applied science applies this stuff, but typically isn't very interesting in terms of the big questions. Contextual behavioral science turns it around and says basically the best science is the science that is done in context, and that you could be asking the big questions at the very same time you're actually improving the quality of life. That expanded agenda is in the very title of this book.

In evolutionary biology, all laboratory science needs to be based on field science. When it isn't, then the laboratory science is in distinct danger of going off the rails because it's not related to the organism in its context. That general idea is pretty well understood by evolutionists, but again, only in the context of genetic evolution—and then not necessarily in the context of applied ecology. I think as soon as evolutionary biologists do become applied in their interests—as with some of our authors—the idea that there can be a positive tradeoff between basic and applied science becomes a very important message.

But CBS, by being so applied, is concentrating on a minority of the population for whom what's happened to them has not worked well. And I wonder how much you've thought about that?

Steve: It's a complex set of issues, and it actually goes back to how the disconnection happened. If you had to pick the biggest thing that ecologists and evolutionary biologists hated about the animal-learning tradition, it was the decontextualization of animal behavior. Putting pigeons and rats in boxes and so on.

Skinner's gamble was that you *might* get high-precision, high-scope principles that would go across tips of evolutionary branches that way. But he realized that he might be mistaken. You can find him saying (I'm paraphrasing here), "It's not that this is arbitrary; I'm not assuming that it is. It's just a strategy. Because if I get immediately into complexity, I'll start focusing on all these features of the specific context, and I might miss what might be higher-scope principles." It turned out, I think, that he got away with it. Schedules of reinforcement apply to boa constrictors and spiders and baboons and human beings in very, very similar ways (up until human beings get language; then they really start changing a lot). Likewise we want principles that will apply to both normal and abnormal behavior, *if possible*.

The Skinnerian tradition breaks from the Watsonian tradition in a paper where Skinner applies behavioral thinking to himself and says, "The issue is not within and without; it's how precise are the contingencies that are controlling my observation as a scientist. If I'm the last person on the planet, if I'm Robinson Crusoe, I could do a science of my behavior and the fact that I can't get public agreement is irrelevant. The issue is can I bring my observation under the control of events in such a way that I save my water enough that I don't die of thirst?" That approach overthrew almost the whole behavioral tradition that Watson had established.

And when I look at evolutionists, I sometimes wonder if they have thought of their *own* behavior the way an evolutionist should? After all, the immediate product of science is just words—so shouldn't that product give me the ability to interact more successfully in and with the world? Where is the selection criterion? Sometimes it doesn't seem as though the evolutionists have thought about it much, although most of our authors are not like that. Of course, scientific goals are up to the scientists—you can't say "My goals should be your goals." It just seems to me that sometimes there has been an automatic assumption in evolutionist circles that basic understanding is what our real scientific goal is, and if it leads to influence in the form of applications, that's just an incidental benefit. Why? That is not a requirement of evolutionary theory per se—there is nothing that says we can't establish our own selection criteria to a degree once language shows up.

David: This apartheid between evolutionary biology and other human-related sciences is because of a taboo. Evolutionary biologists who think clearly about their particular organisms are just getting sweaty palms when they're asked to apply their findings to humans. They just don't go there. And it's not a principle—they simply haven't thought much about it. It's just a boundary that they respect. And they're just not going to barge through that door with the "No Trespassing" sign.

Steve: I think that's exactly right. But here's the thing that I just want to put in the room. When behavior analysis and behavioral modification first came on the cultural scene in the sixties as we began to realize we could actually do some things with these animal-learning principles, within just a few years you had *Clockwork Orange*; you had people in the *New York Times* equating behavioral modification with psychosurgery. And then with Eugenics, and even Nazis. The functional, contextual perspectives that you and I represent, each in our different ways and at different levels, have both been held to account for things like overt racism done with some hand-waving to Mendel.

The behavioral science folks have learned how to move forward through that cultural barrier, and the way we've done this is by hanging on to applied work, where someone is suffering and you can do something about it. That is our face to the culture: if you have a sister with an addiction problem, give us a call; if you have a child who can't learn and is not acquiring a language, give us a call. That formula brought behavioral thinking back into the culture, from the *Clockwork Orange* era fifty years ago to now, when the government is listing behavioral methods among the evidence-based procedures that they want to see implemented in health care systems.

There's not yet something like that on the evolution side, at least in the United States. There is in some very progressive countries such as Norway, and you've been involved with them, but what I'm saying is that I think we have a common fate. To this day, behaviorists are very cautious about talking about things that might be able to be engineered in our schools, societies, businesses. There is that dark shadow out there that you have to face in your side, too, if you want to apply evolutionary ideas. I'm not sure either one of us is going to be able to do what we came here to do in the promotion of human well-being if we don't solve this problem.

David: It's ironic that there are these two histories of scaremongering. One of behaviorism—that it's going to lead to eugenics and Nazis and all that kind of thing. And for evolution, social Darwinism. I've looked at the social Darwinism issue in great detail. I have published with Eric Michael Johnson a special edition of *The View of Life* called "Truth in Reconciliation for Social Darwinism," which actually goes through this history. Basically, these are boogeyman stories, told by intellectuals. Hitler was not influenced by Darwin—quite the contrary! We have good scholarship existing side by side with this crap—with boogeyman stories. The same with behaviorism. At the end of the day, what you have to do is some clear thinking about the application of scientific knowledge of all sorts to public policy—basically, to efforts to make the world better. When does it become insidious and when does it become benign?

Doing nothing won't work. We need to draw upon knowledge in order to do something. Is science useful in this regard? Well, of course it is! So how to we make it benign and keep it from being insidious? It becomes insidious when it involves people manipulating other people. When I impose what I think is best on you, that's insidious. But if *we* agree that this is something that *we* want to do, or if it's just some individuals who agree that they're in a bad way and they want to do something, then of course that is more likely to be benign—not certain to be benign, but more likely to be benign. So we need some clear thinking on egalitarianism; and on making decisions by consensus. What widespread groups of people openly propose to do is likely to be benign, as opposed to various forms of manipulation and exploitation.

Steve: I agree with that. Clearing up the distortions is important. But I don't think it's enough. For one thing, you can certainly find very racist things being said by early evolutionists. Why wouldn't they? The whole culture was racist and it's really not fair to blame them. You can find Darwin talking in judgmental ways about the Maori. And you can find behavioral science misused by behavioral scientists—it happened again very recently in the torture of prisoners by our own government!

Here CBS may have something to offer. One of the things that an applied psychologist does is that if you're going to work with somebody, you're going to have to work with them within their values, their purposes, their goals. And on the ACT side of things, there is a whole technology about how to do that and do it in a way that's not imposing your values on others.

To go back to the point I made earlier, evolutionary theory will never fully enter into the culture in a way that fits what it can bring to the culture, until it's able to show lots of natural applied extensions, and people see and understand that the way that is being done is egalitarian and respectful of their purposes and goals.

We have seen this in the project that we're working on together, Prosocial, which is talked about in the pair of chapters in the Small Groups section of this volume. It mixes a descriptive set of principles that are evolutionarily sensible coming out of the Nobel Prize–winning work of Elinor Ostrom—the core design principles—and because we wanted to apply that to prosocial groups, we added an ACT-based conversation about individual and group values and the things that get in the way of them. I think that gives a little template for how this integrated framework we are seeking might work.

David: I think that it's going to surprise a lot of people how much evolutionary theory shows that for any group to work well, no matter what its size, there has to be a balance of power among its constituents. As soon as you have an imbalance of power, then decisions will be made and cultural evolution will take the form of privileging those with the power, and so balance of power is everything, at all scales. So that's the evolutionary message, not what people associated with social Darwinism.

You don't have to go back to Victorian culture to see how much trust we placed in our leaders. Well, not anymore. Injustices did take place. Not because they follow from behaviorism, not because they follow from social Darwinism, but because they follow from the imbalance of power.

Steve: I agree. And multilevel selection points to it. Behaviorists call it the importance of counter-control: you have to keep the people who are being influenced in a position where they in turn can influence the influencers, or you get this kind of authoritarian result. When ACT first hit the scene and became popular after *Time* wrote it up in 2006, the early criticisms were: "'Acceptance' is just teaching the downtrodden to accept their fate because the people in power want to dominate…" My response was, well, the first randomized trial in the modern era on ACT was done by Frank Bond, who went in to a call center and did work on the psychological flexibility of these workers who were having to work in very difficult environments—very stressful, almost piecework-like.

David: One of the worst jobs you could have.

Steve: It is. And what happened? When you did that individual work, the workers started demanding changes in their work environments! Because what was keeping them from going to the supervisor and saying "This is really not right" was in part fear. Now, sometimes that fear is real, but sometimes those internal barriers show up in environments that would support innovation. "Acceptance" as a word can indeed mean to resign yourself to your situation and your behavior, but that is not what it is about in ACT. It's about accepting your history, focusing on your values, and *changing* your situation and behavior to better foster your own prosperity and development, if that is possible. That's in fact what's happened as we've put some of those methods into Prosocial.

David: So this might be a good high note to end on. If you look at what we're doing, it attempts to be prosocial at all scales. A lot of this began at the individual scale, but it's working up to the worldwide scale. If you take our friend Tony Biglan, who is the one who introduced us to each other, he's currently working on more nurturing forms of capitalism. He's showing that capitalism isn't bad; capitalism is like rapid cultural evolution, but it's taking place in an environment that's causing it to run amok. We have to tune the parameters so that these powerful forces of cultural evolution actually lead to prosocial outcomes, at all scales, from the individual, to the small group, all the way up to the planet Earth.

We've taken a prosocial agenda, an agenda that nobody could disagree with, and we've trained science to that end. When science is trained toward antisocial ends—we stand against that. I think we've taken a moral position and we're harnessing that to the science, which

I think makes it difficult to criticize on the basis that this is likely to lead to an immoral outcome. We're here to make sure that doesn't happen.

Steve: It goes back to that issue of what are the selection criteria, and those are our responsibility. As scientists, I don't think we should be ashamed of saying that we're interested in developing knowledge that has a prosocial impact. We want to play science in the high-integrity way science should be played. But the purpose of playing the game that way is to develop knowledge that can be vetted against prosocial criteria. It is entirely within our prerogative as human beings to say, "This is what we want to do with our careers and lives with this wonderful tool called science." In an abstract sense, science is values-free, but in the lives of scientists and of the consumers of scientific knowledge it is anything but.

Part of what we're trying to work out together, I think, is how to develop a kind of knowledge that walks that line and stays in balance; that does high-quality work at all these levels and different dimensions; and yet can be held to account for the moral status of the enterprise. I think people reading this volume will recognize and appreciate that dynamic as we attempt to bring about an integrated framework that can pass through that level of scrutiny.

David: Well said. You had the last word.

Editor **David Sloan Wilson, PhD,** is president of The Evolution Institute and a SUNY distinguished professor of biology and anthropology at Binghamton University. He applies evolutionary theory to all aspects of humanity in addition to the biological world. His books include *Darwin's Cathedral, Evolution for Everyone, The Neighborhood Project,* and *Does Altruism Exist?*

Editor **Steven C. Hayes, PhD,** is foundation professor in the department of psychology at the University of Nevada, Reno. An author of forty-four books and over 600 scientific articles, his career has focused on an analysis of the nature of human language and cognition, and the application of this to the understanding and alleviation of human suffering and the promotion of human prosperity. He has received several awards, including the Impact of Science on Application Award from the Society for the Advancement of Behavior Analysis, and the Lifetime Achievement Award from the Association for Behavioral and Cognitive Therapies (ABCT).

Foreword writer **Anthony Biglan, PhD,** is senior scientist at the Oregon Research Institute (ORI) in Eugene, OR, and director of the Centers for Community Interventions on Children and Prevention of Problems in Early Adolescence, both at ORI.

Index

leadership: flexible organizations and, 196; organizational styles of, 176–179

learning: associative, 31–32, 39–43; classical vs. operant, 32–39; contextual science of, 25–27; contingency, 286; dialogue on, 48–54; evolutionary psychology and, 20–21; functional approaches to, 16–17; genetics related to, 4, 10; intergenerational transmission of, 249; mechanistic approaches to, 16, 17, 19–21; memory systems distinguished from, 34–35; modern education and, 213–214; respondent vs. operant, 263; self- vs. world-, 36, 37; study of spatial, 17–19; symbolic, 264

learning breadth, 32

learning theory, 4, 6, 7, 10

LeDoux, Joseph, 149

Lewin, Kurt, 171, 172, 182

Lewis, David, 68

lifestyle change, 226–229

light exposure, 213

limited associative learning, 39–40, 41, 42–43

Lipkins, Gina, 86

Locke, John, 68

M

macrosystems, 138

managerial challenges, 174–176

Market Pricing (MP) model, 175, 178

matching law, 225–226, 243

Mayr, Ernst, 246

McEnteggart, Ciara, 55

McMahon, Kibby, 133, 163

Mead, Margaret, 8

Meaningful Roles Intervention, 128

mechanistic approaches to learning: functional approaches vs., 16–17; study of spatial learning and, 17–19

memory, 34, 43, 52

Menard, Louis, 3

Mendel, Gregor, 3

mental health. *See* health and wellness issues

Metaphysical Club, The (Menard), 3

methodological individualism, 245, 247, 250, 251

microsystems, 138

mindfulness, 186, 196, 227, 267, 306

Mischel, Walter, 314

Monestès, Jean-Louis, 283, 312

multidimensional form, 180–181

multidimensional, multilevel (MDML) framework, 60–62

multi-divisional form (M-form), 179–180, 181

multilevel selection (MLS), 173, 250, 251, 252–253, 328

mutual entailment, 58, 61, 62–63, 85, 86, 87

N

natural selection thinking, 6

naturalistic fallacy, 96

near-decomposability principle, 179–180

negative thoughts, 120–121

neuroendocrine structure, 300–301

neuroscience, 251

neuroticism, 305

niche construction, 207–208

nonhierarchical organizations, 202

noticer skills, 113–114

O

obesity and depression, 224

object permanence experiments, 265

O'Connor, Lynn E., 149, 163

On Interpretation (Aristotle), 68

openness: to discomfort, 195; to experience, 268, 316

operant conditioning: behavior variability and, 286; classical conditioning vs., 32–39; consequences of behavior and, 263; evolutionary distinctiveness of, 35, 38–39; functional distinctiveness of, 35–36; taxonomical distinctiveness of, 37–38. *See also* classical conditioning

operant learning, 263

organizational behavior (OB): contextual behavioral science and, 187–188; field of study related to, 187

organizational citizenship behaviors (OCBs), 177

organizational culture, 176, 182

organizational development, 171–182; change related to, 171–172; dialogue on,

200–206; evolutionary perspective on, 171, 172–174; leadership styles and, 176–179; managerial challenges and, 174–176; organizational design and, 179–181

organizational flexibility, 185–197; awareness and, 195–196; effective work design and, 194–195; hexaflex and orgflex models of, 189, 190, 191; leadership and, 196; openness to discomfort and, 195; organizational behavior and, 188; planned action and, 191–192; purpose/goals and, 190–191; situational responsiveness and, 193–194; six characteristics of, 190–196

Origin of Species, The (Darwin), 2

Ostrom, Elinor, 200, 206, 251, 252, 327

P

parenting, 215–216

pathological altruism, 158

patterns of practice, 234

Pavlov, Ivan, 32

Peirce, Charles Sanders, 3, 69

Pepper, Stephen, 134, 137

personality variation: human behavioral change and, 304–306; neuroendocrine structure and, 300–301; psychotherapy effectiveness and, 306; tradeoffs in neural processes and, 301–304

perspective taking, 60; Core Design Principles and, 269–272; group cooperation and, 265–266

phenotypic flexibility, 298, 316

phenotypic plasticity, 125, 298

phenotypic traits, 7, 8, 248

physical activity, 211–212, 229–230, 306

physical health. *See* health and wellness issues

physiology: emotions disrupted by, 151; empathy impacted by, 141

Pinker, Steven, 8

planned action, 191–192

planned organizational change, 171–172, 182

political scientists, 251

polycentric governance, 253, 257, 276

poverty of the stimulus, 82, 89

power distance, 142

pragmatic accuracy, 140

Pragmatism, 3, 320, 321, 323

Pragmatism (James), 3

predication structure, 75, 77–80

prediction-and-influence goal, 188

present-moment awareness, 195, 227

privileged units of analysis, 247–248

proactive personality types, 304

project-definition approach, 192

prosocial emotions, 151, 157

Prosocial project, 254–256, 262, 267, 269–272, 275, 327

prosocial values/goals, 268–269

psychological flexibility, 185–187; Core Design Principles and, 269–272; evolutionary concepts related to, 293; group cooperation and, 267–269; hexaflex representation of, 189, 190; lifestyle change and, 226–229; organizational flexibility and, 189, 196; successful psychotherapy and, 291–293

psychological inflexibility, 284–285, 289

psychopathology: behavioral variability and, 285–288, 290; developmental model of, 94, 95–96, 97; dialogue on behavior change and, 312–318; empathy impacted by, 141

psychotherapy: emotional experience and, 158–160; empathy required in, 143–144; personality variation and, 306; psychological flexibility and, 291–293

puberty-specific changes, 98–100

purpose, organizational, 190–191

Q

quantification, 75, 80

R

Raczaszek-Leonardi, Joanna, 71

rational choice theory, 245

reactive personality types, 304

reafference, 40

recursive affordance, 75, 76–77

reductionism, 245, 247, 250, 251

referential grounding, 71–72

relational frame theory (RFT), 12, 55, 59–64; evolutionary science and, 26, 62–64; human language/cognition and,

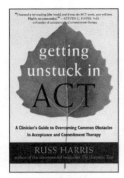

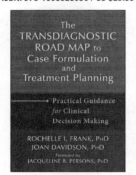